# THE

# HUMAN

# GENOME

## A USER'S GUIDE

# THE

# HUMAN

# GENOME

## A USER'S GUIDE

R. SCOTT HAWLEY

CATHERINE A. MORI

Section of Molecular and Cellular Biology
Division of Biological Sciences
University of California, Davis

*With Commentary by*

JULIA E. RICHARDS

*Departments of Ophthalmology and Epidemiology*
*University of Michigan, Ann Arbor*

San Diego   London   Boston   New York   Sydney   Tokyo   Toronto

*Cover photograph:* A mouse oocyte at anaphase in the first meiotic division. The microtubules are stained to appear green, and the chromosomes to appear red. The image was captured on a confocal microscope. Courtesy of Patricia Hunt of Case Western University.

This book is printed on acid-free paper.

Academic Press
*a division of Harcourt Brace & Company*
525 B Street, Suite 1900, San Diego, California 92101-4495, USA
http://www.apnet.com

Academic Press
24-28 Oval Road, London NW1 7DX, UK
http://www.hbuk.co.uk/ap/

Harcourt/Academic Press
200 Wheeler Road, Burlington, MA 01803
http://www.harcourt-ap.com

Library of Congress Catalog Card Number: 98-85438

International Standard Book Number: 0-12-333460-8

PRINTED IN THE UNITED STATES OF AMERICA
99   00   01   02   03   04   EB   9   8   7   6   5   4   3   2

*To Professors Laurence Sandler and Crellin Pauling,
who understood why I had to write this book, and
to Jeanne, Tara, and Christopher who bore the
burden of its composition —RSH*

*and*

*to Matt, my mother, my father, Claire, John,
Zachary, and Alex —CAM*

# CONTENTS

# 3

## How Genes Work: The Story of How Genes Encode Proteins, or "The Central Dogma of Molecular Biology"

# 4

## How Genes Move: Chromosomes and the Physical Basis of the Laws of Mendel

# 5

## Absent Essentials and Monkey Wrenches: How Mutations Produce a Phenotype

## SECTION II

### How Genes Determine Our Sex

## 6

### Sex and Chromosomes, Sex and Hormones, Sex and . . .

## 7

### Sex Causes Problems: The Inactivation of the Second X Chromosome

# 8

## GENDER IDENTIFICATION AND SEXUAL ORIENTATION

# SECTION III

## WHEN MEIOSIS OF MENDELIAN INHERITANCE FAILS

# 9

## FAILED CHROMOSOME SEGREGATION AND THE ETIOLOGY OF DOWN SYNDROME

# 10

## EXTREME MUTATION: TRIPLET REPEAT SYNDROMES

# 11

## IMPRINTING OR EPIGENETIC CHANGES IN GENES AND CHROMOSOMES

## SECTION IV

## HUMAN GENES

# 12

## INTRODUCTION TO GENE CLONING

# 13

## DNA POLYMORPHISMS AS GENETIC MARKERS IN HUMANS (AND THE MIRACLE OF PCR)

# 14

## HUMAN GENE MAPPING: A GENERAL APPROACH

# 15

## CYSTIC FIBROSIS

# 16

## MAMMOTH GENES: MUSCULAR DYSTROPHY
## AND NEUROFIBROMATOSIS

# 17

## Genes and Cancer

## Section V

## Interactions of Genes and the Environment

# 18

## Multifactorial Inheritance: Toward Finding the Genes for Manic Depression and Schizophrenia

# 19

## The Monoamine Oxidase A Gene and a Genetic Basis for Criminality?

# 20

## GENETICS OF THE HUMAN AIDS VIRUS

# SECTION VI

## PRENATAL DIAGNOSIS

# 21

## METHODS OF PRENATAL DIAGNOSIS

# 22

## POTENTIAL FOR GENE THERAPY

## Epilogue: Fears, Faith, and Fantasies

### Appendix

# PREFACE

The intent of this book is to put genetics in the middle of people's lives, specifically your life. Thus, we will eschew lengthy discussions of the genetics of fruit flies in favor of considerations of issues such as how we become boys and girls and men and women. Indeed, a discussion of the development of human sexual differentiation and sexuality will form the major thread throughout this book. We have chosen to use sexual development as our major emphasis both because of the depth and elegance of our understanding of this process and because of the obvious interest we all share in this topic.

Perhaps most crucially, a focus on sexual development allows us to follow a set of interrelated genetic pathways toward many ends at many levels. We will focus in some detail on the role of the Y chromosome in initiating one pathway of development. We will also spend a good deal of time discussing the role of hormone-controlled differential gene expression in terms of the development of somatic sex. Finally, we will discuss the role of genetics in sexual orientation and sex-role development. These last issues will provide us with a real opportunity to discuss the interface between science and society. Along the way we will endeavor to cover the basic intellectual processes that underlie genetic analysis and to cover a number of other relevant topics such as the genetics of cancer, AIDS, and mental illness.

Our objective is to tell you what we know about the genetics of our species and how we feel about the ways in which genetics does, could, and should impact our lives. Because of that point of view, we will do things you will not normally find in a textbook, much less a science textbook. We will intermingle real and fictional stories to make our points and we will speak to you, when appropriate, in the first person. In doing so we will endeavor to tell you who we are, how we came to write this book, and how deeply we feel about the issues we discuss. We may even argue with each other in print. Such nontraditional communication is enclosed in parentheses and we will always sign our initials to such contributions.

It is not our intention to turn any of you into a geneticist (*although that wouldn't be so awful—RSH*). Our real hope is to impart enough knowledge so that you will be able to bring this subject into your own lives. It is hoped that when you put this book into mothballs you will know when and why you might seek the counsel of a human geneticist or genetic counselor, should you ever need one. It is also hoped that you will have become sophisticated enough to sort out the myths and misconceptions about human heredity that pass for simple truths in the folklore and the press. To the extent that we achieve even a small measure of success with either of these goals, this book will have been well worth the effort.

P.S. We can't begin without thanking some very special people, especially our editors at Academic Press, David Phanco, Garrett Brown, and Dr. Anne Chen, and especially their predecessor, Dr. Jasna Markovac. The fact this book was ever finished is a testament to their tenacity. The real animus for this book began as a series of lecture notes for a course titled Human Genetics that RSH teaches at U.C. Davis. The purpose of that course is to initiate nonscience majors in the beauty and complexity of human genetics and the way in which it has an impact on their lives. We also owe a very real debt to the students of MCB 162 at U.C. Davis, since their intellectual curiosity really drove much of our efforts. We are grateful to Kathy Bayer for so many of the illustrations. RSH especially thanks Dr. Julia Richards (an additional voice in Sections IV–VI) for coming to his aid in time of crisis and for vastly improving Section IV and much of the rest of the book as well. Many thanks are also due to Dr. Cheryl Crumpler for being the conscience of several chapters of this book and for her unflagging support, pinch-hitting, and assistance through much of the writing. RSH is also grateful to Joan Esnayra for many helpful comments. Thanks are also due to Dr. Kenneth Burtis for assistance and for convincing RSH to keep going when stopping seemed like a much better plan; to Kara (the wonder student) Koehler, Heather (it's only good if it's RIGHT!) Collins, Soni (it's a small w, not a capital W) Lacefield, Christina ("how can I help?") Boulton, and Sharon ("try to imagine how YOU would feel if YOUR child was affected with . . .") Murphy for reading large portions of the book and helping to improve it; and to Linda Pike, Alison Kirschner, Cheri Harrington, and Louisa West, who were essential during the early stages of composition. I thank them very much. But most of the burden fell upon Ms. Amy Ring. Amy, thank you for collating, correcting, making figures, catching typos, and retyping the whole thing several times. To Kim McKim, Dawn Milliken, Lisa Messina, Jeff Sekelsky, and Sarah Wayson, thank you for keeping the lab running while RSH was busy writing. Thanks are especially due to Drs. Terry Hassold, Patricia Hunt, and Hunt Willard at CWRU for answering endless questions year after year and to genetic counselor Catherine Downs for helping us increase the accuracy of the sections on deafness and on medical geneticists and genetic counselors. Thanks as well to Drs. Ken Burtis, Carl Marrs, Michael Turelli, Michael Zwick, and Margit Burmeister for helpful suggestions and to Dr. Dan Gusfield for helpful comments on the manuscript. We especially thank Drs. Stephanie Sherman and Terry Hassold for valuable comments on the manuscript. To his dear friend and scientific "big sis," Adelaide T. C. Carpenter, RSH owes a big thank you and a hug. RSH also thanks the muses (J. Stewart, M. C. Carpenter, Meatloaf, and H. Schock) who kept him constant company during too many late nights of writing, and especially Harlan Ellison for living up to RSH's image of him. Finally, to our families and friends we can only say, "Yes, THAT book is finally finished." They have truly borne most of the burden for this project. We thank them for their support.

# THE BASICS OF HEREDITY

This section provides a thumbnail description of what genes are, how they work, and how they are moved around during cell division. We also present several examples of how faulty genes can cause anything from a minor change in hair color to a major change that results in illness or death.

# 1

# THE ANSWER IN A
# NUT SHELL:

## GENES, PROTEINS, AND THE
## MEANING OF LIFE

*There are always those who ask, what is it all about? For those who need to ask, who need points sharply made, who need to know "where it's at," this. . . .*
*—Harlan Ellison\**

*Our genes provide a blueprint for our bodies. In doing so they set some upper and lower limits on our potential. Our interaction with the world and others defines the rest. . . .*
*—RSH*

Our bodies contain billions of cells. In each of those cells is a nucleus that contains all of the information required to make a complete human being. This information exists in the form of 50,000 to 100,000 structures called *genes.* Each gene possesses the ability to encode one *protein,* and proteins provide both form and function within our cells.

Virtually all of the cells in our body contain exactly the same full set of genes. The diversity of cell types and functions reflects the expression of a different pattern of proteins in each cell type and a changing pattern of protein production during development. The ability of cells to exhibit these diverse patterns of protein production reflects the ability of cells to express only certain genes at specific times. Indeed, the development of a human being from conception to death is the result of a complex program of expressing genes in some cell types and not in others at specific times during development.

Many, if not most, of the differences that exist between us reflect the fact that genes can be altered by a process called mutation. The term mutation refers to a startlingly large array of processes that can permanently alter the structure, and thus the information content, of genes. Although mutation occurs at a very low frequency, there are an awful lot of us, we breed well, and we have been breeding for a very long time. Thus, there has been ample opportunity for mutations in each of our genes to occur and in many cases to be spread widely throughout our population. These altered genes may produce an altered protein or produce no protein at all. Although the latter class of defects is frequently the cause of lethality or genetic disease, the former class of mutation provides much of the diversity we see around us.

Accordingly, genes influence our form, appearance, physical abilities and limitations, talents, and much of our behavior as well. Each of us received one complete copy of the "human gene set" from our mom and one set from our dad. Thus each of us carries two copies of each gene. As noted earlier, copics of the same gene may be different (different forms of the same gene are called *alleles*) and thus confer different effects on the organism. A good example is the various alleles of the gene that encodes the protein that colors our eyes. Whether your eyes are blue, brown, green, or hazel is a consequence of the alleles you possess for the genes that determine eye color.

When we make gametes (sperm or eggs), we place only one of our two copies of each gene in each gamete, thus half of our sperm or eggs will carry the copy of a given gene that we inherited from our mother and the other half the copy that we inherited from our father. This trick sees to it that each generation will always have two copies of each gene and it introduces an amusing bit of randomness to the process. Each sperm or egg that we produce consists of a different combination of genes derived from our own mothers and fathers. Thus each new baby is the result of implementing a set of genetic instructions created by two rolls of the genetic dice, one that took place in the father and one that took place in the mother.

So how do genes work? How can differences between various copies of a given gene cause differences in people or cause disease? The answers to these rather fundamental questions lie in understanding what genes are, how they work, and how our cells move them around. That is the business of this book. Most textbooks begin with the formal development of the concept or theory of the gene by an obscure Austrian monk named Gregor Mendel. Indeed, we will feel pedagogically obligated to cover exactly the same ground in Chapter 2. The remaining chapters of this introductory section will then raise the appropriate questions with respect to the physical structure of genes, their manner of expression, and how cells and organisms move them around. However, like most things in life, much of our description of the process by which genes function will make more sense if we tell you the answers or "bottom line" up front.

As shown in Fig. 1.1, a gene is a region of a very long molecule called *DNA* (*deoxyribonucleic acid*) that is itself composed of repeating units called *base pairs*. (A large number of adjacent base pairs make up a gene the way a number of letters make up a word.) As shown in Fig. 1.2, genes function by directing the assembly of other large molecules called proteins. To a first approximation, each of your genes has the capacity to encode a given specific protein. (The sequence of base pairs within each gene determines the type of protein that a given gene produces.)

If genes are the source of cell information, then proteins are the business end of cellular processes. Proteins come in a variety of flavors such as structural proteins, enzymes, hormones, receptors, and DNA-binding proteins. Structural proteins, such as tubulin, keratin, and collagen, are the subunits out of which cells build various kinds of scaffolds and skeletons both inside and outside of cells. Enzymes direct and catalyze a host of biochemical reactions, such as digestion and energy production, that are truly essential for life. Hormones are chemical messengers that carry signals from one cell to another. You can see color and detect smells because of receptor proteins such as color opsins and odor receptors. Your heart or skeletal muscles move because of proteins called actin and myosin. Your body fights off infection with the help of proteins called immunoglobins.

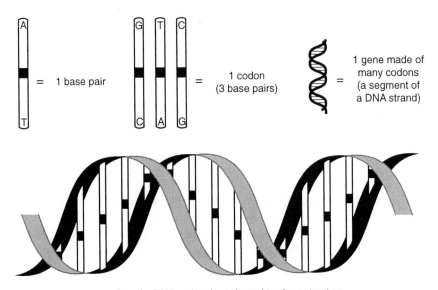

a portion of a DNA molecule, enlarged to show structure

FIGURE 1.1    The relationship between genes and DNA.

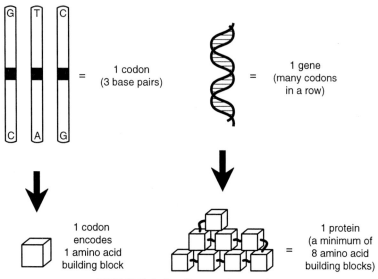

FIGURE 1.2    Genes make proteins.

Even your genes themselves are regulated by proteins called DNA-binding proteins.

Genetic diseases, or inborn errors, result from cases where the DNA blueprint is incomplete, usually because a specific gene is damaged or missing. In such a case, the cells of an individual bearing such a genetic defect will not be able to produce the protein specified by that damaged or missing gene (Fig. 1.3). For example, people who lack functional copies of the genes to make color opsins will be unable to distinguish colors *(like me—RSH)*. In other cases, the gene has been changed so that the protein it produces is actively doing something different than what it usually does, perhaps inhibiting the process in which it normally assists or perhaps doing something even more complex.

As noted earlier, such damaged genes arise as the consequence of the process of *mutation.* Mutations are permanent and inheritable changes in the ability of a gene to encode its protein. Much like typographical errors, which can change the meaning of a word, or even render a sentence as gibberish, such changes in gene structure can have severe effects on the ability of a gene to encode a protein. Some mutations prevent any protein from being produced, some produce a nonfunctional or only partially functional protein, and some produce a faulty or poisonous version of the protein.

Consider a case in which defects in a pair of genes borne by a developing zygote result in a severe defect in development. In this case, both the sperm

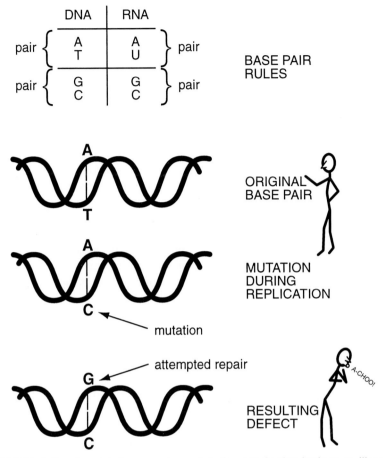

FIGURE 1.3    A mutated gene makes a defective protein that leads to an illness in an individual.

and the egg carried damaged (nonfunctional) copies of a gene required to make a protein called CFTR, which is required for proper function of the cells that make up the lung and pancreas. As a consequence of this genetic deficiency, the resulting child will be unable to produce this essential protein. As a result, this child will develop severe problems in lung and pancreatic function, a disease known as *cystic fibrosis.*

There is a really fundamental point to be made here. It is important to note that it is not the damaged CFTR genes that produce the disease, it is the absence of the protein those genes were supposed to produce. A damaged gene, like the blueprint for a cruise missile, is in and of itself pretty harmless. It is the product of that blueprint (either the cruise missile or the defective or absent protein) that poses a problem. *No one has ever died because of a defect in their genes; they die because of the consequences of that defect in terms of the production, or lack thereof, of an essential protein.*

For much of the last century the science of human genetics has consisted of finding families in which such damaged genes are being passed down from generation to generation, resulting in illness or death, and then attempting to determine the cause of that anomaly in terms of the missing or abnormal protein. As one can imagine, this analysis has had to deal with some rather difficult but intriguing questions, such as how could two otherwise normal individuals produce a child with cystic fibrosis or with other genetic diseases? If the child obtained its genes from its parents, then why don't the parents have cystic fibrosis as well? Are parents whose first child was born with cystic fibrosis more likely to have other children with cystic fibrosis? Could we identify parents at risk?

Research done on fruit flies (yes, fruit flies!) during the first half of this century has provided the answers to most of these questions. As noted earlier, every gene is present in two copies in each of our cells. This reflects the fact that each sperm and each egg carry one copy of each gene. So one gets a copy of the CFTR gene from mom and a copy from dad. Individuals can exist who carry both a functional and a damaged copy of the CFTR gene (see Fig. 1.4). Because they have the one functional CFTR gene, these individuals will be fine. However, when two such carriers of a damaged CFTR gene produce children, there is a chance that each parent will donate a sperm or egg bearing a defective CFTR gene. The result will be a zygote with two nonfunctional CFTR genes that will develop into a child with cystic fibrosis. However, such parents can also produce offspring with two normal copies of the CFTR gene, or with one good and one damaged copy.

As the century proceeded, research in bacteria provided the crucial understanding of the mechanisms by which genes encode proteins and demonstrated that genes were made of DNA. These studies have also provided us with a set of tools, called cloning, that allow us to specifically isolate and analyze the genes in which we are interested. Thus, we can obtain a bottle filled with copies of the CFTR gene, isolated away from the other 50,000 to 100,000 genes in the *human genome* (the complete collection of human genes, which is present in each of our cells, is referred to as the human genome), and precisely determine the order of base pairs that make up that gene. We can ask with real precision when and where that gene is expressed. Research in many organisms has also given us an ever-sharpening picture of what specific proteins actually do in the cell and how various biological processes actually occur.

As we write this, our colleagues are focused on a means to identify the genetic defects that influence the whole host of human ailments and characteristics. Yes, we want to understand the genetic defects that result in muscular dystrophy or diabetes, but we also want to know if there

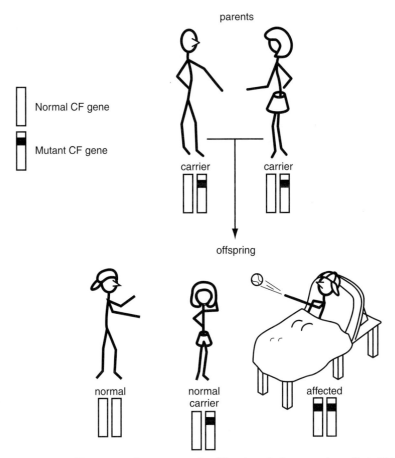

FIGURE 1.4   Two parents heterozygous for CF and producing normal or affected kids.

are genetic components that underlie our behavior, our personality, our creativity, our intelligence, and our sexuality. Progress on this massive endeavor has been both impressive and a bit frightening. Although our crystal ball is still clouded, it seems likely that in the near future genetic testing of a newborn will tell us not only the diseases to which it may become susceptible but much about its future personality and appearance as well.

One can see pluses and minuses here. Yes, a real understanding of the genetics of mental illnesses may soon provide clinicians with invaluable diagnostic tools and researchers with a deeper understanding of the biological basis of diseases such as schizophrenia and manic depression, which may enable us to successfully treat or cure these diseases. However, the same research will likely allow us to identify people at risk for these illnesses both before and after birth. Is there an obvious benefit to being able to tell parents that their beautiful newborn will likely develop schizophrenia two decades hence, without an available treatment? Might parents be likely to terminate such a pregnancy *in utero* even if such treatments did exist?

We also now possess the first rough-hewn tools to attempt gene replacement or "gene therapy." Can we cure cystic fibrosis by permanently replacing the damaged gene with a dose from our bottle of functional CFTR genes? Assuming that our tools get better, the door is open to begin to change the hereditary makeup of future generations. So perhaps the answer to the ethical conundrum raised by schizophrenia in the last paragraph is: "Don't worry, we'll just cure it by putting good genes in." Who will decide what the good genes are and who will get them?

We will return to each of these issues in time. Before we can really discuss them again, we need to have a more detailed knowledge of genes themselves. And so we begin. . . .

# 2

# MENDEL AND THE CONCEPT

# OF THE GENE

*In the beginning . . .*

We suspect that people have been curious about how heredity works ever since they figured out what caused it. (Which is a story in and of itself actually. Think about it for a minute or two; how do you suppose our ancestors drew the connection between sex and babies? Remember this was substantially before cable and MTV.)

It is important to note that our current sophistication in these matters is of fairly recent origin. We know that "like begets like" (Lucretius) or, in the words of Frank Stahl, "the fundamental observation of genetics is that dogs have puppies, cats have kittens, people have babies, and nothing else ever happens." This seemingly obvious knowledge that children will be like their parents would have been heresy to the ancient Greeks who wrote extensively on the progeny of matings between members of different species, such as between swans and sheep.

We also know that children share similarities with both of their parents. For long periods in our history, people imagined that children were the offspring of only one parent (either the mother or the father). There were schools of thought in which children were preformed only in their mothers; the father was thought to provide only a "vital spark" (much like jump-starting a dead battery). However, the early microscopists, all of whom were men, imagined that babies were preformed in the father and sailed their little sperm down the vaginal canal into awaiting uterine incubators. Indeed there are drawings of these tiny preformed men (called homunculi) piloting little sperm. Curiously, these drawings never show little preformed women.

These myths persisted despite the realization by farmers that offspring appeared to be a mixture of both of their parents. Face it, if you mate a big horse to a small horse, for the most part you get a medium-sized horse. This philosophy of heredity, known as blending, took a long time for us to incorporate into our views of our own heredity. Some texts claim that this understanding was forced upon our ancestors by the age of commerce and exploration and the resulting matings of people of light and dark skin coloration. These matings gave rise to children of intermediate coloration (regardless of which parent was light or dark in skin color). We are troubled by this explanation for two reasons. First, such matings must have been observed throughout human history. Second, skin color is just one of the hundreds (thousands) of noticeable human traits. Surely someone must have noticed the blending of other features. Nonetheless, by the mid-19th century, most people were willing to accept the concept that the traits observed in children were some mix of those observed in both parents and in both sets of grandparents.

As silly as it seems today, blending really was a very reasonable way to look at things. As stated earlier, if a large horse is mated with a small horse, odds are the colt will grow into a medium-sized horse. The same can be said for some traits in humans. People imagined that there was some kind of substance, such as blood, that blended in the offspring to produce a mixture of traits in the child. (Note the term "blood relative," which implies a shared ancestry, not a relative by marriage.)

Still there were some surprises that blending did not explain: blue-eyed kids born to brown-eyed parents, blond children of raven-haired moms and dads, kids who are taller than either parent, and so on. Blending, although a useful way to understand some traits like height and weight, just did not explain everything.

It was into this rather curious intellectual environment that Gregor Mendel was born. During his lifetime this man's intellect would boldly go where no person's mind had gone before. Like Galileo, Newton, Freud, and Einstein, Mendel's vision would change the course of human understanding. That vision comes from doing one unbelievably simple experiment.

*(Now we both know that all of you have heard about good old uncle Gregor since Montessori school. But please don't tune out here. We're going to show you how to understand pure genius, that magic moment in human cognition when it all becomes so painfully and beautifully obvious and clear. We can do that if you stay with us. Really we can. We promise. . . . RSH and CAM)*

## WHAT MENDEL DID

Mendel began with a specimen, the pea plant, which was simple to cultivate. He chose to study the inheritance of several simple and obvious traits, such as green vs. yellow peas. These were simple "yes-or-no" traits and not quantitative traits such as weight.

He also began with pure-breeding populations of plants. That is, he had a bunch of plants with yellow seeds that produced only plants with yellow seeds when bred to each other. Similarly, he had a bunch of plants with green seeds that produced only plants with green seeds when bred to each other. The experiment of concern to us is diagramed in Fig. 2.1.

Please note that in the first generation, when Mendel crossed plants with green seeds to plants with yellow seeds, all he saw in the progeny were plants with seeds identical in color to those of the green-seeded parent. NONE OF THE PLANTS HAD SEEDS OF AN INTERMEDIATE COLOR. It didn't matter which way the cross was made (i.e., green males crossed to yellow females or vice versa), all the offspring had green seeds. So much for the theories of uniparental inheritance. In addition, a real adherent to blending would have postulated that the progeny of the first generation should have been yellowish-green, not true green. Mendel's observations were simply incompatible with any sort of blending hypothesis.

In the subsequent generation, green-seeded plants crossed to themselves or each other produced both yellow- and green-seeded plants. Bang! Blending is dead. If you try to use the blending hypothesis to explain two green parents making a yellow offspring, you will realize that it just doesn't work. No way at all.

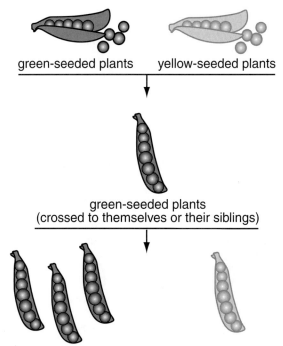

green-seeded plants     yellow-seeded plants

green-seeded plants
(crossed to themselves or their siblings)

75% green-seeded plants     25% yellow-seeded plants

FIGURE 2.1    The results of crossing green-seeded plants to yellow-seeded plants.

Please note that there is nothing unique to pea plants here. Fig. 2.2 displays a similar thought experiment for the inheritance of the human trait albinism. The same rules apply.

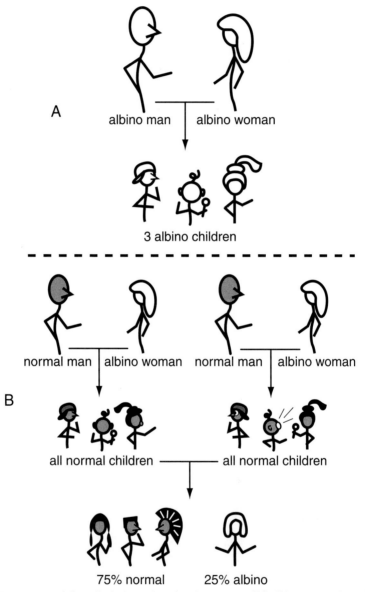

FIGURE 2.2    A hypothetical set of matings between an (A) albino man and an albino woman; (B) two albino women and two normal men; and (C) the marriage of their children.

In order to make sense of all this Mendel had to do three essential things. First he had to separate the genes that encoded a given trait (the *genotype*) from the physical manifestation of the trait itself (the *phenotype*). Mendel argued that there were discrete units of heredity, called *genes,* that were immutable and were thus passed down unchanged from generation to generation. He also argued that these genes specified the appearance of specific traits but were not the traits themselves. This insight gave rise to Mendel's first law: the purity and constancy of the gene. In simple terms, Mendel's first law says that you will pass on your genes, which you received from your parents, to your children in a precise and faithful fashion.

In order to explain differences in traits, Mendel supposed that genes could take different forms, called *alleles,* that specified different expressions of the trait. For example, Mendel claimed that there was a gene that gave seed color and two different forms or alleles of that gene: one specifying green color and one specifying yellow color. We will refer to those alleles that specify green color as *G* alleles and those that specific yellow color as *g* alleles. All individuals expressing green seed color must receive at least one copy of the G allele from their parents. Similarly, plants with yellow seeds must lack the *G* allele, having received only the *g* alleles from their parents.

In addition, an individual must also be able to carry genetic information for a trait it does not express. (We know that the green-seeded progeny produced by the first generation carried the information to produce yellow seeds because they were able to pass it on to the yellow-seeded progeny in the next generation.) Similarly, the normally pigmented kids of the marriages between normally pigmented and albino people described in Fig. 2.2 must carry the albino gene. They are then capable of passing that gene onto their children. Thus some plants with green seeds must carry both the *G* and the *g* alleles. The term *heterozygous* is used to denote this state (e.g., *Gg*) as opposed to *homozygous,* a term used to denote individuals carrying two identical alleles (*GG* or *gg*). *(Realize that heterozygotes are formed upon joining of an egg and sperm that carry different alleles of a gene, and homozygotes are formed by egg and sperm that carry the same alleles of a gene. — RSH)*

Mendel's second insight was that this pattern of inheritance could only be explained if green (*G*) alleles could mask the expression of the yellow (*g*) alleles, such that individuals getting a green allele from mom and a yellow allele from dad would be just as green as those that got green alleles, and only green alleles, from both parents. According to Mendel, those individuals who carried both green and yellow alleles will be green in appearance. As noted earlier, such *Gg* individuals are called *heterozygotes* because they carry two different alleles, whereas *GG* and *gg* individuals, who carry pairs of identical alleles, are called *homozygotes.* To denote this difference between the ability of alleles to determine a phenotype, Mendel

introduced the terms *dominant* and *recessive*. In this case, the *G* gene allele
is said to be dominant and the *g* allele recessive because *Gg* plants were
green in appearance.

Mendel's third insight was to assume that although there is but one gene
for each trait, each of us receives one copy of that gene from each parent.
We all thus carry two copies of that gene, one derived from mom and one
from dad. Similarly, our sperm or our eggs will carry only one of those two
genes. This is Mendel's second law: the law of the gene. You receive only
one copy of each gene from each parent and transmit one and only one in
each gamete.

So far so good. But how do we explain the fact that in the second
generation there are so many more green offspring than there are yellow
offspring? The answer lies in the statement made earlier: "We all carry
two copies of that gene, one derived from mom and one from dad. Similarly,
our sperm or our eggs will carry only one of those two genes." So an
individual of genotype *Gg* will produce both *G*- and *g*-bearing gametes
with equal frequency (Fig. 2.3). As shown later, if two such individuals

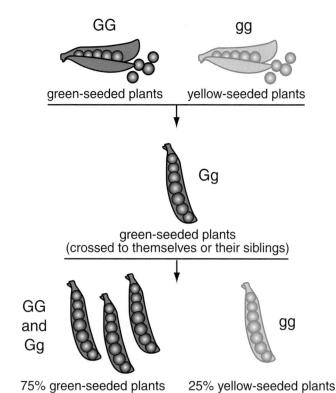

FIGURE 2.3    The genetics of crossing green-seeded plants with yellow-seeded plants.

mate, the odds of producing a *gg* offspring are only 1 in 4, or 25%. This is because only half of the sperm carry a *g* allele and only half of the eggs carry a *g* allele. So the odds of a *g*-bearing sperm getting intimate with a *g*-bearing egg are one-half times one-half or one-fourth.

As shown in Fig. 2.4, this can be diagramed by drawing a Punnett square. A Punnett square can be created by drawing the possible maternal alleles down the side of the box and the possible paternal alleles across the top. Each box in the grid then represents the combination of that paternal and that maternal allele. *(I am always amazed at the reluctance of people to use Punnett squares. I've been doing genetics professionally since January 1973 and I use them all the time. They save time and make it easy to see how a cross will come out.—RSH)*

By postulating the existence of these particles called genes, Mendel provided a basis to understand heredity. Recall that we wondered in the previous chapter how two normal parents could produce one or more children with cystic fibrosis. It is easy; the parents are both heterozygous for the normal and mutant alleles of the CFTR gene. Thus approximately one in four of their kids will be homozygous for the mutation that prevents the production of the CFTR protein. It is these children who will develop cystic fibrosis. Still, the question remains: What are these genes really and how, mechanically, does the Mendelian inheritance function? How can each sperm or egg be made to carry one and only one copy of each pair of genes?

We can drop the formalisms here. As shown in the next chapter, genes are not the weightless intellectual pebbles that Mendel envisioned. Rather they are composed of a molecule called DNA (Fig. 2.5). DNA is made up of pairs of molecules called bases arranged in a linear array, much as a novel, such as *War and Peace,* is made up of a rather long string of letters.

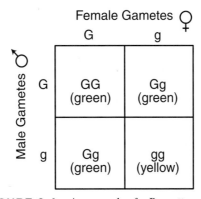

FIGURE 2.4    An example of a Punnett square.

FIGURE 2.5    An artist's conception of DNA.

Beating this analogy to death, let us equate the information encoded on all of the DNA molecules in a cell with a novel. We can then suggest that genes are to the sum of DNA molecules in the cell as individual sentences are to the novel. Each sentence (gene) is made up of many letters (base pairs) and many sentences are required to complete a novel (Fig. 2.6).

Each DNA molecule then includes many genes. For example, human beings posses somewhere on the order of 50,000 to 100,000 genes arranged

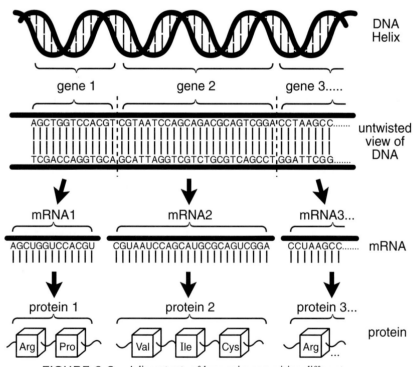

FIGURE 2.6    Adjacent sets of base pairs comprising different genes.

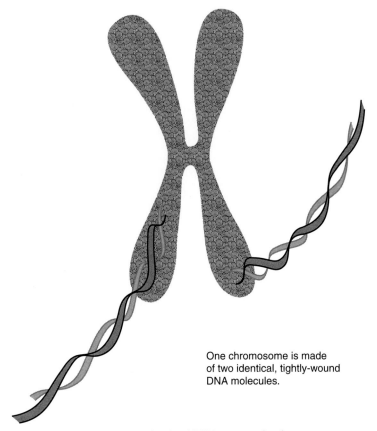

One chromosome is made
of two identical, tightly-wound
DNA molecules.

FIGURE 2.7   A molecule of DNA as part of a chromosome.

on 46 DNA molecules (the average human DNA molecule is 65,000,000 bp in length for a total DNA content of 6,000,000,000 bp). Chapter 3 will describe in detail how genes actually work; which is to say how they actually encode proteins.

Genes are not assorted by some random lottery-like device as brother Gregor would have liked us to believe. Individual DNA molecules in the cell are maintained as *chromosomes* that are composed of DNA and a great many types of proteins (Fig. 2.7). These chromosomal proteins act to package the DNA, to protect it from damage, to move it around during cell division, and to control its expression during cell division. As detailed in Chapter 4, the physical basis of Mendelian inheritance is the behavior of these chromosomes during the process of gamete formation.

# 3

## HOW GENES WORK:

### THE STORY OF HOW GENES ENCODE PROTEINS, OR "THE CENTRAL DOGMA OF MOLECULAR BIOLOGY"

As noted previously, genes are composed of DNA and their function is to encode proteins. (Evidence for both of these assertions is provided in Boxes 3.1 and 3.2.) The question is then: How can a DNA molecule provide the information required to encode a protein? We begin with an examination of the structure of the DNA molecule itself.

#### STRUCTURE OF DNA

DNA is an abbreviation for a zipper-like molecule called deoxyribonucleic acid. It is composed of two separate chains (molecules) that are wrapped around each other to form a structure rather similar to a spiral staircase. Like the two handrails of that staircase, the two chains are connected to each other by step-like structures called base pairs.

As shown in Fig. 3.1, each of the two chains that comprise the DNA molecule is composed of a very long repeating array of only four different subunits called *nucleotides* containing the bases adenine (A), thymine (T), cytosine (C), and guanine (G). Each nucleotide is composed of one base linked to a sugar. The sugar molecules of adjacent bases are linked together by phosphate molecules and the bases project away from the back bone like teeth in a zipper. Because the sugar molecules are asymmetric, each sugar–phosphate backbone has a polarity. The ends of such a chain are denoted as 5' and 3', corresponding to the portion of the sugar molecule that is exposed at each end in reference to this polarity.

The DNA molecules that make up our genes consist of two individual nucleotide chains wrapped around each other to form a double helix.

## BOX 3.1 EVIDENCE THAT DNA IS THE HEREDITARY MATERIAL

It took a surprisingly long time for most biologists to accept the fact that DNA was indeed the hereditary material. They were finally convinced by four lines of evidence. First, the types of radiation that can cause mutation are more efficiently absorbed by DNA than they are by other cellular constituents, such as proteins. Second, it was demonstrated by bacteriologists that when viruses (which are composed of DNA wrapped in protein coat) infect bacteria it is only the DNA of the virus that enters the bacterial cell, as the protein is left outside the cell and sloughed off. Thus upon entering the cell, the DNA has all of the information needed to cause the cell to synthesize new viruses (i.e., to direct the synthesis of both viral DNA and viral proteins). Third, it was demonstrated that unlike other molecules in the cell, the amount of DNA in gametes was precisely half of the amount of DNA found in other cells of the body. (Think about this for a moment; it is exactly what is required of the hereditary material based on the properties of Mendelian inheritance.)

The most important line of evidence came from the study of the genetics of the bacterium that causes pneumonia in mice (*Streptococcus pneumoniae*). Back before your parents were born, Frederick Griffith demonstrated that a harmless variant of this bacteria (form R) could be transformed into a virulent form (form S) that is capable of causing disease just by transplanting DNA from S to R form cells. (This is where we get the term *transformation* used by modern molecular biologists to describe the generalized phenomenon of being able to transform a cell's characteristic by putting DNA into the cell.)

The term "form R" refers to bacteria whose colonies appear to have a rough surface, whereas "form S" colonies have a glistening smooth surface. If R cells are injected into a mouse, nothing happens, the mouse's immune system just destroys them. However, injected S cells can evade the mouse immune system and kill the mouse. The difference between S and R cells is a cell surface protein that both protects the cell from the mouse immune system and makes colonies of these cells appear smooth and refractile. R cells lack the ability to make this protein, presumably as a consequence of a mutation in the corresponding gene.

Griffith showed this by simply growing R form cells in the presence of heat-killed S form cells. Griffith reasoned that R form cells had to be taking up something from dead S cells that transformed them into S cells *(sounds kind of like the plot to a bad horror/vampire movie: harmless R cells eat the rotting carcasses of the dead S cells and in doing so become deadly themselves).*

Several decades later workers at Rockefeller University (Professors Avery, MacLeod, and McCarty) determined that the ability of dead S cells to transform R cells into new S cells resided solely in the DNA of S cells. Thus, the transforming agent (or transforming principle as it was then called) was DNA. By taking DNA from S cells and putting it into R cells, one could stably and heritably alter the character of the recipient R cells. Unlike the first three lines of evidence, which seemed to convince only those who already wanted to believe in DNA, the experiments of Avery and colleagues convinced the entire scientific community.

## BOX 3.2 EVIDENCE THAT EACH GENE ENCODES A PROTEIN

Several lines of evidence show that the products of genes are proteins and that (with a few rather arcane exceptions) each gene encodes a single protein. The most important of these is the work of Beadle and Tatum on the genetics of fungus *Neurospora crassa*. These workers isolated a number of mutations that prevented this fungus from being able to synthesize an essential amino acid called arginine. Now it takes a number of specific chemical steps (like steps on an assembly line) for this bug to make arginine. Each of these steps is carried out (catalyzed) by a specific enzyme. (Recall that enzymes are a class of proteins that execute specific chemical reactions.)

What Beadle and Tatum showed was that their rather large collection of mutations fell into a number of different genes. (We will discuss how one shows that two mutations occurred in the same gene in Chapter 5; it is called a complementation test.) Most importantly, all of the mutations in a given gene knocked out the same enzymatic step and mutations in different genes knocked out different enzymatic steps. Thus each gene appeared to be responsible for synthesizing a single enzyme. This crucial insight was called the "one gene–one enzyme" hypothesis and was a major milestone in the development of modern molecular genetics.

The next crucial discovery came from the analysis of the abnormal hemoglobin protein that is produced by individuals carrying the mutation that causes sickle cell anemia. In 1957, Vernon Ingram demonstrated that the sickle cell mutation causes the production of a hemoglobin that carries a specific amino acid substitution, resulting in a nonfunctional protein. This is to say that a specific mutation was associated with a specific amino acid substitution.

Some years later Charles Yanofsky would demonstrate that each of his large collection of mutations in the gene that encodes the tryptophan synthetase enzyme in the bacterium *E. coli* was associated with a change at a specific site in the protein. More crucially, the order of amino acid substitutions within a gene corresponds precisely to the order of those mutations within the gene. This is not surprising because, as described *ad nauseam* in this text, the first 3 bp of the gene encode the first amino acid of the protein, the second 3 bp encode the second amino acid, and so on. Thus, if we change one of the second sets of 3 bp we will specifically change the second amino acid of the protein. This experiment exemplifies the experiments that underlie another crucial concept, namely "the colinearity of the gene and protein."

As noted in this chapter, eukaryotic genes are not precisely colinear with their proteins. The coding regions (exons) of most eukaryotic genes are interrupted by introns that must be spliced out before translation can occur. The existence of splicing has allowed some eukaryotic genes to evolve in such a way that a given gene can be spliced in multiple different ways, each of which produces a different mature mRNA molecule, and thus a different protein. Transcripts might differ, for example, in the inclusion of one or more exons or in the choice of several different terminal exons. This process is known as *differential splicing*. Organisms can use differential splicing to produce different forms of a given protein either in different cell types or at different times in development. So when we say that a single gene makes a single protein, you must realize that there are some genes in eukaryotes capable of producing several different versions of that protein.

These two nucleotide chains have opposite polarity and are oriented such that the bases project inward toward each other. The sugar–phosphate backbones are on the outside of the helix like banisters on a spiral staircase. *Most crucially, this structure requires that the bases from each chain be paired and the only allowable pairings are G=C and A=T.* A simplified drawing of this structure is provided in Fig. 3.2.

This structure of DNA was deduced by James Watson and Francis Crick in 1953. It was an accomplishment filled with gossip, brilliance, and even a bit of intrigue. It was also a fundamental leap forward that deserved and won a Nobel prize. For those of you who want to know a fuller version of the story of this discovery we recommend two books: *The Double Helix* by James Watson and *The Eighth Day of Creation* by Horace Freeman Judson.

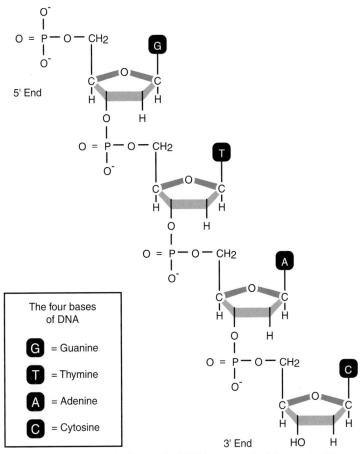

FIGURE 3.1    A single strand of DNA composed of four nucleotides.

## REPLICATION OF DNA

In addition to providing us with the chemical structure, this discovery also provided a molecular basis for heredity that easily explained the mechanism by which a gene replicates itself. Two rather prominent features of the DNA molecule, namely that the strands have opposite polarity and bases must be paired and those pairings must be A=T and G=C, allow us to visualize the faithful replication of DNA as nothing more than simply separating the two complementary strands and resynthesizing two new complementary strands (see Fig. 3.3). The replication of each of the original strands of the helix results in the production of two identical daughter DNA molecules each containing

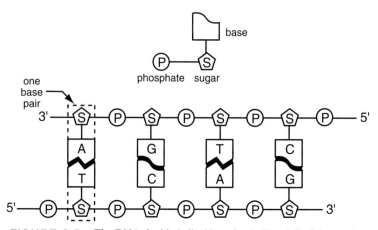

FIGURE 3.2    The DNA double helix. Note the A–T and G–C base pairs.

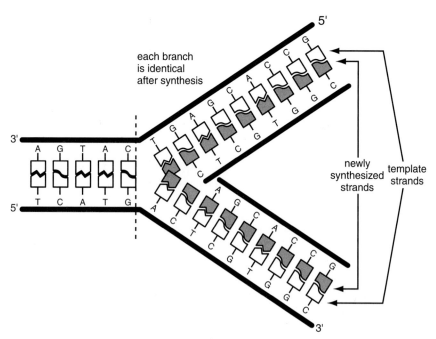

FIGURE 3.3    The replication of DNA.

one original and one newly replicated strand. Thus the daughter molecule with the original "Watson" strand will carry a newly replicated "Crick" strand and vice versa. This process is known as *semiconservative* DNA replication.

DNA replication is mediated by enzymes called DNA *polymerases*. Before each cell division these enzymes must replicate the entire human genome *exactly* once. The fidelity of this replication process is sufficiently precise that errors occur at frequency of less than one error in every 10,000,000,000 replicated bases. Realize that replication is a processive process in which new bases, complementary to the base on the original or *template* strand, are added sequentially to the 3' end of the replicating strand. As shown in Fig. 3.3, replication is initiated at a large number of sites in the genome, called *origins of replication,* and proceeds out from those sites in both directions along the DNA molecule.

## HOW GENES ENCODE PROTEINS

The sequence of base pairs in a gene determines the protein product that gene can encode. This specificity lies in the order of base pairs, sometimes called the *sequence* of the DNA, and changes in that order is the molecular basis of mutation.

However, a discussion of the mechanism of gene expression requires a brief discussion of the structure of our cells. We are *eukaryotes,* i.e., most of our genes in our cells are sequestered into a specific structure called a *nucleus* (see Fig. 3.4). (The exception to this rule is that our mitochondria, the tiny energy-producing organelles found in the cytoplasm of our cells, also contain small circular chromosomes that carry a number of genes required for mitochondrial function.) Although most of our cell's DNA is in the nucleus, the sites of protein assembly, called *ribosomes,* are located out in the remainder of the cell called *cytoplasm* (Fig. 3.5).

The first rule for running a healthy eukaryotic cell is "never let your DNA go wandering out into the cytoplasm, very bad things could happen to it out there and you might not get it back!" To solve the problem of getting information to the ribosomes the genes composing the nuclear DNA make portable and expendable copies of themselves in the form of a chemically and functionally different nucleic acid called *messenger RNA* (or mRNA), which is described below.

Thus, the order of information transfer goes DNA → mRNA → PROTEIN. Although there can be a lot of variation in the lifetime of a given mRNA or protein molecule, depending on its function, the DNA molecules serve as a permanent reservoir of information.

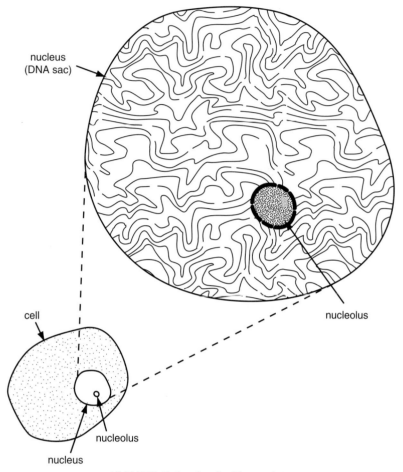

nucleus
(DNA sac)

cell

nucleolus

nucleolus

nucleus

FIGURE 3.4    A cell with a nucleus.

## PRODUCTION OF MESSENGER RNA: TRANSCRIPTION

mRNA molecules are one class of another type of nucleic acid, known as RNA (or ribonucleic acid). RNA differs from DNA in that the sugar molecule in the backbone is ribose instead of deoxyribose and that the base uracil (U) is used instead of thymine (T). Whether or not you know what the difference is between ribose and deoxyribose, the important point here is that a ribose backbone "looks" significantly different from a deoxyribose backbone to the machinery of your cells and completely different sets of

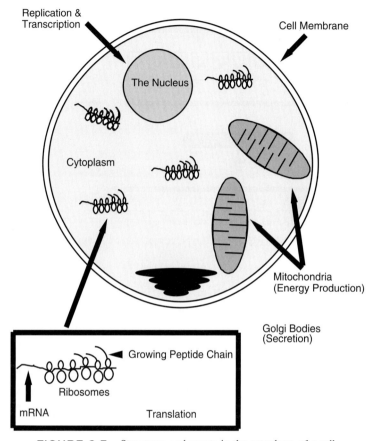

**FIGURE 3.5**  Structures and events in the cytoplasm of a cell.

enzymes and other biological processes are associated with the handling of the two kinds of molecules, DNA and RNA.

Messenger RNA molecules are synthesized from DNA molecules by a process called *transcription*. During transcription only one strand of the DNA corresponding to that gene (known as the template strand) is copied into a messenger RNA molecule (see Fig. 3.6). Obviously, each gene will produce a mRNA molecule that is complementary to the bases that compose the template strand of that gene. Unlike nuclear DNA molecules, most mRNA molecules are quite short lived. They go out to the cytoplasm, direct the synthesis of a protein and then are destroyed.

Transcription is similar to DNA replication in that the newly synthesized RNA molecule has a polarity opposite to that of the DNA strand from which it was derived. Transcription also depends on complementary base

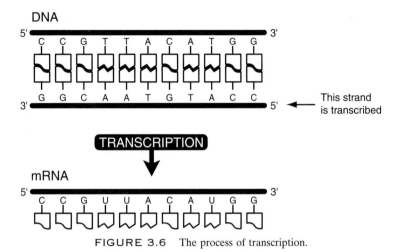

**FIGURE 3.6**    The process of transcription.

pairings, such as A = **U**, T = **A**, G = **C**, C = **G** (ribonucleotides are indicated in bold). Transcription differs from replication in that only one strand of the DNA molecule is transcribed. Thus the resulting RNA molecules are single stranded. Moreover, RNA molecules are relatively short compared to the huge chromosome-length DNA molecules that encoded them. The amount of transcription at any given gene can be carefully controlled by the cell. This control is the molecular basis of gene regulation and will be discussed in detail in a later chapter. The single-stranded RNA copy of a DNA molecule retains all of the information present in the DNA strand from which it was derived (this includes the base-pairing properties, in that U pairs with A).

The resulting mRNA molecules are transported out of the nucleus and into the cytoplasm. Once in the cytoplasm the mRNA molecules are loaded onto ribosomes where protein synthesis occurs by a process called *translation*. Ribosomes are composed of several types of RNA molecules called *ribosomal RNA* molecules (or rRNAs) and a large number of proteins called, not surprisingly, *ribosomal proteins*. Unlike mRNA molecules, rRNA molecules play no role in encoding the amino acid sequence of proteins. Rather they serve as integral components of the ribosome itself.

## MRNA MOLECULES DIRECT THE SYNTHESIS OF SPECIFIC PROTEINS: TRANSLATION

Like DNA, proteins are linear arrays of a set of subunits called amino acids that are joined together by structures called chemical bonds. However, unlike DNA, which is composed of only four types of nucleotides, there are 20 different amino acids. A list of amino acids is provided in Table 3.1. The rules by which the base sequence of the mRNA molecule is translated into the primary amino acid sequence of a protein are referred to as the *genetic code.*

The function of a protein within the cell is determined by the order and number of the amino acids that make it up. The order of amino acids determines how the protein will fold up, and its final three-dimensional structure determines its function. As noted earlier, some proteins, known as *structural proteins,* form scaffolds or skeletons within the cells on which other proteins can be arranged or transported. Other proteins catalyze specific biochemical reactions within the cell and are called enzymes. In order to perform either set of functions the order of amino acids within a protein must be specified

**TABLE 3.1**    A Codon Usage Table

| Amino acid | Codons |
|---|---|
| Alanine (Ala) | GCU, GCC, GCA, GCG |
| Arginine (Arg) | CGU, CGC, CGA, CGG, AGA, AGG |
| Asparagine (Asn) | AAU, AAC |
| Aspartic acid (Asp) | GAU, GAC |
| Cysteine (Cys) | UGU, UGC |
| Glycine (Gly) | GGU, GGC, GGA, GGG |
| Glutamic acid (Glu) | GAA, GAG |
| Glutamine (Gln) | CAA, CAG |
| Histidine (His) | CAU, CAC |
| Isoleucine (Ile) | AUU, AUC, AUA |
| Leucine (Leu) | CUU, CUC, CUA, CUG, UUA, UUG |
| Lysine (Lys) | AAA, AAG |
| Methionine (Met) | AUG |
| Phenylalanine (Phe) | UUU, UUC |
| Proline (Pro) | CCU, CCC, CCA, CCG |
| Serine (Ser) | UCU, UCC, UCA, UCG, AGU, AGC |
| Threonine (Thr) | ACU, ACC, ACA, ACG |
| Tryptophan (Trp) | UGG |
| Tyrosine (Tyr) | UAU, UAC |
| Valine (Val) | GUU, GUC, GUA, GUG |
| STOP | UAA, UAG, UGA |

precisely. Even the substitution of one amino acid can often completely inhibit the ability of a protein to function.

Basically the nucleotides of the mRNA molecule are read by the ribosome in such a way that each set of three nucleotides, called a *codon*, can specify a single amino acid. Thus the first three bases of the mRNA will encode the first amino acid of the protein, the second three bases the second amino acid, and so on. *Each three nucleotides, called a codon, specify the incorporation of one and only one amino acid.*

Obviously, as there are 64 different possible types of codons (each of the three bases can be A, U, C, or G, thus the total number of possible codon is 4 × 4 × 4 = 64) and only 20 different types of amino acids, some codons must code for the same amino acid (thus the code is said to be degenerate); however, no codon encodes more than one amino acid. A list of the codons and the amino acids they encode is provided in Table 3.1. Please note that three of the codons do not specify the incorporation of any amino acid. These codons, which are found at the end of the coding sequence contained on the mRNA (yes, there actually is a long tail of RNA beyond the end of the coding sequence), tell a ribosome to stop translating the message and release the assembled protein and are appropriately called *stop codons*.

The mRNA is read, or *translated,* from one end to the other, beginning at the 5′ end and proceeding one codon at a time toward the 3′ end (see Fig. 3.7). For our purposes, translation usually begins at the first *start codon* (AUG) encountered by the ribosome as it reads along the message. This codon directs the addition of the amino acid methionine (Met). As each successive codon is read, the ribosome incorporates the indicated amino acid into the growing protein. Translation stops when the ribosome encounters one of the three stop codons (UAA, UGA, or UAG) that do not specify the incorporation of an amino acid.

## HOW MUTATIONS ALTER SEQUENCE OF PROTEINS

We hope we have convinced you by now that the translation process has a defined reading frame based on a triplet code. Thus, a mutation in a gene that changed the sequence of the base pairs (say an A → G mutation) would also change the codon in which that base pair resides. As a consequence the altered codon might specify the incorporation of a different amino acid. Mutations that direct incorrect substitutions are referred to as *missense mutations.* An example of a missense mutation is shown below (but see also Fig. 3.8).

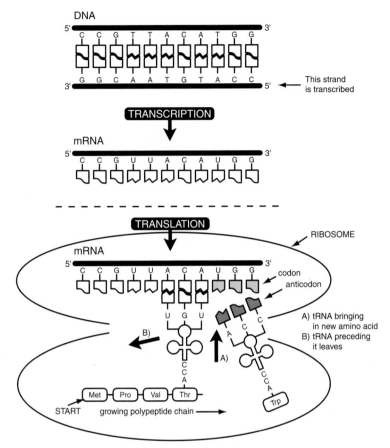

**FIGURE 3.7** The process of translation.

| Part of the normal gene | Same part of the mutant gene |
| --- | --- |
| ATTAGGT**A**CGTATGTGAT<br>TAATCCA**T**GCATACACTA | ATTAGGT**G**CGTATGTGAT<br>TAATCCA**C**GCATACACTA |
| *Part of the mRNA from the normal gene* | *Part of the mRNA from the mutant gene* |
| AUUAGGUACGUAUGUGAU<br>read as<br>AUU/AGG/UAC/GUA/UGU/GAU | AUUAGGUGCGUAUGUGAU<br>read as<br>AUU/AGG/UGC/GUA/UGU/GAU |
| *Part of the protein encoded by the normal<br>   gene* | *Part of the protein encoded by the mutant<br>   gene* |
| Ile–Arg–**Tyr**–Val–Cys–Asp | Ile–Arg–**Cys**–Val–Cys–Asp |

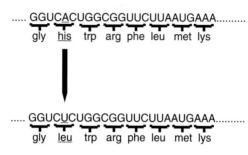

A MISSENSE MUTATION

..... GGUCACUGGCGGUUCUUAAUGAAA.........
gly  his  trp  arg  phe  leu  met  lys

..... GGUCUCUGGCGGUUCUUAAUGAAA.........
gly  leu  trp  arg  phe  leu  met  lys

A single base change in the mRNA,
due to a base change mutation in the
DNA, results in the incoproration of a
different amino acid.

FIGURE 3.8    A missense mutation.

When we consider why changing an amino acid might cause any problems, we need to keep in mind that the functions of a protein depend on its folded shape, which isn't just a long line of amino acids like a rope but in fact is a twisted, folded, complicated structure. Each of the amino acids has a different set of biochemical properties that help determine the final structure and function of the protein. Examples of why certain types of amino acid replacements might (or might not) be deleterious are listed below.

- Some of the amino acids are small. Others are large and will fill up a lot of space in their immediate vicinity. Putting a large amino acid into a space normally filled with a small amino acid can interfere with functions taking place in that part of the protein.
- Some amino acids have chemical properties, such as positive or negative charges, that will cause them to be attracted to or repelled from other amino acids in the protein chain. Putting a negatively charged amino acid in the place of an uncharged amino acid can change how charged amino acids in other parts of the protein interact with the substituted amino acid.
- Some amino acids are more likely to be found in portions of the protein that bend or fold, so replacing an amino acid that causes bending in the protein chain with an amino acid that does not cause or allow bending can change the shape of the molecule.
- And finally, because some amino acids have very similar properties of size and charge and a tendency to bend, some amino acid substitutions might not cause much of a change in the final shape or function of the protein.

Note that in the example just given, a single mutation caused the incorporation of a cysteine (Cys) instead of a tyrosine (Tyr) at position 3. What is that change going to do to the developing protein? That depends on how important the tyrosine residue was. If the tyrosine residue was in a crucial portion of the protein (e.g., in the catalytic active site of a protein), the resulting protein would be completely inactive. Cysteines also have the ability to form strong chemical bonds with other cysteine residues. Such bonds may play a crucial role in creating or maintaining that protein's three-dimensional structure. The loss of such a cysteine might well then create an inactive or partially active protein or it might be that the cysteine residue was in a fairly nonessential part of this protein. In that case this mutation might have little or no effect on protein function.

Suppose this protein made up part of the set of pigments that give hair its color. The mutational change might just alter the way in which the pigment protein absorbed and emitted light. In doing so this mutant allele might result in a different hair color. It all depends on the nature of the protein and on the nature and position of the amino acid substitution.

We should also note that some mutation will convert a codon into one of the three stop codons. Such mutations will cause the ribosome to stop adding amino acids to the growing protein, resulting in the release of a truncated and usually inactive protein. This type of mutation is referred to as a *nonsense mutation*. An example of a nonsense mutation may be seen in Fig. 3.9.

Because the mRNA is read in codons of three bases, mutations that cause the insertion of a single nucleotide, or the deletion of a single base pair for that matter, would throw the ribosome out of frame. As shown in Fig. 3.10, a deletion of the first base in one of the codons would cause the ribosome to read a new codon consisting of the second and third bases of that codon followed by the first base of the next codon. Not only that, but every codon after that would be misread in the same way, resulting in a kind of molecular

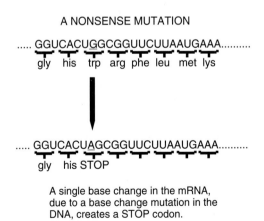

A NONSENSE MUTATION

..... GGUCACUGGCGGUUCUUAAUGAAA.........
  gly   his   trp   arg  phe  leu  met  lys

..... GGUCACUAGCGGUUCUUAAUGAAA.........
  gly   his STOP

A single base change in the mRNA,
due to a base change mutation in the
DNA, creates a STOP codon.

FIGURE 3.9    A nonsense mutant.

## A FRAMESHIFT MUTATION

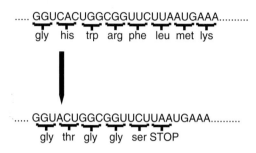

..... GGUCACUGGCGGUUCUUAAUGAAA.........
gly his trp arg phe leu met lys

..... GGUACUGGCGGUUCUUAAUGAAA.........
gly thr gly gly ser STOP

A single base deletion in the mRNA,
due to a base pair deletion mutation in the
DNA, creates a 'gibberish' sequence
including a STOP codon.

FIGURE 3.10   A frameshift mutation.

gibberish as the ribosome adds a completely different set of amino acids to
the growing chain. If a stop codon turns up in one of these new out-of-frame
codons, then the amino acid chain will not only differ in sequence from the
sequence in the correct chain, but it is also likely to be a different length of
amino acid chain from the original protein. Such mutations are thus referred
to as *frame-shift mutations*. However, the insertion of any three bases (or
multiple of three) will specify the incorporation of an additional amino acid,
but it will not throw off the reading frame. Similarly, deletions of three bases
will remove an amino acid but will not mess up the reading frame.

So the code is read three bases at time. Fine. But just how does that work?

## TRANSLATION REQUIRES AN ADAPTER
## MOLECULE CALLED тRNA

The codons in an mRNA molecule cannot and do not directly recognize
the amino acids whose incorporation they direct. This process instead depends on a class of small RNAs called *tRNAs* that serve as adapters. It
turns out that there is a specific tRNA molecule for each possible codon.
The exceptions to this rule are the three stop codons; no tRNA molecules
exist that can read these codons. A set of three bases at one end of each
type of tRNA molecule is called the *anticodon* and can base pair with the
complementary codon in the mRNA, whereas the free 3' end of the tRNA
is attached to the appropriate amino acid (see Fig. 3.7). Thus tRNA molecules are really the translators here. They have a codon-reading anticodon
at one end and the appropriate amino acid at the other.

A ribosome ratchets along the mRNA reading it codon by codon. At each codon it searches for a tRNA molecule whose anticodon is complementary to that codon. That tRNA brings along the appropriate amino acid, which is then incorporated into the growing polypeptide chain. Once this is done the spent tRNA molecule is released (to eventually be recharged within the cell by adding back the appropriate amino acid) and the ribosome ratchets onto the next codon. This process continues until the ribosome reaches a *stop codon,* for which there is no corresponding tRNA, and the ribosome releases both the mRNA molecule (perhaps to be translated again) and the completed protein.

The important concept here is that the tRNA molecule functions as an interpreter. It reads codons (words) in the mRNA molecule and translates them into amino acids. This interpreting permits a string of codons to specify a string of amino acids. Moreover, it allows us to infer from the sequence of base pairs with a given gene the amino acid sequence of the protein encoded by that gene. [Obviously, because the code is degenerate (i.e., several codons specify the same amino acids), we cannot go backward from the amino acid sequence to the DNA sequence.]

## GENE REGULATION

### VIRTUALLY EVERY CELL IN YOUR BODY
### CONTAINS A COMPLETE COMPLEMENT OF GENES

Not all genes are or should be on in every tissue. Moreover, a given cell or type of cell may need to express some genes at some times and not at others. Think about this for a moment. You don't need eye pigment proteins in your skin cells or the ability to produce hair in the cells that make up your tongue. Instead each cell in your body expresses only a distinct subset of its genes at any one time. For example, although all cells need to produce the enzymes necessary for basic cell maintenance and metabolism, only red blood cells need to produce hemoglobins and only the cells of the retina in the eye need to produce the light-sensitive proteins, such as rhodopsins, that permit us to see. This is not to say that liver cells lack the genes for hemoglobin or rhodopsin; rather liver cells simply do not express these genes (i.e., they do not make RNA transcripts from those genes). Instead the liver cell expresses its own specific repertoire of genes needed to make the proteins that carry out the specific functions of the liver.

To achieve this diversity of protein function during normal development, different cells express different sets of genes in a precisely regulated fashion. The effect is similar to the ability of the same orchestra (genome) to play a multitude of different symphonies (cell-type-specific gene expression) by playing different instruments (genes) in a different pattern. Did we beat that metaphor to death or what?

## THE CONCEPT OF A PROMOTER

Gene regulation occurs primarily at the level of transcription. A given cell transcribes only a specific set of genes and not others. Regulation can also be imposed by modulating how often an active gene is transcribed (i.e., how many RNA copies of that gene are made in a given interval of time).

As shown in Fig. 3.11, transcription begins by the binding of a very large enzyme called the *RNA polymerase* to a site upstream of the gene called the *promoter,* which is some number of bases away from the beginning of the coding region of the gene. Once the polymerase is bound to the pro-

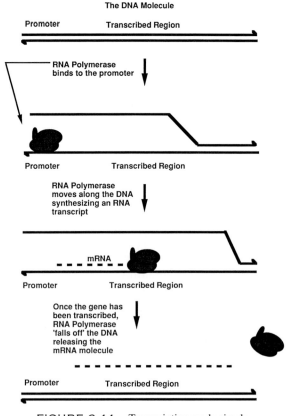

FIGURE 3.11    Transcription made simple.

moter, it moves along the DNA, making a single mRNA copy from only one strand of the DNA double helix. When the RNA polymerase reaches the end of the DNA that comprises the gene, it detaches from the DNA and releases the newly made mRNA molecule. The mRNA molecule is then chemically modified in several fashions by a set of events referred to as RNA processing before it is exported from the nucleus to the cytoplasm where it is translated.

Most organisms control gene expression either by regulating the ability of the RNA polymerase to obtain access to the promoter or by regulating the ability of the RNA polymerase to begin transcribing the gene. We begin by considering an example of gene regulation in the bacteria *Escherichia coli.* This example deals with the regulation of a gene, called the *lacZ gene,* that encodes a protein called *β-galactosidase,* which is essential for using the sugar lactose to produce energy (i.e., *lactose metabolism*). It is important to understand that the β-galactosidase enzyme is only useful to the cell when it is growing in an environment that contains lactose. There is no point in the bacterium transcribing the lacZ gene, and thus producing this protein, unless there is lactose in its growth medium.

*E. coli* controls the expression of the lacZ gene by a specific regulatory protein called a *repressor.* This protein binds to the LacZ gene at a site between the promoter and the beginning of the coding sequence. This repressor binding site is called the *operator.* By binding to the operator, the repressor protein physically blocks the polymerase from transcribing the gene (Fig. 3.12A).

Regulation of the lacZ gene results from the fact that whenever lactose is in the medium and taken up by the cell, it inactivates the repressor protein. A small number of lactose molecules bind to the available repressor proteins, which makes them unable to bind to the operator (Fig. 3.12B). The RNA polymerase molecule is now free to proceed down the lacZ gene and produce a message that will encode β-galactosidase.

The lactose molecules in the cell are then metabolized by the newly produced β-galactosidase enzyme. As the concentration of lactose in the cell diminishes, there are fewer and fewer lactose molecules that can bind to the repressor (they are all quickly being eaten by β-galactosidase). Soon there is no lactose left in the cell and the repressor molecules return to their normal function, namely binding to the operator and turning off the transcription of the lacZ gene. This cessation in lacZ transcription causes the cell to stop producing β-galactosidase.

The regulation of the lacZ gene can then be summarized as follows:

add lactose → inactivate repressor → transcribe lacZ
consume lactose → reactivate repressor → stop transcription

A    In the absence of Lactose the Repressor Binds to the Operator, Blocking
     RNA Polymerase

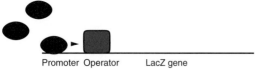

Promoter Operator        LacZ gene

B    In the presence of Lactose, the Repressor Binds to Lactose and NOT to the Operator,
     Allowing the Gene to be Transcribed

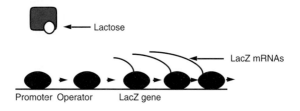

Promoter Operator        LacZ gene

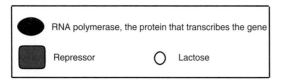

FIGURE 3.12    Regulation of the transcription of the lacZ gene in *E. coli.*

Try to realize that this is some ways more of a rheostat (or "dimmer"
switch) than a simple on/off switch. The cell can precisely adjust the amount
of lacZ transcription to the amount of lactose available.

## THINGS ARE MORE COMPLEX IN
## HIGHER ORGANISMS

Gene regulation in complex organisms like ourselves also depends primarily on regulating gene expression at the level of transcription. In eukaryotes, a variety of different kinds of regulatory sequences can be found
either upstream or downstream from the gene, and sometimes even within
the gene, but are more commonly found upstream from the promoter. Most
genes are controlled by a number of specific DNA sequences or *regulatory
elements* (RE) in the promoter region that act like "on" switches, "off"
switches, "dimmer" switches, or boosters of expression. A specific protein
called a *transcription factor* binds to each type of RE and acts to regulate

the process of transcription (Fig. 3.13). REs may function as negative or positive regulators of the gene. Indeed, there are cases in which one type of transcription factor binding to a given regulatory element upregulates transcription while the binding of other factors downregulates the expression of that same gene. Regulatory elements may have large or small effects on gene function and often function in combinatorial patterns.

Each different class of regulatory element, and there are many classes of REs, confers a specific temporal and tissue-specific pattern of transcription on the gene or genes that it regulates. This to say that some REs are tissue specific — they will activate their gene(s) only in a given tissue or set of tissues. There are, for example, REs that only function in white blood cells or in nerve cells. In all other cell types they are silent. A gene that is only active in one cell type might be regulated by only a single tissue-specific RE. Most genes, however, are active at numerous times during development and in more than one tissue. These genes are usually associated with multiple REs, each of which can control that gene's expression at a given time or in a given tissue. There are also cases where expression in a single tissue requires the function of two or more different REs. This pattern of gene regulation allows for a combinatorial control of gene expression in which the same gene can be regulated by different REs in different tissues. It allows for groups of genes that share a common RE or set of REs to be coordinately regulated.

As noted earlier, RE sequences function by binding specific proteins called transcription factors. Transcription factors are DNA-binding proteins

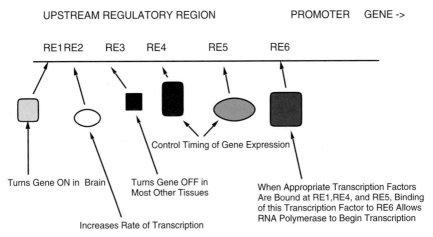

FIGURE 3.13 Enhancers and transcription factors in eukaryotic cells. A schematic diagram of the upstream regulatory region for a brain specific transcript is provided.

that possess the ability to bind only to very specific DNA sequences. Some of them will bind only if other specific transcription factors are also present. There are a very large number of different kinds of transcription factors, with colorful descriptors such as zinc-finger proteins and leucine zippers.

As was true for the lactose repressor proteins, transcription factors can also be activated or inactivated by binding to other molecules. For example, some hormone receptors act as transcription factors that function only when bound to a hormone. This allows the cell to activate a specific set of genes in the presence of the hormone and to leave them inactive in its absence.

As shown in Fig. 3.14, eukaryotes also differ from prokaryotes in that the coding sequences along the gene, and within the initial transcript (called *exons*), are interspersed with noncoding sequences (called *introns*). Once the initial transcript is produced, the introns are spliced out to form the completed and translatable message called an mRNA. The function of introns is not well understood, although in some cases regulatory elements have been found to lie within introns. However, because introns can be quite large and numerous, some genes can be much larger than the final mRNA molecule might suggest. This is especially true for the gene DMD, which, when mutated, produces muscular dystrophy. This gene is approximately 2.5 million bp in length, has over 70 introns, and encodes a message some 17,000 base pairs in length. We shall return to the DMD gene in Chapter 16 (Box 3.3).

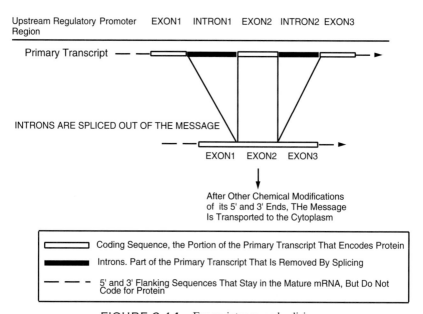

FIGURE 3.14    Exons, introns, and splicing.

## BOX 3.3 JUST WHAT DOES THE HUMAN GENOME CONSIST OF?

We talk about genes in this chapter in a way that should tell you that in eukaryotes we find noncoding DNA not only between genes but also within genes (in the introns). What may not be clear is just how much noncoding DNA there is between and within. Since the genes have not yet all been cloned and sequenced, we still do not know just how many there really are. But most people guess that the upper limit for the total number of human genes is 100,000. The average size of an mRNA is around 1500 base pairs, although, as we shall see in Chapter 15, some genes can be much much larger. Simple multiplication, then, argues that only 150,000,000 base pairs of the 3,000,000,000 base pairs in the human haploid genome gets turned into mRNA. What is the rest of all that DNA doing? Are there really 2,850,000,000 base pairs in the human genome that do not end up becoming mRNA? Perhaps not; the answers are not all "in" yet. However, it is clear that a large fraction of the DNA spread along the human chromosomes is located within introns or between genes.

What is all of that DNA doing, and what kinds of sequences are there if they are not coding sequence or some other part of the mRNA? Some of these sequences are what we call single copy, occurring once in a haploid genome. Many of the genes in the human genome are single copy. Other sequences that are present more than once and less than a million times are called moderately repetitive. And sequences present more than a million times are called highly repetitive. Regions of hundreds of thousands of based pairs of repeated sequence are often interspersed with long stretches of single-copy sequences.

Some of the biggest clusters of repeated sequences are near the centromeres, the sites on the chromosomes that allow them to move (see Chapter 4). Other sequences that are repeated are genes such as ribosomal RNA genes which occur in clusters on the short arms of chromosomes 13, 14, 15, 21, and 22 of the human genome (rRNA can sometimes make up 80% of the RNA in a cell). One of the most famous repeat sequences in human DNA is called the Alu repeat. About half a million of the 300 basepair Alu sequences can be found throughout the human genome, and some rare mutations in human genetic diseases turn out to be the result of insertion of a copy of an Alu repeat into the coding sequence of a human gene.

Many of the repeated sequences are varying lengths of simple sequences such as

CACACACACACACACACA
or
CAGCAGCAGCAGCAGCAGCAGCAGCAG.

The simple sequence repeats are usually in between the genes, in introns, or in parts of the mRNA that do not get translated into protein sequence. However, these simple sequences, especially the ones based on unit lengths that are multiples of three, are sometimes found in coding sequences and some very severe human genetic diseases have been found to result when simple sequence repeats get substantially longer than normal (see Chapter 10).

There is still a lot that we do not know about all of the "extra" DNA between and within the genes. It has been argued that some of it is junk DNA, serving no useful purpose to the organism. It has also been suggested that Alu sequences could serve as origins of replication, focal points at which copying of the chromosomal DNA would begin. A currently popular view is that some blocks of repeat sequences may affect the overall structure of the DNA over large regions and thus control expression of genes in adjoining regions. However, none of these theories have been proven so we must all continue waiting for the answer to the question, "What is all of that DNA doing?"

## SUMMARY

This chapter tried to present a brief sketch of both the physical structure of genes and the manner in which genes encode proteins. We have told you that one strand of DNA is copied (transcribed) into a complementary RNA molecule. In eukaryotes, that RNA is then spliced to remove the introns and is otherwise chemically modified at its ends to make the mature mRNA molecule. That mRNA molecule is then transported to the ribosome where, with the help of tRNA adapter molecules, the mRNA is translated into a protein. Each set of three bases specifies a specific amino acid. We have also given you a glimpse of how the process of transcription can be controlled. The next section presents a discussion of how these genes are transported during cell division and during gamete formation.

# 4

# How Genes Move:

## Chromosomes and the Physical

## Basis of the Laws of Mendel

The DNA within the nucleus is arranged in structures called chromosomes. Each chromosome, prior to its replication, is composed of only one long DNA molecule that contains thousands or tens of thousands of genes arranged in a linear array. After replication, and before cell division, the chromosome consists of two identical DNA molecules called *sister chromatids* (Fig. 4.1).

Human beings carry 46 chromosomes per cell (Fig. 4.2). These chromosomes are in reality 23 pairs of identical chromosomes. Forty-four of these chromosomes are truly members of 22 sets of identical pairs and are called *autosomes*. All of us have two copies of each autosome. The remaining two chromosomes are referred to as *sex chromosomes,* specifically the X and Y chromosomes. This pair of chromosomes differs in males and females, such that males carry an X and a Y and females carry two X chromosomes. We will return to the function and behavior of the sex chromosome pair in Chapter 6.

One member of each pair of chromosomes is provided by the individual's mother and one by the individual's father. Thus each sperm or each egg carries only 23 chromosomes: one member of each pair. Note that this is simply a restatement of Mendel's second law (i.e., each individual possesses two copies of each chromosome). During sperm or egg production, one, and only one, of those two chromosomes is destined to be packaged in each gamete. A typical cell with two copies of each chromosome is said to be *diploid* and a cell with only a single copy of each chromosome is said to be *haploid.*

To begin our explanations of these phenomena we start with a discussion

A DNA MOLECULE

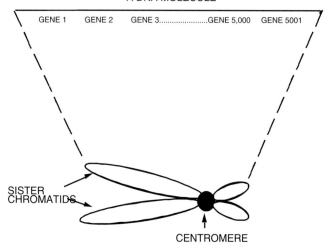

A CHROMOSOME AT THE START OF MITOSIS IS COMPRISED OF TWO
SISTER CHROMATIDS. EACH SISTER CHROMATID CONSISTS OF A SINGLE
LONG DNA MOLECULE. GENES ARE ARRANGED ALONG THAT DNA MOLECULE
LIKE BEADS ON A STRING.

FIGURE 4.1    A chromosome.

of a basic process of cell division that creates identical cells. We have all
repeatedly used this process, called *mitosis,* to develop into large animals
from the single-celled zygotes that resulted from *that* sperm finding *that* egg.

## THE BASIC PROCESS OF CELL DIVISION
## IS *MITOSIS*

Let us begin by imaging a very simple organism consisting of only a few
cells whose genes are arranged on only a single pair of identical chromosomes. One member of this pair of chromosomes came from its mother
and the other member came from its father *( please don't worry about how
such an organism can have a full and meaningful sexual life; trust me, it
can—RSH ).* During most of the cell's life these DNA molecules are loosely
entwined with each other in the cell nucleus, going about the gentle business
of running various aspects of metabolism and growth. During this time,
the chromosomes are not visible as separate entities; rather the nucleus
looks like an old brillo pad *(if you have never seen a brillo pad, clean your
kitchen now!—RSH and CAM ).*

Suppose that one cell in this organism needs to divide in order to form
some necessary structure consisting of two or more cells. The process of
cell division is known as mitosis and involves the completion of a series of

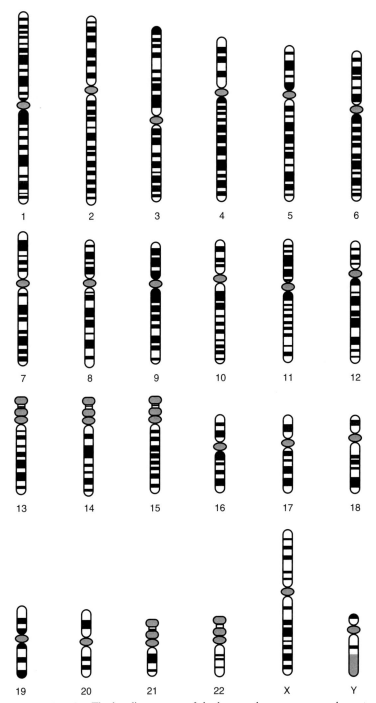

FIGURE 4.2    The banding pattern of the human chromosome complement.

cellular events known as a *cell cycle*. At the beginning of the cell cycle there are two DNA molecules and thus two chromosomes. The DNA molecule in each chromosome is replicated once during each cell division to form a pair of *sister chromatids* (each sister chromatid possesses a complete copy of the DNA molecule). As shown in Figs. 4.1 and 4.2, these sister chromatids are held together by a specific site on the chromosome called the *centromere*. The centromere is the motor apparatus of a chromosome that both controls sister chromatid cohesion and allows chromosome movement during cell division.

Once replication is complete the chromosomes begin to condense and become visible under the microscope as distinct entities. The process is called *prophase*. Although all chromosomes possess a centromere, they do differ in terms of the position of that centromere. There are chromosomes with the centromere in the middle (metacentric), toward one end (acrocentric), or at the end (telocentric). Chromosomes can often be distinguished from each other by their length, by their centromere position, and by a staining process called *banding* (see Fig. 4.2) that produces light and dark bands rather like the bar codes read by automated check-out systems in a grocery store. Banding is a modern-day alchemy in which chromosomes that have been affixed to the surface of a glass microscope slide are stained with dyes that bind to some regions along the length of each chromosome but not to others. By virtue of this technique all 23 different human chromosomes can be differentiated. Of course both members of a pair of autosomes (known as homologous chromosomes) will have the same length, centromere position, and banding pattern.

As prophase continues, the membrane around the nucleus disappears. The cell also assembles a scaffold called a spindle, made of microtubules, on which chromosome movement will occur. Microtubules are like train tracks along which the centromeres can pull the chromosomes. By *metaphase* the centromeres of each sister chromatid are connected by microtubules to both poles of the spindle, such that one sister chromatid is oriented toward each pole (see Fig. 4.3). As a result of these attachments to the poles of the spindle, chromosomes have lined up in a single row along the equator of the spindle (also known as the *metaphase plate*).

The next step in cell division is known as *anaphase* (see Fig. 4.4). All of a sudden the two sister chromatids completely split and the two chromatids move rapidly to opposite poles. We are just now beginning to figure out how this works. Suffice it to say that there are proteins at the centromere that function as motors and pull the chromatids toward the poles of the spindle along microtubule tracks.

The phase of the cell cycle that occurs once the chromatids have reached the poles of the spindle is called *telophase*. At this point, actual cell division, called cytokinesis, occurs (Fig. 4.5). There are now two daughter cells identical in genotype and DNA content to the original cell.

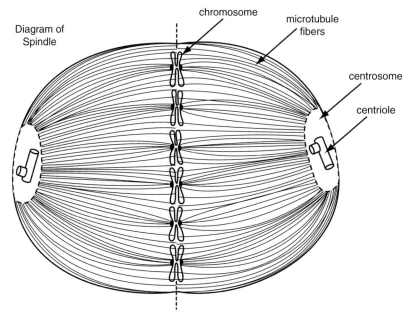

**FIGURE 4.3**   A schematic diagram of chromosomes on the spindle at metaphase.

It is this process of mitosis that allows individuals to develop from a single-celled *zygote* (the product of sperm and egg fusion) to a complex organism with "gazillions" of cells, all of which are genetically the same. In humans this process is somewhat more complex than in the simple animal we have just considered. For example, in a human mitosis there are 46 chromosomes at the metaphase plate, but the basic pattern of events in the cell cycle is the same.

So far so good. But suppose this fictitious little critter gets the urge to mate. Now it needs to make a gamete. In order to do things the Mendelian way, it needs to produce a cell with only one copy of each chromosome pair or, in this particular case, one chromosome. This requires a rather different process of cell division known as *meiosis* (see Fig. 4.6).

## MEIOSIS MADE SIMPLE

Meiosis actually encompasses two cell division events, cleverly called the first and second meiotic division. The early events of the meiotic cycle, such as DNA replication, are the same as those observed in the mitotic cycle. However, during first meiotic prophase, homologous chromosomes

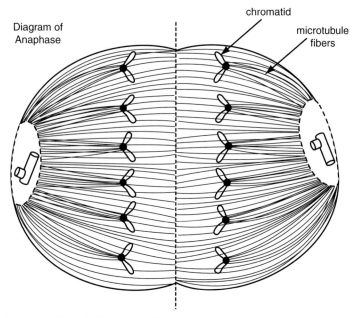

FIGURE 4.4    The spindle presented in Fig. 4.3 advanced to anaphase. The sister chromatids split apart and move to opposite poles.

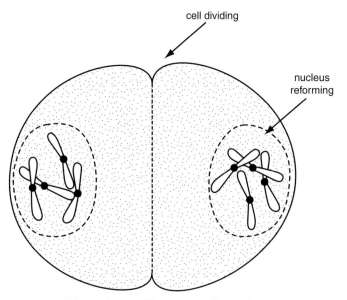

FIGURE 4.5    The same spindle at telophase.

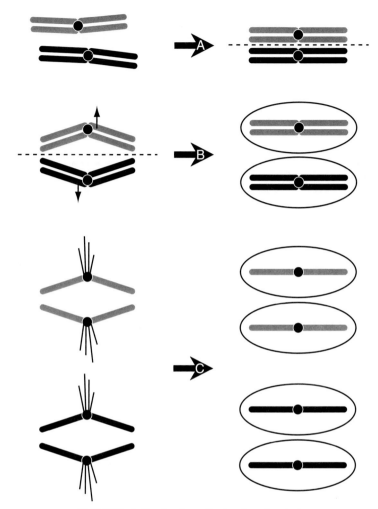

FIGURE 4.6    A schematic drawing of meiosis.

pair along their entire length; this does not normally occur in mitotic cells. These paired chromosomes, known as *bivalents,* are allowed to exchange large regions of homologous DNA by a process known as *recombination, exchange,* or *crossing over* (see Figs. 4.7 and 4.8). These recombination events serve to link homologous chromosomes together and ensure their segregation from each other at the first meiotic division.

   *(Please make sure that you understand this point! The function of meiotic recombination is to ensure homologous chromosome segregation. Meiotic recombination does this by taking advantage of the fact that sister chromatids*

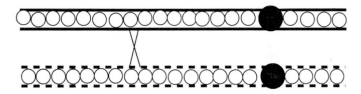

TWO HOMOLOGOUS CHROMOSOMES UNDERGOING A RECOMBINATION EVENT

● CENTROMERE   ○ PROTEINS THAT HOLD SISTER CHROMATIDS TOGETHER   ▬ ▬ ▬ ▬ CHROMATIDS

**FIGURE 4.7**   Meiotic recombination between two paired chromosomes.

*are held tightly together up until the beginning of the first meiotic cell division (Fig. 4.7). This sister chromatid cohesion is facilitated by a set of proteins that function to hold the two sisters tightly together. As the first meiotic division begins, sister chromatid cohesion is released along the arms of the chromosomes and maintained near the centromeres (Fig. 4.8). Thus, meiotic recombination uses sister chromatid cohesion to hold homologs together until the onset of the first meiotic division. In the absence of such exchange*

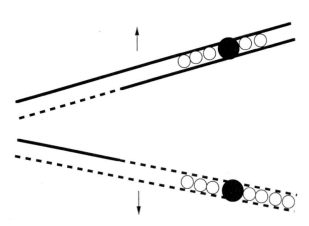

TWO HOMOLOGOUS CHROMOSOMES THAT HAVE UNDERGONE A RECOMBINATION EVENT
AS THE FIRST MEIOTIC DIVISION BEGINS

● CENTROMERE   ○ PROTEINS THAT HOLD SISTER CHROMATIDS TOGETHER   ▬ ▬ ▬ ▬ CHROMATIDS

**FIGURE 4.8**   The segregation of two homologous chromosomes that have recombined to opposite poles. Note the change in the pattern of sister chromatid cohesion.

*or recombination events, the homologs will often fail to go to opposite poles at meiosis I. Thus the function of recombination is to ensure chromosome segregation. I don't care what your professor told you about recombination and generating genetic diversity. They don't work on meiosis. I do!—RSH)*

Prophase of meiosis I is frequently divided into five substages: *leptotene* (the early stages of meiotic chromosome pairing and alignment), *zygotene* (the chromosomes are now aligned along their length), *pachytene* (the homologous chromosomes are intimately synapsed and more tightly condensed), and *diplotene/diakenesis* (the chromosomes now seem to become further condensed and may appear to begin to repel each other). However, we find that such terms are often confusing and not applicable to a given meiotic process (e.g., human female meiosis does not display a classical diplotene/diakenesis; see later). For these reasons we tend to think of meiotic prophase as consisting of three functional steps:

match them → lock them → move them
(pairing) (recombination) (alignment on the metaphase plate)

At first meiotic metaphase, it is bivalents and not individual chromosomes that line up on the metaphase plate. During the first division of meiosis each centromere is attached to only one pole by microtubule fibers (Figs. 4.7 and 4.9 and Box 4.2). In each bivalent the two homologous centromeres are attached to opposite poles. This implies that at this point in the meiotic cycle the centromere is doing something fundamentally different from what it does during mitosis (namely, the mitotic centromere is functionally differentiated into two sister centromeres, which can attach to two opposite poles, whereas the meiotic centromere can attach to only a single pole).

At first meiotic anaphase, the two homologs that comprise each bivalent separate and move to opposite poles of the spindle (Figs. 4.8 and 4.10, see also Box 4.3). This is the crucial meiotic event (why?) and explains Mendel's observation that only one copy of a given pair of alleles will be included

FIGURE 4.9    The arrangement of a bivalent on the metaphase plate. A schematic drawing of a centromere recombination event and microtubules. The directions of force are diagramed.

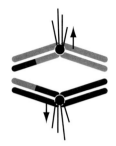

FIGURE 4.10    Anaphase I.

in a gamete (the other allele just went to the opposite pole at anaphase I). At this point each homolog still consists of two chromatids.

The chromosomes reach the pole at telophase and cytokinesis ensues. The result is a single diploid daughter cell that has divided to produce two haploid daughter cells, each with only half the normal number of chromosomes (Fig. 4.11). *The two daughter cells then proceed directly to prophase of the second meiotic division without undergoing DNA replication.* The remainder of this division follows the rules of mitosis. Each chromosome lines up on the metaphase plate with one sister chromatid oriented toward each pole. At the second meiotic anaphase the sister chromatids split and one sister chromatid moves toward each pole of the spindle (Fig. 4.12). After telophase, cytokinesis occurs (Fig. 4.13). The result is four haploid gametes carrying unreplicated chromosomes (i.e., each chromosome is composed of a single chromatid). Fusion of two such gametes will produce a diploid cell that is ready to begin the mitotic cycle and thus develop into a complex organism (Fig. 4.14).

We can now consider the meiotic process in terms of two or more pairs of genes. At least at first blush, there are two cases we need to consider: (1) when two genes map onto different pairs of chromosomes and (2) when both pairs of genes map onto the same pair of homologous chromosomes.

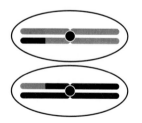

FIGURE 4.11    Telophase I.

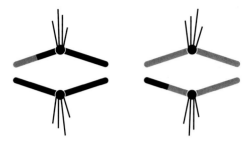

FIGURE 4.12    The sister chromatids split apart and move toward opposite poles.

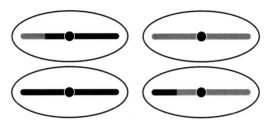

FIGURE 4.13    Telophase II.

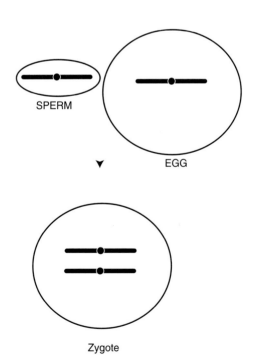

SPERM

EGG

Zygote

FIGURE 4.14    Fertilization.

Because the first case is both simpler and more common, we will consider it first.

## GENE PAIRS LOCATED ON DIFFERENT CHROMOSOMES SEGREGATE AT RANDOM

Mendel asserted that an individual of the genotype *AaBb,* where *A* and *a* are alleles of one gene and *B* and *b* are alleles of a *different* gene, will produce four types of gametes (*AB, Ab, aB,* and *ab*) with equal frequency. Note that the gamete carrying the *A* allele is as likely to carry the *b* allele as it is to carry the *B* allele. The same is true for gametes carrying the *a* allele. Mendel stated that the two gene pairs segregate independently such that there is no preference for a gamete to carry a particular combination of alleles. He referred to this rule of segregation as *independent assortment.* As shown in Fig. 4.15, independent assortment results from the fact that two bivalents will orient at random on the metaphase plate with respect to each other.

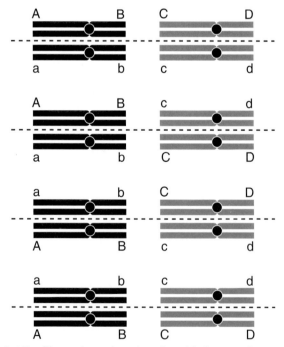

FIGURE 4.15   The random orientation of two bivalents on the metaphase plate.

Independent assortment gives rise to those icky ratios (such as 9:3:3:1) that we use to torture the pre-meds in most genetics courses. Imagine a mating of two individuals of the genotype *AaBb*. As described earlier, both of these individuals will produce four types of gametes (*AB, Ab, aB,* and *ab*) with equal frequency. As shown in Fig. 4.16, using a Punnett square, such a mating will produce 16 types of zygotes, some of which will be genetically identical.

Assuming that *A* and *B* are fully dominant, the progeny ratios are 9(*A_B_*):3(*A_bb*):3(*aaB_*):1(*aabb*). (The symbol *A_* indicates that the individual's genotype is either *AA* or *Aa*. Because the *A* allele is dominant, the two genotypes result in progeny of the same phenotype and you cannot tell whether the second allele is *A* or *a*.) That is to say that three individuals will express the phenotype caused by the recessive *a* allele, three individuals will express the phenotype caused by the recessive *b* allele, and only one individual will express the phenotypes caused by both *a* and *b* alleles. (You should, however, note that when considered separately the two loci behave exactly as expected for a single allelic pair, e.g., three-fourths of the progeny are *A_* and one-fourth are *aa*.)

This reasoning can be extended to individuals of the genotype *AaBbCc* or even of the genotype *AaBbCcDdEe . . . . YyZz*. We won't, but we could if we wanted to.

## RECOMBINATION AND PAIRS OF GENES THAT MAP ON THE SAME CHROMOSOME

As shown in Fig. 4.17, the rule of independent assortment does not apply when two genes, *R* and *S,* map on the same chromosome. Indeed, in the absence of crossing over, one would predict that an individual of the genotype *RrSs*, where *R* and *S* alleles are on one homolog and the *r* and *s* alleles are on the other, would only produce *RS* or *rs* gametes. This exception to Mendel's rule of independent assortment is called *linkage.*

MALE GAMETES ♂

|  |  | AB | Ab | aB | ab |
|---|---|---|---|---|---|
| | AB | AABB | AABb | AaBB | AaBb |
| FEMALE GAMETES ♀ | Ab | AABb | AAbb | AaBb | Aabb |
| | aB | AaBB | AaBb | aaBB | aaBb |
| | ab | AaBb | Aabb | aaBb | aabb |

**FIGURE 4.16**    A Punnett square showing the 16 types of zygotes.

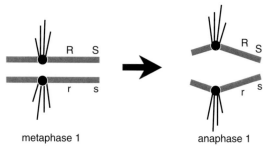

metaphase 1                    anaphase 1

FIGURE 4.17    Two genes on the same chromosome at metaphase and anaphase I.

As shown in Fig. 4.18, *Rs* or *rS* gametes can only be produced when crossing over occurs between two genes. Recombination or crossing over events are, however, relatively frequent during human meiosis. (Note there are two synonyms for crossing over, namely recombination and exchange. All three terms will be used interchangeably.) There is usually at least one such recombination event per bivalent. In the case of large chromosomes, recombination may be more frequent. This is especially true for large metacentric chromosomes in which recombination events will likely occur on both arms of the bivalent.

Recombination events can involve any two nonsister chromatids and can occur at any point along the length of a chromosome (except in the immediate vicinity of the centromere). This means that recombination events in your meiotic cells can only involve one chromatid from the chromosome you got from your mother and one chromatid from the chromo-

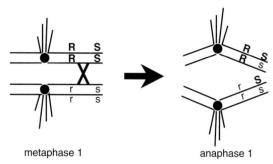

metaphase 1                    anaphase 1

FIGURE 4.18    Two genes on the same chromosome at metaphase and anaphase I with a crossover.

some that you got from your father. Recombination events do not normally occur between the two sister chromatids on the chromosome you received from your mother or between the two sister chromatids on the chromosome that you got from your father.

Because recombination can occur at most sites along the chromosome, the probability that a recombination event will occur between two genes is dependent on the distance between those two genes on the chromosome. Thus if two genes map at opposite ends of a chromosome, the probability of a recombination event occurring between them is high. However, as one considers genes that lie close to each other, the probability that an exchange will occur between them is diminished. Figure 4.19 shows six exchange events involving a bivalent marked with *Ee, Ff,* and *Gg.* All the exchange events fell between the *E* and the *G* genes. Five fell between *E* and *F,* but only one of the exchanges fell between *F* and *G.* Thus the frequency of exchange events between any two markers is proportional to the physical distance between them (see Box 4.1).

Take a minute to think about the differences between mitosis and meiosis (these are reviewed and discussed more fully in Boxes 4.2 and 4.3). Meiosis serves the vital function of reducing the number of chromosomes, and thus the number of copies of each gene pair, from two back down to one. The fusion of two such gametes at fertilization can then return the cell to a diploid state.

## MEIOSIS IS EXECUTED QUITE DIFFERENTLY IN HUMAN MALES AND FEMALES

Given the importance of this process, it is surprising that it takes place in such different fashions and at different times in men and women, but it does and these differences are truly impressive.

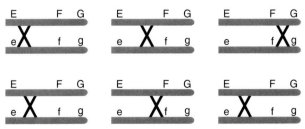

FIGURE 4.19    The same bivalent marked with *Ee, Ff,* and *Gg* with crossovers in six different positions.

---

BOX 4.1 MAPPING MADE SIMPLE

We can estimate the distance between genes on the same chromosome by determining the frequency of recombination events. The frequency of crossing over between the *R* and *S* genes can be assessed by determining the fraction of *Rs* and *rS* recombinant chromatids among the gametes.

This method is a major tool used by geneticists to determine the relative distance between genes and the relative order of groups of genes. We will return to it later.

A note for those of you who are mathematically inclined: The frequency of crossing over, or map distance, is calculated as

recombination frequency $= (Rs + sR)/(RS + sr + Rs + sR)$.

By multiplying this fraction by 100, we convert it to map units (or Morgan units), which is the standard term for expressing recombination frequencies. Because each crossing over event involves only two nonsister chromatids, the frequency of actual recombination events per bivalent is twice the calculated recombination frequency. This is because each crossing over event produces two recombinant and two nonrecombinant chromatids.

A note for those of you who are not mathematically inclined: Ignore that last paragraph.

---

BOX 4.2 DIFFERENCES BETWEEN MEIOTIC
AND MITOTIC CENTROMERES

1. Mitotic centromeres are functionally split into two halves and thus can attach to microtubules emanating from both poles [each half, or sister, centromere is connected by microtubules to one pole (see Fig. 4.25)].

2. At the first meiotic division the centromere of each chromosome is undivided. Thus each centromere can attach to only one pole. At the second meiotic division the centromeres are functionally split and behave like mitotic centromeres (see Fig. 4.26).

3. In the anaphase of mitosis (and of the second meiotic division), the two half centromeres separate, allowing the two chromatids to move to opposite poles (see Fig. 4.27).

4. The centromeres do not split at the first meiotic anaphase. Thus both sister chromatids move to the same pole (see Fig. 4.28).

## BOX 4.3 A SUMMARY OF THE DIFFERENCES BETWEEN MEIOSIS AND MITOSIS

1. Mitosis results in a cell producing two identical daughter cells. Meiosis results in a cell producing four gametes, each with one-half as many chromosomes as the original parent cell. The fact that meiosis pairs off the chromosomes, such that each gamete receives only one copy of each chromosome, explains Mendel's laws.

2. Mitosis takes place whenever cells divide. In an adult human being, mitosis takes place most frequently in the cells that comprise the skin, the bone marrow, and the inside of the gut. Meiosis only takes place in ovaries and testes. The purpose of meiosis is to produce sperm or eggs. Once an egg is fertilized by a sperm it becomes a zygote. A zygote increases its cell number by successive mitotic divisions.

3. Chromosomes only pair and recombine during the first meiotic division. They do not pair or recombine during the second meiotic division or during mitosis.

4. Sister chromatids *do not* separate from each other at the first meiotic division. Instead, paired homologs separate from each other. The centromeres do not split, and thus the sister chromatids do not separate from each other until the second meiotic division. At mitosis each chromosome lines up on the metaphase plate. The two sister chromatids separate from each other and then go to opposite poles.

5. In humans, recombination is essential to ensure proper segregation. If two homologous chromosomes do not recombine with each other for any reason, they will segregate at random (and thus go to the same pole, i.e., nondisjoin, in one-half the cases). Many of you may suffer from the misconception that if two chromosomes fail to recombine they will always nondisjoin. This is not true; they simply go at random. Recombination also serves to create new combinations of alleles for genes on the same chromosome.

6. Nondisjunction can occur at either meiotic division, although in humans it seems to most commonly occur at the first meiotic division. Most, but not all, nondisjunction at the first meiotic division results from a failure of chromosomes to pair and/or recombine. A failure of nondisjunction at the first meiotic division does not cause nondisjunction at the second meiotic division. Nondisjunction at the second meiotic division is quite rare. When it does occur, it results from the two sister chromatids separating from each other and then going to the same pole (i.e., they separate but do not segregate from each other). Nondisjunction can also occur in either parent. However, some classes of nondisjunction result in sperm that cannot function.

7. One can often determine in which parent and at which division nondisjunction occurred. This requires visible differences, or *polymorphisms,* in banding or alleles between the chromosomes carried by the two parents. For example, consider a case of trisomy 21 in which the parent in which the nondisjunction occurred was heterozygous for markers on chromosome 21. When both chromosomes from the same parent are the same, with respect to that marker, then nondisjunction occurred at the second meiotic division. When both parental alleles are present, then nondisjunction occurred at the first meiotic division.

8. Human embryos are pretty insistent on having two and only two copies of each chromosome. Monosomies, cases where an individual possesses only a single copy of a given pair of chromosomes (such as just one copy of chromosome 1), are not tolerated well for any pair of chromosomes. (Before you claim that human males are monosomic for the X chromosome, please note that XO human embryos, which truly are monosomic for the sex chromosomes, spontaneously abort in greater than 99.9% of the cases. We will return to this point in the next section.) Trisomies of chromosomes 13 and 18 sometimes get to birth but usually die soon after. Only trisomy 21 and trisomy for the X are really consistent with life.

Perhaps the most notable sex difference is in the timing of meiosis. In human females, meiosis begins during fetal development. All of the oocytes a human female will possess in her lifetime are produced while she is still *in utero.* These oocytes all begin the meiotic process during fetal development but arrest at the end of *pachytene* (the period of early meiosis during which chromosomes are observed to be fully synapsed along their length and are known to have completed meiotic recombination). Thus, all of the meiotic recombination a human female will ever do is completed before she is born. These arrested oocytes remain quiescent until the girl enters puberty. At that point a few oocytes are allowed to begin the maturation process during each menstrual cycle. This maturation consists only of progression to metaphase I. Usually only one oocyte is ovulated, allowing it to proceed to complete the first meiotic division. Because completion of the second meiotic division is triggered only by fertilization, the number of completed meioses experienced by a human female roughly equals the number of conceptions. Thus, in human females, recombination events must ensure chromosome segregation events that will occur decades later! Chapter 9 discusses the fact that this long delay in completing meiosis in females may well underlie the observation that the frequency of birth defects due to meiotic errors increases dramatically with advancing maternal age after age 35.

Male meiosis begins at puberty and continues uninterrupted throughout the life of the male. Male meiotic cells, known as *spermatocytes,* are continuously reproducing cells known as *stem cells.* Thus, unlike oogenesis, where all the oocytes exist at birth, spermatocytes are constantly produced throughout the life of the male. Once a spermatocyte initiates the meiotic process, it takes less than 75 days to produce mature sperm. (Compare this with the process of oogenesis that must span decades!) Thus, in human males the meiotic process is basically free running with cells usually progressing through the meiotic process in an uninterrupted fashion. It is thus perhaps not surprising that geneticists have observed subtle differences in the patterns of recombination in the two sexes. Perhaps the temporal differences in meiotic prophase and the different requirements for ensuring chromosome segregation (see later) have imposed different pressures on the evolution of recombination patterns in males and females.

These differences in biology of oogenesis and spermatogenesis result in some rather large differences in the number of meiotic cells and of the number of gametes produced by the two sexes. Each females carries some 2–3 million oocytes at birth, but usually less than 400 of these oocytes will eventually mature during her life. However, the production of spermatocytes and the subsequent process of male meiosis occur at a rate sufficient to produce the roughly 200,000,000 sperm present in each ejaculate (approximately 1,000,000,000,000 sperm during the life of the average male). The most important numerical difference is this: each female meiosis produces only a single oocyte, the remaining products of meiosis become nonfunctional cells called *polar bodies.* However, each male meiosis produces four functional sperm. Several workers have noted that the large number of stem cell divisions required to continue this process of sperm production throughout a male's life should provide more opportunities for newly arising mutations in the male germline than in the female. Indeed some data suggest that the mutation rate may be higher in sperm than in eggs and may increase with advancing male age. Thus we see that genetic errors increase in both eggs and sperm with advancing age, but the kind of errors that occur are substantially different: errors in getting the right number of chromosomes into a cell in females and errors in correct sequence of a gene in males.

The actual molecular mechanisms that ensure meiotic segregation appear to be different as well. In human males the meiotic spindle is organized by cytoplasmic structures called *centrosomes.* The chromosomes then attach to the developing spindle. In females, the chromosomes themselves bind to the microtubules and build the spindle from the inside out without the assistance of centrosomes. Moreover, whereas human female meiosis includes frequent preprogrammed stops, and selection appears to act at multiple points in the process, male meiosis appears to run uninterrupted once initiated. However, as noted in Chapter 9, there do appear to be multiple checkpoints or control points in male meiosis that allow a spermatocyte that has made errors in meiosis to abort the meiotic process. Whether such checkpoints exist in female meiosis is a hotly debated issue.

It may be surprising to realize that meiosis is so different in the two sexes, but try to think about what the organism needs to accomplish. A sperm and an egg are very different cells. A sperm is basically a genetic torpedo. It has a payload (23 chromosomes), a motor, and a rudder. Its function is to survive for a day or so and to swim to an egg. Once the sperm nucleus (called a *pronucleus*) is delivered to the cytoplasm of the egg, the rest of the sperm cells are destroyed. An egg, however, must possess all the supplies and determinants required to support embryonic development until the embryo can attach to the endometrium of the uterus and access the mother's blood supply. These two roles call for very different cellular machinery, and the process of human reproduction requires that a vast excess of sperm be produced for every egg, as the probability of any one sperm finding the one egg is just too low!

## THE CHROMOSOME THEORY OF HEREDITY

Up until now we have blithely assumed that genes map onto chromosomes, which actually took our intellectual ancestors some effort to prove. Although the proof is one of the more elegant examples of genetics, it unfortunately did not involve the study of human subjects. Indeed the experimental organism was the common fruit fly, *Drosophila melanogaster.* Although we are loathe to deviate from our central focus, namely ourselves, this proof is critical to ensuring that the relationship between Mendelian inheritance and meiotic chromosome behavior is really understood. Moreover, in addition to its heuristic importance, this inquiry, which centers on how chromosomes determine sex in flies, also serves as a useful transition to our next major topic: how chromosomes determine sex in humans. With that apology, we turn to the work of Calvin Bridges in 1912.

By the first decade of this century, there were two lines of evidence that genes were carried by chromosomes. First, Theodor Boveri had shown that sea urchin embryos, which were missing one or more chromosomes, developed abnormally and that the pathology of that improper development differed according to which chromosomes were missing from the genome. Second, it was clear that the behavior of chromosomes during meiosis paralleled and explained the behavior of genes. But this was only correlative evidence. What was needed was proof.

The proof came from a rather egregious exception to Mendel's laws. T. H. Morgan found a mutation in fruit flies called white (*w*), which causes the eye color to be pure white rather than the brick red color that is characteristic of normal, or *wild-type,* flies. As shown in Fig. 4.20, crosses involving pure-breeding red and white stocks (the white stocks were homozygous for the *w* mutation) displayed some rather odd results. We will denote the allele that produces a white eye color allele as *w,* and the allele that produces a red eye color as *W.*

One could imagine that CROSS A1 was really *ww* males crossed to *WW* females and that *W* is dominant. If you make that assumption, then CROSS A2 is a cross of *Ww* males to *Ww* females. So far so good. One then expects one-fourth of the progeny to be white (*ww*) and three-fourths to be red (*WW* or *Ww,* often abbreviated *W___*). That is exactly what you see. *However, note that all of the white-eyed progeny of CROSS A2 are males.* You cannot wiggle out of this by saying that the white-eyed trait can only be

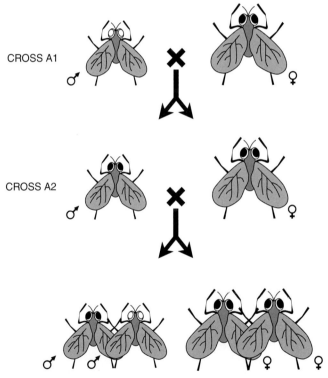

**FIGURE 4.20**    CROSS A1: White males were crossed to red females. This cross produced only red-eyed sons and daughters. CROSS A2: Red sons were crossed with red daughters. All the daughters were red-eyed, one-half of the sons were red-eyed, and one-half of the sons were white-eyed.

expressed in males because there are white-eyed females in your original breeding stock. Something is terribly wrong. As shown in Fig. 4.21, matters get worse when we do the cross backwards.

To figure this out, Morgan's student, Calvin Bridges, had to hypothesize that sons did not receive a copy of the *W* gene from their father. They instead carried only a single copy of the *W* gene (either the *w* or *W* allele) that they inherited from their mother. He further argued that females carried two copies of the *W* gene; thus females receive a copy of the *W* gene from both mom and dad. Please read this paragraph several times until you are sure, really sure, that you understand it. This is crucial.

To explain this rather curious exception to Mendel's laws, Bridges turned to a difference between the chromosomes of male and female flies. Flies have four pairs of chromosomes numbered 1–4. Chromosomes 2, 3, and 4 are identical in both sexes, but the first pair looked different in males and females (recall that there is also a pair of sex chromosomes in humans). In males there is an acrocentric chromosome (known as the X chromosome) that pairs with a metacentric chromosome (called the Y chromosome). In females there are only two X chromosomes.

Bridges reasoned that the X and Y chromosomes were sex-determining (i.e., that males are males because they carry both a X and a Y chromosome

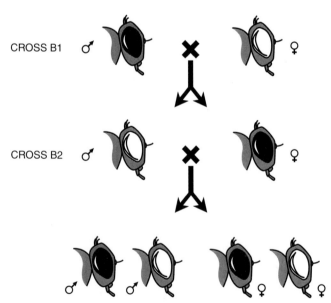

FIGURE 4.21    CROSS B1: Red males were crossed to white females. This cross produced only white-eyed sons and red-eyed daughters. CROSS B2: White sons were crossed to red daughters. One-half of the daughters were red-eyed, one-half of the daughters were white-eyed, one-half of the sons were red-eyed, and one-half of the sons were white-eyed.

and females are female because they carry two X chromosomes). He also reasoned that by placing the *W* gene on the X chromosome he could explain its inheritance. Bridges's proof of this hypothesis proved that genes did indeed reside on chromosomes and that the behavior of chromosomes during meiosis provided a physical basis for the laws of Mendel.

In males the X and Y chromosomes segregate away from each other at meiosis 1. Thus males produce either X- or Y-bearing sperm. Females produce only X-bearing eggs. According to Bridges, the fusion of Y-bearing sperm with X-bearing eggs makes sons whereas the fusion of X-bearing sperm with X-bearing eggs makes daughters (see Fig. 4.22). Note that sons get their X chromosome only from their mother and their Y chromosome from their father. By placing the *W* gene on the X chromosome, Bridges found a way to explain the fact that males carried only one copy of the *W* gene that they received from their mother. We can now rediagram CROSS A and CROSS B in a sensible fashion in Fig. 4.23.

This was a truly elegant piece of science. Bridges had correlated a usual pattern of inheritance with an unusual pattern of chromosome segregation, but it was still a correlation. Nothing more, nothing less. To prove his hypothesis he would need to demonstrate that errors in chromosome segregation caused errors in gene transmission. Fortunately, the experiment he was already doing provided him with exactly the exceptional progeny that such an experiment needed.

## FAILED MEIOTIC SEGREGATION (*NONDISJUNCTION*) AS PROOF OF THE CHROMOSOME THEORY OF HEREDITY

The proof that genes map on chromosomes came from rare exceptional progeny that came out of CROSS B1: red-eyed (*WY*) males crossed to

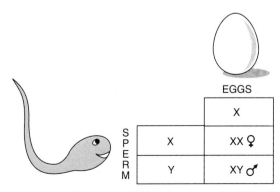

FIGURE 4.22    The fusion of Y-bearing sperm with X-bearing eggs.

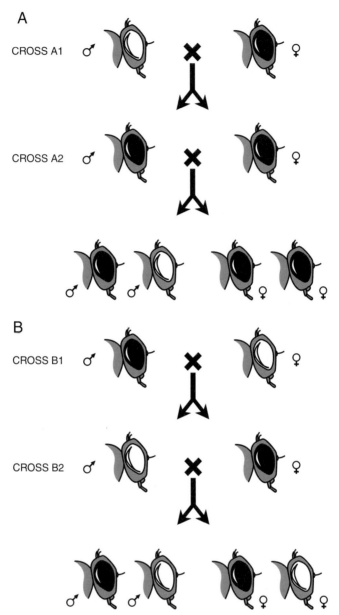

FIGURE 4.23 CROSS A1: White-eyed males $(wY)$ were crossed to red-eyed $(WW)$ females. This cross produced only red-eyed $(WY)$ sons and red-eyed $(Ww)$ daughters. CROSS A2: Red-eyed sons $(WY)$ were crossed to red-eyed daughters $(Ww)$. All daughters were red-eyed $(W_-)$, one-half of the sons were red-eyed $(WY)$, and one-half of the sons were white-eyed $(wY)$. Similarly, in CROSS B1, red-eyed $(WY)$ males were crossed to white-eyed $(ww)$ females. This cross produced only white-eyed $(wY)$ sons and red-eyed $(Ww)$ daughters. CROSS B2: White-eyed sons $(wY)$ were crossed to red-eyed $(Ww)$ daughters. One-half of the daughters were red-eyed $(W_-)$, one-half of the daughters were white-eyed $(ww)$, one-half of the sons were red-eyed $(WY)$, and one-half of the sons were white-eyed $(wY)$.

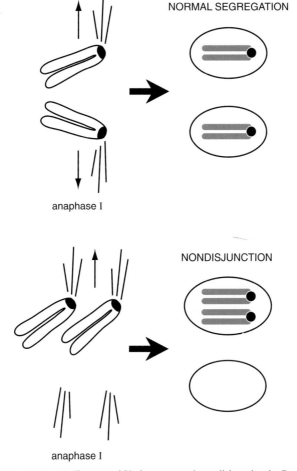

FIGURE 4.24    A diagram of X chromosomal nondisjunction in *Drosophila*.

white-eyed (*ww*) females. At low frequency (~1 in a 1,000), Bridges found white-eyed daughters and red-eyed sons. Bridges realized, and cytological studies confirmed, that these white-eyed females carried two X chromosomes and a Y chromosome (*wwY*) and that the males carried but a single X chromosome (*W*).

As shown in Fig. 4.24, Bridges realized that such flies could arise if two X chromosomes had failed to segregate from each other at the first meiotic division in the mother. Assuming a normal meiosis II, such an event produces both eggs with two X chromosomes (diplo-X) and eggs with no X chromosomes (nullo-X). Fertilization of a diplo-X (*ww*) egg with a Y-bearing sperm creates white-eyed (*wwY*) exceptional females, whereas

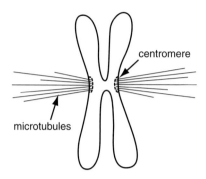

metaphase

FIGURE 4.25    A mitotic centromere at metaphase.

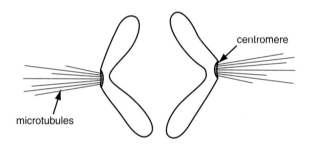

anaphase I

FIGURE 4.26    A mitotic centromere at anaphase.

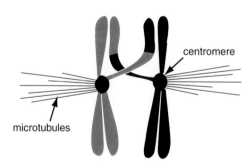

metaphase I

FIGURE 4.27    A meiotic centromere at metaphase I.

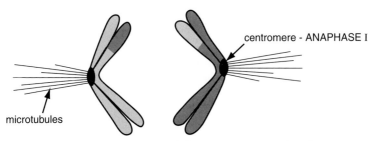

FIGURE 4.28 A meiotic centromere at anaphase I.

fertilization of the nullo-X egg with an X-bearing sperm creates the exceptional red-eyed flies (*W*). (Fertilization of diplo-X eggs with X-bearing sperm creates *XXX* zygotes that are lethal. Similarly, zygotes arising from the fertilization of nullo-X eggs by Y-bearing sperm die because they lack an X chromosome.)

In this experiment, Bridges showed that meiotic nondisjunction of the X chromosomes in the mother had resulted in nondisjunction of the carried X chromosome (i.e., a female had passed on both copies of the *w* allele to her XXY daughter). This was proof.

Bridges had also shown that sex in flies was determined by the number of X chromosomes an individual carried. Two X chromosomes made one a female, whether or not they carried a Y chromosome. Similarly, individuals with just one X chromosome were male, whether or not they carried a Y chromosome. This finding that chromosomes, and the genes that they carried, could determine something as important as the sex of an individual convinced even the most skeptical critics of the correctness of the chromosome theory and of the central importance of genetics.

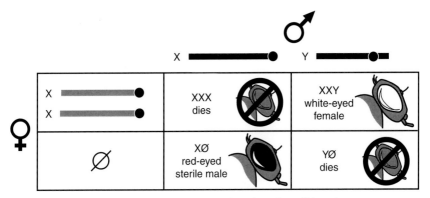

FIGURE 4.29 Bridges's explanation of nondisjunction.

Bridges's paper on nondisjunction was the first paper published in the prestigious journal *Genetics*. The reference is easy to remember: Bridges (1916), *Genetics*, volume 1, page 1. In a real sense this paper began the science of modern genetics. *(Indeed the issues raised in this paper still form the basis of much of my last 20 years of research. When pushed too far by my friends, or worse yet by my parents, to explain "just what the heck I do all the time in that laboratory," I simply state that I am solving problems that were at the center of scientific interest in 1929.—RSH)*

As it turns out, nondisjunction is a major cause of birth defects in humans, most importantly of Down syndrome. We will return to this issue in a later chapter.

# 5

## ABSENT ESSENTIALS AND

## MONKEY WRENCHES:

### HOW MUTATIONS PRODUCE

### A PHENOTYPE

*Nature, like a Greek oracle, is both eminently fair and cruel.
It will obligingly answer any question you ask; and will do
so with the fullest measure of truth. But the oracle will only
answer the question you actually ask, not the question you
think you are asking. One must therefore learn to be cautious. . . .*

*—D.R. Parker*

The last four chapters defined a gene, determined its structure, described
its function, and followed its movement during cell division. We have
discussed various allelic forms of a gene, mutations if you will, in terms of
their effects on phenotypes and used them as markers in genetic crosses.
Now we wish to answer a few more precise questions: How do different
types of mutations segregate in families, just how do mutational changes
in a gene bring about differences in a phenotype, and why do some forms
of a gene have a greater effect on phenotype than others?

#### PATTERNS OF INHERITANCE

*(When my daughter was born, my wife and I were immediately surrounded
by relatives, many of whom I had never met. Indeed, in some cases I wasn't
aware of their existence. But some of them brought to my daughter's crib
the most extraordinary bits of genetic folklore. I remember one of them
staring at my daughter Tara and saying, "She has her grandfather's eyes,*

*but then girls always get their grandfather's eyes." Wait a minute; I'm a geneticist, and a passably good one at that, and this was big news to me. Well, such wisdom kept coming for the next few days, to the point that I started writing it down. Much of it was just folklore, but some had good basis in fact. The point of this story is that people have long known that traits move through families in patterns, patterns that we now call "modes of inheritance." Those patterns tell us much about the genetics of a trait. That is why we discuss them below.—RSH)*

Most texts define mutations in terms of their behavior in pedigrees. Does the mutation appear to be dominant or recessive? Does it show the segregation patterns expected for a normal Mendelian gene, and thus presumably lie on an autosome, or does it show sex-linked inheritance?

For example, Fig. 5.1 displays a pedigree for an *autosomal dominant* mutation that causes a degenerative neurological disease called Huntington disease. In such pedigrees, males are represented by a square and females by a circle. Because the average age of onset for this disease is 35 years, many people who develop the disease have produced their own children before they become affected. Note that this pedigree is characterized by five basic features:

1. All affected children have at least one affected parent.
2. One-half of children of each affected individual are affected themselves if they are old enough to have developed the disease.
3. Except for very rare cases of new mutations, unaffected parents do not produce affected offspring.

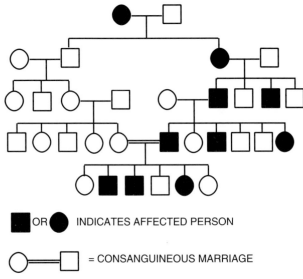

■ OR ● INDICATES AFFECTED PERSON

○══□ = CONSANGUINEOUS MARRIAGE

FIGURE 5.1   A pedigree for Huntington disease.

4. There are an approximately equal number of affected males and females.
5. Both affected males and females can transmit the disease to their children.

In addition to Huntington disease, which we will discuss again in Chapter 10, other autosomal dominant disorders you might have heard of include neurofibromatosis (discussed in Chapter 16), Lou Gehrig disease (amyotropic lateral sclerosis or ALS), and Marfan syndrome, which is believed to have affected Abraham Lincoln.

A second major type of inheritance describes the behavior of recessive mutations that map to the autosomes (i.e., *autosomal recessive* mutations). One such case (discussed in detail earlier) is the mutation in the CFTR gene that causes cystic fibrosis. A pedigree for Tay-Sachs disease, which shows a similar pattern of inheritance, is presented in Fig. 5.2. Note that this pedigree is characterized by four basic features:

1. Affected children can be, and in this case always are, derived from unaffected parents (homozygotes for the mutation that causes Tay-Sachs disease die as very young children). Obviously both parents must be heterozygotes to produce an affected child.
2. One-quarter of the children produced by the mating of two heterozygotes will be affected.

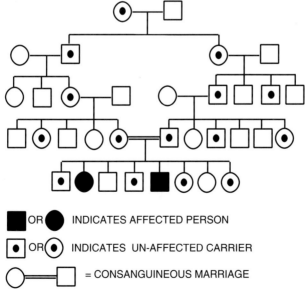

FIGURE 5.2    A pedigree for Tay-Sachs disease.

3. There are an approximately equal number of affected males and females.
4. The matings of affected individuals with unrelated and unaffected individuals usually produce normal children. (Obviously, if the unaffected mate was a carrier, affected children might be produced, but is usually quite rare. The rate at which this happens is directly related to the frequency of carriers in the population, which is low for many, but not all, autosomal recessive disorders.)

Other common diseases that show autosomal recessive inheritance include Lesch-Nyhan syndrome, pituitary dwarfism, and phenylketonuria (PKU, for which tests are done on newborn babies in many states).

Mutations on the X chromosome show a third and fourth pattern of inheritance, namely *sex-linked dominant* and *sex-linked recessive,* or collectively *sex-linked traits.* Obviously, X chromosomal genes are different from the autosomal genes because a male only carries one copy of the X chromosome, which he inherited from his mother. Sex-linked dominant traits are quite rare. An example of a sex-linked dominant mutation is a disease called vitamin D–resistant rickets, which causes extremely short stature. Please now look at Fig. 5.3. In the second generation, the male can only pass this trait to his daughters, not to his sons. In the next generation, one of these affected women marries a normal male and passes it onto a few of her sons, a few of her daughters, but some are unaffected (why?). *(The best way to understand sex linkage is to draw it out yourself. Label all*

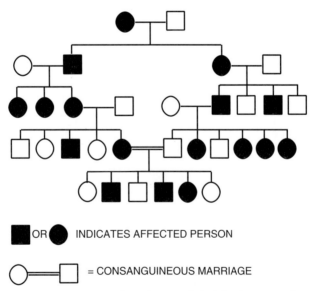

■ OR ● INDICATES AFFECTED PERSON

○══□ = CONSANGUINEOUS MARRIAGE

FIGURE 5.3    A pedigree for a sex-linked dominant mutation.

*the males with an XY and all of the females with an XX. To differentiate
the X chromosomes carrying the disease and those that do not carry the
disease, use a different color of ink for the two X chromosomes.—CAM)*
This disease is dominant and thus the presence of this gene on only one
chromosome is all it takes to be affected.

The critical and distinguishing features of sex-linked dominant inheritance are:

1. Affected fathers cannot pass the trait on to their sons.
2. All of the daughters of an affected male will be affected.
3. One-half of the offspring of heterozygous women will be affected
   and there will be an equal number of affected daughters and sons
   (obviously, as is true for an autosomal dominant as well, all of the
   offspring of a female homozygous for a sex-linked dominant will
   be affected).

Figure 5.4 displays an example of sex-linked recessive inheritance. The
recessive trait exhibited in this diagram is color blindness. *(Being partially
color-blind, this pattern of inheritance has special meaning to me.—RSH)*
This disorder is the result of mutation in one of the color opsin genes that
allows us to see colors. Starting with the first generation, notice that neither
the father nor the mother is affected. However, the mother is a carrier
(denoted by the dot in the center of the symbol). Moving to the next

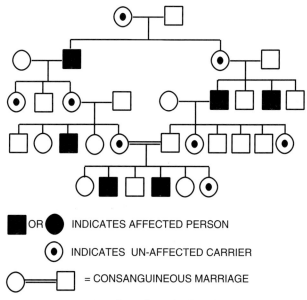

**FIGURE 5.4**   A pedigree for a sex-linked recessive mutation.

generation of Fig. 5.4, the carrier daughter marries a normal male. This woman produces two color-blind sons and two sons who are not affected. Two of the sons are affected because they inherited the X chromosome carrying the mutant (nonfunctional) allele of the color opsin gene. Because there is no homologous X chromosome with a normal copy of the color-blind gene, these males will not be able to see colors properly. The two unaffected sons received the other X from their mother, the one bearing the normal allele of the color opsin gene. Now consider the case of the other mating in the second generation, the one in which a color-blind male married an unaffected (noncarrier) female. Their daughters will not be color blind because they have two X's, one of which came from their mother and thus must carry a normal allele of the color-blind gene, but they must be carriers. In the third generation, an unaffected male marries one of the carrier females. Together, these two individuals produce an affected son, an unaffected son, one carrier daughter, and two noncarrier daughters. *(Again, to figure out the genotypes of everyone on this pedigree I suggest drawing this out to fully comprehend sex linkage. Trust me, professors just LOVE to play the pedigree game on exams.—CAM)*
    The critical and distinguishing features of sex-linked recessive inheritance are:

1. Affected males will be much more common than affected females.
2. Affected fathers cannot pass the trait onto their sons.
3. Assuming that his mate is not a carrier, none of the daughters of an affected male will be affected.
4. One-half of the sons of heterozygous women will be affected. As long as that woman marries a normal male, none of her daughters will be affected.

Sex-linked recessive traits include red-green color-blindness and the kind of hemophilia (blood-clotting disorder) that history has reported among some of the European royal families.
    Some of the older text books also refer to a pattern of inheritance exhibited by genes present only on the Y chromosome, so-called *holandric* traits. A common such example is a man with large tufts of hair sprouting from his ears. It turns out that other than the TDF gene, which determines sex (see the Chapter 6), there do not appear to be many genes that are truly specific to Y chromosomes. The so-called holandric traits reported in earlier texts presumably represent examples of *sex-limited* traits, which are traits resulting from mutations that can only be expressed in one sex or the other. For some traits the limitation is obvious (e.g., a defect in the coiling of the vas deferens or in the structure of the fallopian tube) but for other traits such as a sex-limited disorder that causes hand malformations, but does not affect primary or secondary sexual characteristics, the sex-limited nature of the disease is not one that leaves us saying, "Of course!"

Finally, we need to at least mention the phenomenon of *mitochondrial inheritance*. Mitochondria are the small organelles in the cytoplasm of our cells that carry out most of our energy production. These organelles, which apparently arose in the cells of our early eukaryotic ancestors as bacterial symbionts, carry their own small circular chromosomes. Most mitochondria carry approximately 10 copies of this 16,600-bp chromosome. These small chromosomes encode 13 proteins, 22 tRNAs, and 2 rRNAs. (If you don't remember what tRNAs and rRNAs are, look back at the section in Chapter 3 on translation.) At fertilization *all* of the mitochondria contributed to the zygote come from the mother. Thus, if a woman carries a given mutation in all of her mitochondrial chromosomes, that mutation will be passed on to *all* of her offspring. (Sperm do not contribute mitochondria to the zygote and thus her affected sons cannot pass along the trait. Therefore, the father's mitochondrial genotype is irrelevant.)

Mitochondrial inheritance of a given trait is easy to spot because *all* of the children of an affected mother will be affected, and *none* of the children of an affected father will be affected. A sample pedigree is shown in Fig. 5.5. A number of inherited human diseases display mitochondrial inheritance. These include *mitochondrial myopathy, progressive external opthalmoplegia,* and *MERFF* or *myoclonus epilepsy* with *ragged-red fibers*. Curiously, all of the disorders just listed appear to be due to mutations in one or another of the tRNA genes; the most reasonable interpretation of their effects is that they impair the translation of various mitochondrial proteins.

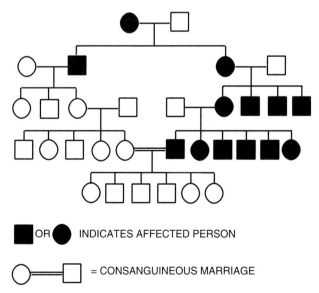

■ OR ● INDICATES AFFECTED PERSON

○══□ = CONSANGUINEOUS MARRIAGE

**FIGURE 5.5** A pedigree for a mutation in the mitochondrial genome.

Some evidence also shows that mutations in one of the rRNA-encoding genes may be responsible for a type of congential deafness. Moreover, mutations in the mitochondrial genome have been implicated in some forms of *Alzheimer* and *Parkinson* diseases and in some aspects of aging. However, some of the observed effects may actually be due to changes in the mitochondrial genome in cells throughout the body of an individual over the course of a lifetime. (Again we will remind you that there can be other forms of these diseases. Not all forms of epilepsy or Alzheimer disease are found in families that show mitochondrial inheritance.)

So, fine, we can follow the segregation of a trait in a pedigree and define the pattern of inheritance, but just what does that mean? How do mutations exert an effect?

## THE RELATIONSHIP BETWEEN THE
## NATURE OF THE MUTATION AND THE
## RESULTING PHENOTPYE

Defining mutations simply in terms of how they behave in pedigrees is all fine and good if one's serious goal is passing similarly phrased questions on a midterm. Sadly, this approach misses the point, at least in our view, because it confuses the final organismal phenotype with the nature of the actual cellular defect produced by a given mutation.

It is much more appropriate to divide mutations into only two classes: those that fail to produce a protein required by the cell (a "loss of function" or an absent essential) and those that produce an abnormal or poisonous protein that disrupts an essential cellular function (a monkey wrench). Classically, people have described such functions in terms of convenient, but often misleading, terms such as dominant or recessive. As shown below, there are no simple correlations between the nature of a mutation and its behavior in pedigrees (i.e., whether it is dominant or recessive).

### A CASE WHERE A LOSS-OF-FUNCTION MUTATION
### BEHAVES AS A RECESSIVE

As discussed in Chapter 1, there is a protein called CFTR that is required for proper function of cells in the lungs. This protein is an enzyme that facilitates the transport of salts and fluids across the membranes of several tissues, most notably the lungs and pancreas. The failure of proper salt transport in the lung cells causes a thickening and hardening of the mucus, which in turns leads to inflammation and the possibility of chronic respiratory infections. These repetitive infections can cause the individual to die of severe respiratory failure.

Now suppose there is a child homozygous for a mutation in the gene that encodes the CFTR protein. This mutation prevents the proper encoding of this protein. Indeed, it is a nonsense mutation that causes translation to be prematurely terminated after only a few amino acids have been incorporated into the growing protein. It should be obvious to you that this child will be unable to make any functional CFTR protein. As noted previously, this child will then suffer from a disease called cystic fibrosis, which has the symptoms described earlier. We want to be extremely clear here: This individual's illness does not occur because he/she is homozygous for a mutation in the CFTR gene but rather he/she is ill only because of an absence of the CFTR protein.

Now consider an individual who carries both the normal (wild type) functional allele of the CFTR gene and the mutant allele. Most texts and even some professors will glibly tell you that everything will be fine in this case simply because the CFTR mutation is a "loss-of-function" mutation and that loss-of-function mutations are *always* recessive (and thus the normal allele is dominant). What we *think* they mean by such statements is that one normal copy of a gene per cell can produce enough good protein, even in the presence of a nonfunctional mutant copy, to prevent the nonfunctional copy from impairing the function of the cell.

Please realize that buried in this explanation for recessivity is the idea that the cells of most individuals, which carry two normal copies of the CFTR gene, produce at least twice as much CFTR protein as they really need! The CFTR protein is such an efficient enzyme that one-half the normal levels of this enzyme can apparently manage the necessary levels of salt and fluid transport. Hence the oft-stated assertion that loss-of-function mutations will be recessive and that recessive mutations are normally loss-of-function alleles. After all, a normal allele on the other homolog should be able to compensate for a mutant allele, shouldn't it? Isn't that the main attraction of being diploid?

Although this explanation for recessivity is frequently correct, there are several other, quite clinically significant, examples of loss-of-function mutations that create severe phenotypes even in the presence of a normal allele. In most cases, these mutations define genes whose products are required as structural proteins rather than as enzymes that catalyze biological reactions.

### A CASE WHERE A LOSS-OF-FUNCTION MUTATION BEHAVES AS A DOMINANT

Most structural proteins, such as type I collagen, are required in simply huge amounts. Moreoever, the demand for these building blocks of biological structures is usually so high that one-half is really insufficient. For example, loss-of-function alleles of the type I collagen gene are lethal when homozygous. However, even when heterozygous with a normal allele, the presence of one such nonfunctional allele results in a mild form of *osteogenesis imperfecta,* a disease that causes brittle bones and early onset deafness (presumably as a consequence of the degeneration of the bones of the inner ear). Because such a mutation can cause a disease even in the presence of a normal allele, it will be classified as a dominant.

Another example, the case of the retinoblastoma (Rb) gene, is even more vexing because it involves a gene that encodes an enzyme that serves to prevent cells in the human retina from developing into a tumor. Even one good copy of this gene in a given cell is sufficient to prevent tumorigenesis. In this sense one might wish to describe mutations that knock out the Rb gene as *recessive.*

However, it turns out that mutations also occur in somatic cells, albeit at very low fequencies (approximately 1 in every 100,000 cells). Accordingly, 1 in every 100,000 cells will suffer a mutational loss of the normal RB gene (see Fig. 5.6). Although this event may seem rare, there are more than a million cells that comprise the human retina, and thus, on average, this event will occur some 10 or so times in each eye, resulting in full-blown expression of the disease. As a conscquence, every individual who inherits a loss-of-function Rb mutation will express the disorder. For this reason, Rb mutations are often considered to be *dominant,* at least with respect to their effects on pedigrees.

Similarly, Chapter 7 will describe the evidence that a serious disorder called Turner Syndrome results from the presence of only a single copy of a gene called rps 4 (again a structural protein; in this case a component of the ribosome) that must be present in two copies for normal development. In the presence of only one copy, normal growth and development are either impeded or completely prevented.

Okay, okay, so some loss-of-function mutations are going to look dominant and some are going to look recessive in pedigrees. But aren't mutations that create disruptive or poisonous proteins always going to be dominant? Unfortunately, as discussed later, the answer is simply no, they won't always be dominant.

## CONSIDER THE CASE OF THE PROVERBIAL MONKEY WRENCH, A POISONOUS PROTEIN THAT ACTS AS A DOMINANT

Like ballet dancers and bank robbers, very few proteins truly act alone! Instead, many function either as dimers, an associated pair of identical proteins, or as parts of large macromolecular assemblies that are composed of many proteins! Imagine then a mutation that produces a protein that can be assembled into such a larger structure but when doing so cripples

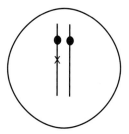

A Somatic Cell With One Mutant and
One Normal Copy of the Rb Gene

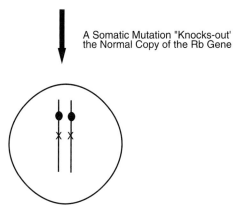

A Somatic Mutation "Knocks-out'
the Normal Copy of the Rb Gene

The Cell Will No Longer Be Able to Produce the Rb Protein,
And Thus Will Initiate Tumor Formation

FIGURE 5.6   A somatic mutation causes the loss of the normal gene in a cell heterozygous for a mutation at the Rb gene.

the structure (kind of like the proverbial weak link in a chain). Such mutations can and do create a cellular disorder/defect, even in the presence of a normal gene product.

A classic example of such a disorder is Huntington disease (see earlier discussion). The mutation that underlies this disorder produces a defective protein that impairs the functioning of the nervous system. The phenotype of individuals whose cells carry this protein is the same regardless of whether their cells are carrying one copy of the defective mutation and one normal allele or two copies of the mutant allele. Huntington disease is also a classic example of a so-called dominant mutation in terms of its effect on pedigrees in that the mutation produces a phenotype regardless of whether a normal allele is also present.

## CASES WHERE MONKEY-WRENCH PROTEINS
## BEHAVE AS RECESSIVES IN A PEDIGREE

More vexingly, there are also some mutations that can create phenotypes even in the presence of a normal allele under some genetic or environmental conditions, but which are masked by the normal allele under others. This point is made most clearly by considering the case of sickle cell anemia, a terribly severe blood disorder that can cause an early and unpleasant death. The genetic basis of this disease is homozygosity for a mutation in the gene that encodes the adult form of hemoglobin. In this case the mutation is referred to as a *missense mutation* because it causes the wrong amino acid to be incorporated at a specific position during translation (specifically a valine is inserted instead of glutamic acid). The consequence of this wrongly inserted amino acid is a hemoglobin molecule that tends to precipitate within the red blood cells. The red blood cells become deformed and block the capillaries.

The incidence of this disease is the highest in black populations and approaches 1 in 25 births in some parts of equatorial Africa. (The incidence of sickle cell anemia among American blacks is approximately 1/500.) Under normal conditions, heterozygotes are fine (although they may exhibit some symptoms at very high altitudes where the oxygen pressure is low). Indeed, in at least one case they are better than fine: Heterozygotes for the sickle cell anemia trait are largely protected against malaria! Thus the sickle cell mutation is recessive in one sense (sickle cell disease) but dominant in another respect (resistance to malaria).

Thus, there would appear to be some standard correlations, that loss of function usually will be recessive and monkey wrenches usually will be dominant simply because that is often what happens. In reality, there are no absolute correlations between the actual nature of mutations, in terms of their effect on gene function, and their phenotype when heterozygous with a normal allele. Any attempts at such simple interpretations will be thwarted by the fact that the relationship between a given form of a gene and its phenotype depends on the nature of the encoded protein, its biological function, the cell type in which it acts, and the environmental factors that influence expression.

For these reasons we prefer to couple the terms dominant and recessive with a description of mutations in terms of the gene's ability to synthesize functional or poisonous proteins. Thus we will try to use terms such as "recessive loss-of-function allele" or "dominant poisonous allele" to tell you both how the mutation acts in a pedigree and what it actually does in terms of protein production.

In addition to avoiding the sort of confusion described previously, this approach also allows us to deal with a more serious set of complexities that arise when different mutations in the same gene produce different phenotypes or when mutations in different genes give rise to identical phenotypes.

## NOT ALL MUTATIONS IN THE SAME GENE CREATE
## THE SAME PHENOTYPE

In the next section we will discuss the fact that some types of mutations in a gene that encodes the testosterone receptor can cause genetic males to develop as females. Other less severe types of mutations in the same gene do not cause sex reversal but do cause a neuromuscular disorder called spinal and bulbar atrophy (see Chapter 6). We will also discuss the different disorders that can result from mutations in the Duchenne muscular dystrophy gene. Not all types of mutations completely prevent protein production or completely prevent function. Thus mutations with different effects on protein structure and function can produce very different phenotypes.

However, some mutations in two different genes can produce the same phenotype.

### MUTATIONS IN DIFFERENT GENES CAN PRODUCE
### IDENTICAL PHENOTYPES

We can cite many cases where the same phenotype is created by mutations in different genes. For example, earlier in this chapter we described the inheritance of color-blindness as a sex-linked recessive. However, other kinds of color-blindness or hemophilia can result from changes in genes located on autosomes. In fact, there are many things that we think of as one disease, lumped together under a common name. There are actually separate genetic disorders that result in symptoms so similar that doctors have called them the same disease. Thus, in one family, cataracts may be inherited as an autosomal dominant disorder, but in another family cataracts resulting from a different biochemical cause may be inherited as an autosomal recessive trait. Thus, different forms of albinism may be autosomal recessive or sex-linked recessive, depending on which albinism family you consider. Osteoporosis may be autosomal dominant in some families and autosomal recessive in others because individuals in the two families really have different genetic disorders, despite the similarity of symptoms and the shared disease name.

A good example of this principle is the genetics of congential deafness. Congenital deafness is often due to loss-of-function autosomal mutations that have no impact on hearing when heterozygous with a normal allele. So when two deaf individuals (deaf as a consequence of a genetic defect) marry, do you expect deaf children? The answer is usually no because although both parents may be homozygous for loss-of-function alleles, in all probability they are homozygous for loss-of-function alleles in different genes.

Imagine that there are at least two such genes (indeed there are dozens), cleverly called gene 1 and gene 2. So let's say that Sean is homozygous for a loss-of-function allele in gene 1 and Mary is homozygous for a loss-of-function allele in gene 2. Because both genes encode products that are absolutely required for hearing, both will be deaf. However, realize that their children will likely inherit a normal allele of gene 2 from Sean and a normal allele of gene 1 from Mary. Accordingly, the children will likely have normal hearing.

Suppose, however, that both Sean and Mary were deaf due to homozygosity for mutations in gene 1. In this case *all* of their children would be affected. Whether you know it or not, you have just done a complementation test (see Fig. 5.7). *(This is one of the most powerful bits of magic in my little cloth bag of geneticist tricks.—RSH)* Its use is more fully described in Box 5.1.

Actually, this problem of mutations in different genes producing the same phenotype can often tell us a good deal about how biological processes really work. In many cases it may take several enzymatic steps to produce a given product or result. Such a pathway will thus require the products

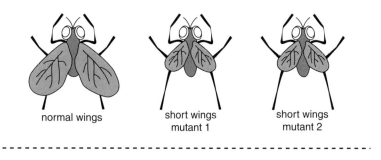

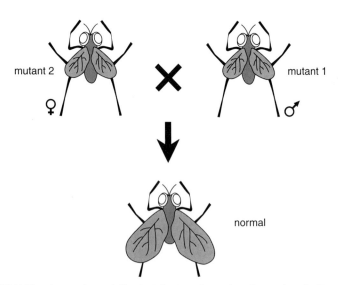

FIGURE 5.7    A complementation test for two short-winged mutations in *Drosophila*.

## BOX 5.1 DO TWO MUTATIONS THAT PRODUCE THE SAME PHENOTYPE DEFINE THE SAME GENE?

Suppose you recovered two different autosomal recessive mutations in the same organism. In both cases the mutations had no effect in heterozygotes but produced a strong phenotype as homozygotes. Suppose that you were also capable of producing pure breeding strains of homozygotes for each of the two mutations. How could you determine if the two mutations had occurred in the same gene?

To address the question let us consider the case of two recessive mutations in fruit flies, both of which result in flies with very short wings. Now suppose both mutations map in the same gene (a gene we will call $A$). Let us denote the two mutations $a1$ and $a2$. Strain 1 flies are $a1/a1$, strain 2 flies are $a2/a2$, and normal wild-type flies are $A/A$. If we cross $a1/a1$ males from strain 1 to $a2/a2$ females from strain 2 (or vice versa), we will get $a1/a2$ progeny. The crucial point here is that although there are two different mutant alleles ($a1$ and $a2$) in this genome there are no normal ($A$) alleles. Thus no normal $A$ product will be produced and the flies will have short wings.

Now suppose that the two mutations are in different genes ($B$ and $C$), both of which make products required to build a wing of normal length. Suppose strain 1 is homozygous for a recessive loss-of-function mutation ($b$) in gene $B$ but carries two normal alleles of gene $C$. We would denote its genotype as $bbCC$. Similarly, suppose strain 2 is homozygous for a recessive loss-of-function mutation ($c$) in gene $C$ but carries two normal alleles of gene $B$. We would denote its genotype as $BBcc$. Now let us cross strain 1 males to strain 2 females. The resulting progeny are of genotype $BbCc$ (i.e., they are heterozygous at both genes). Note that these progeny carry one normal function allele for each gene ($B$ and $C$). Thus, because $B$ and $C$ alleles are dominant, these flies will have wings of normal length.

What you have just read is a description of a *complementation test*. If the phenotype of the double heterozygote (mutant 1/mutant 2) is mutant then the two mutations define the same gene. Conversely, if the double mutant is normal the two mutations are in different genes. This is basically the same experiment described for deafness in the text. If two individuals marry, both of whom are deaf because of homozygosity for recessive mutation in the same gene, then all of their children will be deaf. However, if they are each homozygous for mutations at different genes then their children will most likely be normal. Complementation tests are one of the most powerful tools in a geneticist's intellectual arsenal.

of several (many) different genes. (Imagine an automobile assembly line in which each station is staffed by a given protein.) Being homozygous for a mutation in any one of these genes will stop the entire process. (No cars will get made!) If all we can assay is the final end product, then mutations in many genes whose products function in this process will have an identical phenotype. (Realize that other types of mutations may not prevent production of a car, but they may get you a car of a different color or a car without a horn or a backseat!)

## THE ABILITY OF SOME MUTATIONS TO CREATE A PHENOTYPE MAY DEPEND ON OTHER GENES IN THE ORGANISM OR ON ENVIRONMENTAL FACTORS

Sometimes even the same mutation, or set of mutations, does not produce an identical phenotype in two individuals. This problem is significant enough that we have a term for it, *penetrance*. Mutations that are fully penetrant will exert a similar effect in every individual, usually without being affected by environmental conditions or by the other genetic variations possessed by that individual (the so-called *genetic background*). Most of the traits that have been studied have high penetrance, precisely because such traits are far easier to study. Nonetheless, examples of weaker penetrance are well known and will be discussed more fully in the chapter on multifactorial inheritance.

### SUMMARY

The relationship between a mutation and its phenotype is obviously complex. To grasp the full complexity of that relationship and the manifest ways in which genes can exert their effects, we need to consider the ways in which functional genes can direct a biological process. One of the best understood systems is the process by which our genes determine our sex. The next section will give you your first real look at how all of this biology can actually direct the assembly of a complex critter like ourselves.

# How Genes Determine
# Our Sex

This section describes the genetic mechanisms by which sex is determined. Our major focus will be the demonstration that a given phenotype, in this case maleness or femaleness, represents the result of several genetic regulatory systems that may act by strikingly different mechanisms.

# 6

## SEX AND CHROMOSOMES, SEX AND HORMONES, SEX AND . . .

*. . . What it comes down to is men wear pants and women
wear skirts . . .*
—*Corporate Myth*

*What are the biological factors that make an individual
a man or a woman? Also, what is the connection between
biology and behavior?*
—*CAM*

Sex is a complicated subject. If a friend of yours says the word "sex" to you and nothing else what do you think? Let's try it: *Sex.* What went through your mind? In this chapter and the following chapters when the word "sex" is used, it will be referring to several different things: *gonadal sex, somatic sex, sex identification,* and *sexual orientation.*

If you are a male, your testes define your gonadal sex. Somatic sex characteristics are broken into *primary* and *secondary.* Male primary somatic sex characteristics are the penis and the scrotum. Secondary characteristics include facial/chest/increased body hair, pelvic build (lack of rounded hips), upper body muscular build, and the ability to generate muscle mass at a faster rate than the female. If you are a female, your ovaries define your gonadal sex. Your primary sex characteristics are your vagina, uterus, fallopian tubes, clitoris, labia, cervix, and the ability to bear children. Your secondary sex characteristics are your relative lack of body hair, thicker hair on your head (in some cases), rounded hips/figure, a decreased ability to generate muscle mass at a fast rate, decreased upper body strength, breasts, ability to nurse children, a menstrual cycle, and increased body fat composition. Sexual identification and sexual orientation define our sex roles and our choices in sexual partners.

The question we want to explore revolves around the degree to which each of these components of sex in human beings is genetically determined. Just how do our genes determine our sex and to what extent do genes determine our sexual behaviors? This chapter discusses the mechanisms that underlie the development of primary and secondary sexual characteristics. Chapter 7 will consider in detail some of the peculiar properties of the sex chromosomes in humans. Finally, we will return to the issue of the role of genes in establishing sex roles or sexual orientation in Chapter 8.

For right now, however, we need to start at the very beginning. As stated in Chapter 4, even the earliest fly geneticists understood that biological sex was determined by a pair of chromosomes called the sex chromosomes.

## GONADS AND CHROMOSOMES

All genetically normal individuals possess 23 pairs of chromosomes; 22 pairs are called *autosomes* and 1 pair is called *sex chromosomes*. A normal female possesses two X chromosomes, whereas a normal male possesses one X and one Y chromosome. The finding of XXY males and XO females (to be discussed further later) convinced geneticists that sex in humans was determined solely by the presence or absence of a Y chromosome (Fig. 6.1). However, it didn't take long to find variant Y chromosomes that were missing quite a bit of material but were still capable of determining maleness. All that seemed to matter in terms of being able to determine maleness was a small region on the short arm of the Y chromosome. These data demonstrated clearly that it is not simply the presence of the Y chromosome that creates a

| XXY | XO |
|---|---|
| Klinefelter Syndrome MALE | Turner Syndrome FEMALE |

FIGURE 6.1    Abnormal karyotypes and sexual differentiation.

normal male but rather a small amount of genetic material now known to be a single gene located on the Y chromosome called the *testis determining factor* (TDF). The TDF gene initiates the development of male genitals.

## EVIDENCE THAT TDF DETERMINES GONADAL SEX

Before we can convince you that it is just one gene (TDF) and not the entire Y chromosome that is male determining, it is necessary to understand the mechanism by which the sex chromosomes pair and subsequently recombine during meiosis. Sex chromosome recombination and segregation in females are actually fairly straightforward to understand. Because the two X chromosomes are homologous, when meiotic crossing over occurs in the female germ cells, the crossover or recombination events can occur at a very large number of sites along the length of the X chromosome. In other words, because X chromosomes are homologous, the pairing region includes the entire length of the chromosomes and is not limited.

However, the fact that X and Y chromosomes are not homologs makes matters more complicated in males. Logically, of course, the two chromosomes are different simply because one chromosome is an X and the other is a Y. However, as shown in Fig. 6.2A, there is a limited region at the tips of the short arm of both the X and the Y chromosome in which the X and Y chromsomes are homologous and can recombine with each other. This region is indicated by a region of Watson–Crick base pairing. Along this limited region, which is scientifically termed the *pseudoautosomal region* (or PAR), are a small number of genes that are shared by both X and Y chromosomes. (There are in fact several such regions of homology between X and Y chromosomes, but crossing over is usually limited to the one just described.) The male meiotic system allows recombination to take place only within this limited region of homology. This paired region of X and Y chromosomes is confined to a specific region within the meiotic nucleus called the *sex vesicle*. The sex vesicle is made up of a specific assembly of proteins and acts as the mechanism to ensure proper pairing and crossing over during meiosis I.

Although the crossing-over event is amazingly precise (see Figs. 6.2B and 6.2C), you may now be thinking, "With all the billions of meiotic events occurring every day in each male, it only seems logical that a mistake in crossing-over will occur at some frequency." (*Good thought; hold onto it.—RSH*) Turn now to Fig. 6.2C and notice the location of the TDF gene; it is not within the pseudoautosomal region but is very close to it. Look further at Fig. 6.2D and observe an abnormal and unequal crossing-over event. (Such events are rare — 1 in 10,000 meioses — but there

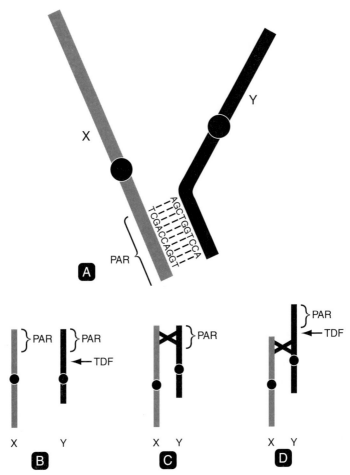

FIGURE 6.2    Sex chromosome pairing. (A) Rough drawing. (B) Position of the PAR and TDF. (C) Normal crossing over. (D) Unequal exchange!

are an awful lot of people!) As a consequence of this unequal exchange, the TDF gene is translocated onto the X chromosome from the Y chromosome creating a chromosome that we will call X(TDF). Suppose a sperm bearing that chromosome fertilized an X-bearing egg? The resulting zygote will carry two X chromosomes, one of which bears the newly acquired TDF gene; is this child a male or a female? Remember what was said earlier: It is ONLY the presence of the TDF gene that determines male gonadal sex.

Indeed, this child will develop as an XY male, but will suffer from testicular atrophy or, more simply put, small testes and thus sterility. Why sterility? This is because it is not possible to have two X's present in the male germline; the mere presence of another X chromosome acts almost like a poison to the germ cells and kills them during meiosis. Hence, this individual is unable to produce healthy and happy living sperm. Nevertheless, regardless of the two XX's present, THIS INDIVIDUAL IS A MALE!

Realize that the same unusual meiotic event also created a Y chromosome from which the TDF gene has been deleted [denoted *Y(del TDF)*]. If a sperm carrying this chromosome fertilized an egg bearing a normal X, a *XY(del TDF)* zygote would result. This *XY(del TDF)* zygote will develop as a female. Internally she will possess a cervix, uterus, and a normal vagina with the exception of small and shriveled ovaries. This female is also sterile because her eggs will die due to the absence of a second X chromosome (see Chapter 7). Again, let me reiterate that although this individual possesses an X and a Y chromosome, THIS PERSON IS STILL A FEMALE! Moreover, we have subsequently found several XY females that differ from normal males only by point mutations within the TDF gene.

To prove to you that the TDF gene alone is responsible for male gonadal sex, Chapter 12 will summarize an experiment that has been done in England. These experimenters used some rather clever tricks of DNA manipulation to insert a mouse TDF gene, and just the TDF gene, into the genomes of XX mouse embryos. (Mice determine sex exactly in the same manner as do humans.) These female mouse embryos, which had been transformed with the TDF gene, developed into healthy but sterile male mice.

So the only thing that matters is the TDF gene, but what does it do and how does it do it? We will explore that question in the following paragraphs.

## THE TDF GENE CAUSES THE INDIFFERENT GONADS TO DEVELOP AS TESTES

When you were first conceived, you began life with a pair of indifferent gonads. The term indifferent gonads is self-explanatory: The fetus' organs are literally "indifferent" to becoming either ovaries or testicles, depending on whether the fetus' germ cells carry either XX or XY chromosomes. The presence of the TDF gene during the seventh to eighth week of fetal development causes the indifferent gonad to develop as a male. Note that the TDF gene acts only during this time during development (it is inactive during the remainder of development). Moreover, it acts only in a specific fraction of cells in the indifferent gonad. The expression of TDF in those cells is, however, sufficient to induce the indifferent gonads to become testicles and initiate male development (see below). If the TDF gene is not expressed, the cells of the indifferent gonads will follow a separate path and develop as ovaries.

*(That's it. The TDF gene turns on for a brief period of time in a minor fraction of fetal cells and is then done until the next generation. I am reminded of Shakespeare's comment about each of us being a bit player who comes on the stage, speaks but a few lines, and is then finished. Surely that describes the role of the TDF gene in human fetal development. Nonetheless whether those few lines are spoken or not determines the course of that fetus' life, male or female.—RSH)*

## FROM THE GONADS TO SOMATIC
## SEX CHARACTERISTICS

Unlike gonadal sex, somatic sex develops independently of the presence or absence of the TDF gene. It is determined by the hormones that are produced by the developing gonads. You began life with two sets of reproductive "plumbing" (as RSH calls it): the Müllerian ducts (female reproductive tract: uterus, primitive fallopian tubes) and the Wölffian ducts (male reproductive tract: vas deferens, seminal vesicles). You also possess a small bud of tissue called a genital tubercule that will form either a penis or a clitoris. This is to say that nature's first choice is both!

If you are a male, you possess a normal Y chromosome (or at least a TDF gene) and within the 8th week of development your indifferent gonads became testicles and began secreting androgen (testosterone) and the Müllerian inhibitory factor (MIF). The MIF causes the regression of the Müllerian ducts. If you are a female, the indifferent gonads become ovaries and produce estrogen. During the 13th week of development, the Wölffian ducts degenerate and the Müllerian ducts develop. The relative levels of estrogen or testosterone also determine the developmental of the primary sexual characteristics. The high levels of testosterone produced by the testicles cause the genital tubercule to develop into a penis. In the presence of high levels of estrogen, the same tissues will form a clitoris.

## ADRENAL HYPERPLASIA AND
## AMBIGUOUS GENITALIA

It turns out that the cells of the adrenal cortex (a part of your adrenal gland) also produce low levels of both estrogen and testosterone. Sometimes as a consequence of overactivity of the adrenal gland during development, or of a defect in hormone synthesis, high levels of either estrogen or testosterone can be produced by the adrenal cortex. Thus a developing female fetus could be exposed to high levels of both testosterone (from the adrenal cortex) and estrogen (from the ovaries). The result is a mixture or confusion of developmental processes, resulting in a newborn whose genitals seem to be "a little of both." These cases of ambiguous genitalia are quite disturbing to parents and physician alike and often are resolved by surgical correction. Treatment of these infants requires genetic evaluation to determine the sex chromosome composition and the presence or absence of ovaries and testis, surgical evaluation to determine the treatment most likely to produce a functional adult, and psychological evaluation and counseling of the parents.

The frequency of such births (approximately 1/10,000) requires that we mention this disorder. We also mention ambiguous genitalia because it vividly makes our point that genitals and other external features of sex are determined by hormonal messengers and not by chromosomes. We are also well aware of the use of surgery and hormone treatment in the sexual reassignment of adult transsexual patients. Both of these cases should focus your attention on the fact that the only step in primary or secondary sexual differentiation that is controlled by genes is the choice of testicles or ovaries. The rest is determined by the hormones produced by testicles or ovaries.

## HOW HORMONES WORK

As puberty begins, these hormones will also determine the development of secondary sexual characteristics. The high levels of testosterone flowing through a male's body are responsible for his physically masculine appearance, whereas the high levels of estrogen flowing through a female's body are responsible for her physically feminine appearance. A general definition of a hormone is a chemical messenger that is produced by one cell type, released into the bloodstream, and received by a target cell with the intention of altering this target cell's pattern of gene expression. The type of hormone concentrated on here is the *steroid hormone*. Steroid hormones include testosterone and estrogen: testosterone is excreted in the testes and the adrenal cortex in the male, whereas estrogen is excreted by the ovaries and the adrenal cortex in the female. Actually both sexes produce both hormones. However, there is much more testosterone than estrogen in males and much more estrogen than testosterone in females.

When hormones are excreted into the blood stream, they are transported to the target cell where they are needed to carry out their purpose which is altering gene expression in that target cell. These target cells usually have receptors that wait for the needed hormone to float on by. When the receptors detect the presence of the hormone, they bind to the hormone and carry it through the plasma membrane of the cell to the awaiting nucleus. Once inside the nucleus, the hormone and the receptor complex bind to DNA regulatory elements and promote gene expression. The protein products of these testosterone- or estrogen-induced genes actually allow the cells and organs to execute sexual differentiation.

## MUTATIONS IN THE GENE THAT ENCODES THE ANDROGEN RECEPTOR

Now imagine if a steroid hormone receptor in your body was not there or was not functional. Your hormones would continue to flow throughout your body, but would never be detected for cell utilization and gene expression. For example, the testosterone receptor is encoded by a gene on the X chromosome, known as the androgen receptor (AR) gene. Loss-of-function alleles of the AR gene are referred to as AIS mutations. Because these mutations prevent the production of any functional testosterone receptor, the resulting phenotype of XY individuals is a disorder known as androgen insensitivity syndrome (abbreviated AIS and sometimes known as testicular feminization) which is seen in approximately 1 in 20,000 live births.

In XY embryos bearing the AIS mutation, the indifferent gonads develop as testes and the Müllerian ducts regress in the presence of MIF. However, the cells of this embryo cannot sense testosterone (the receptor has been mutationally destroyed). Instead, the somatic cells respond to the low level of estrogen secreted by the adrenal cortex of both sexes and develops along the female pathway. Consequently the child at birth appears as a perfectly normal female. However her vagina ends in a blind duct. There is neither a cervix nor a uterus. Fallopian tubes are also absent. Instead of fallopian tubes there are two fully developed but undescended testes pumping out high levels of testosterone. These females are externally normal throughout childhood, puberty, and adult development with the exception of a scarcity of underarm and pubic hair. Obviously they will neither menstruate nor be able to bear children. Given that such women are detected as teenage girls there is a serious issue in terms of how much information should be provided during diagnosis and counseling, how it should be provided, and who should receive the information.

As noted above, this female also possesses a fully developed set of testes that are located internally above where the scrotum would normally be. Throughout this female's life, her testes excrete high levels of testosterone that flow throughout her body without end. One quick note about the internally located testes. It is essential that such a female have her testes removed by the time she is 31 or 32 years old because of an increased risk of testicular cancer. The testes are not in a scrotum and consequently are kept at a higher temperature than normal, thus greatly increasing the rate of cancer.

AIS females exhibiting androgen insensitivity syndrome are usually taller than the average woman. They are also often considered to be quite attractive by contemporary standards. Psychologically, they are usually stable and live normal lives as women and, when they so choose, as mothers of adopted or stepchildren.

## X-LINKED SPINAL BULBAR ATROPHY AND
## ANOTHER ROLE OF THE TFM GENE

It turns out that the AR gene also plays a later role in life, namely in the maintenance of skeletal muscle. X-linked spinal bulbar atrophy is categorized as a *dystrophy.* Many of you are familiar with muscular dystrophy, which is actually a general name for many diseases that are characterized by muscle degeneration or "wasting." All of the muscular dystrophies will be discussed further in a later chapter.

Like the AIS mutation, mutations for X-linked spinal bulbar atrophy disease alter the function of the AR gene, the gene that codes for the testosterone receptors. Within the coding region of the AR gene a specific CAG triplet repeats itself 21 times in normal individuals. However, in affected individuals, the repeat occurs anywhere from 40 to 52 times. The greater the number of repeats, the earlier the onset and the more severe the symptoms. *(We will revisit the issue of triplet repeat expansion in more detail in Chapter 10. Until then, just trust us that it happens.—RSH )* X-linked spinal bulbar atrophy patients experience a progressive muscle weakness in the back as well as increasing pain in the back that eventually becomes excruciatingly uncomfortable. (A fictional description of an encounter between two individuals bearing these two rather different types of mutations in the testosterone receptor gene is presented in Box 6.1.) Unfortunately, there is no cure and there is no answer as to why the triplet is amplified. Another scientific mystery is how testosterone receptors and muscle weakening in the back are interrelated.

The reason we discussed AIS and X-linked spinal bulbar atrophy in this chapter should be quite obvious now. The malfunction of alleles in the AR gene results in complete somatic feminization, and amplification of the CAG triplet in the AR gene results in atrophy of the spinal–skeletal muscu-

### BOX 6.1 SPINAL BULBAR ATROPHY AND AIS: A FICTIONAL ENCOUNTER

Steve could remember Grandpa's bad back. In his family, Grandpa's back was famous for keeping Grandpa out of action. When Steve started having back trouble he was not ready to admit that he might be heading down the same painful road that had kept Grandpa bedridden during his visits to his grandparents' farm. Steve tried exercise and aspirin and finally, when the pain got to be too much, he gave up and went to see a doctor, all the while trying to tell himself that he surely had just strained something. He was surprised at some of the questions the doctor asked, not just about himself but also about his family and whether it was his maternal grandfather that had the back problem. He was surprised when the doctor suggested that they should run some genetic tests.

On his return visit to the doctor, he was given a diagnosis that he didn't really understand at first: X-linked spinal bulbar atrophy, something he had never heard of before. The doctor said, "You have a mutation, a change in the genetic information in one of your genes. This gene is responsible for creating a protein that plays an important role in the spinal–skeletal musculature. Because of a very small change in this gene, you are experiencing back pain. Unfortunately, we cannot make the mutation go away, we can only offer pain control and physical therapy.

He didn't want to believe it, but then he recalled his grandfather and the way the grandkids used to have to run and fetch things for him when his back was "out." And this business of there being no cure — well that fit in with his recollections of Grandpa, too.

A gene! He had certainly read enough about genes recently, as the newspaper headlines seemed to constantly be full of announcements about genes that had been found. "So, doctor, what kind of gene are we talking about? What does it do and what is wrong with it?"

"The gene you are dealing with is the AR gene. It encodes for . . . . "

Steve had a fit. He said, "THE AR GENE? The testosterone receptor gene? I've heard of that one. In fact, a friend of mine met this gorgeous model and I thought he was so lucky to be dating her until he found out that she had this AR mutation. The way I understand it she's actually a guy or something! She's got a Y chromosome even! You'd NEVER believe it by looking at her! She seems happy enough with her life — married my friend in fact — but this is a disaster! Am I going to end up looking like her? Frankly, doc, I am pretty firmly committed to staying male!"

Steve was quite rattled at this point, but the doctor calmed him down.

"No, Steve, you are not about to turn into a girl. Even though the gene is the same, the kind of change that has occurred in the gene is quite different. It is likely that your friend's wife ended up developing as a female because her body does not make the androgen receptor. She did not start out as a guy and then turn into a girl later. The difference between the sex predicted by her chromosome complement and the sex she turned out to be actually started out before she was even born."

"You developed as a male, so your androgen receptor is able to perform some of its key functions. In your case, your body is making the AR protein, the androgen receptor, but the protein has had some extra amino acids added to it so that it does not function perfectly. When we look at your family history we can see other cases of back pain that might have been caused by this same mutation."

"Oh, right, Grandpa," said Steve.

"Yes, Grandpa," said the doctor. "In addition to our discussions of therapies and medications that we will use to help you cope with your back pain, I would like you to also consider going to talk to the medical genetics people here in the clinic. The genetic counselor you would talk to can help explain more about this mutation and its origins, and can also help you and your family decide what other kinds of information or testing you would like. Since it looks like this might run in your family, you will need to decide what you want to do about telling your relatives. I expect that some of them are also going to have questions and concerns, especially if any of them are suffering from back problems!"

lature. The key point: Different mutations in the same gene can result in very different phenotypes.

## A SUMMARY OF THE BASIC MECHANISM OF SEX DETERMINATION

Sex determination in humans requires at least four elements: a Y chromosomal signal; a sensing mechanism in the indifferent gonad to respond to the Y chromosomal signal; a hormonal signal produced by the gonads (androgens or estrogens); and a set of sensors, androgen and estrogen receptors, in the somatic tissues. With this background in place, let us now consider the effects of altering the number and structure of the sex chromosomes.

## CHANGES IN THE NUMBER OF SEX CHROMOSOMES (OR SEX CHROMOSOME ANEUPLOIDY)

The term aneuploidy refers to an abnormal chromosome number. One vivid example of human aneuploidy is *Klinefelter syndrome.* Klinefelter males possess an XXY genotype and occur at a frequency of 1/1000 live births (see Fig. 6.3 for the meiotic nondisjunction events that produce Klinefelter males). Klinefelter males often have small testes, are sterile,

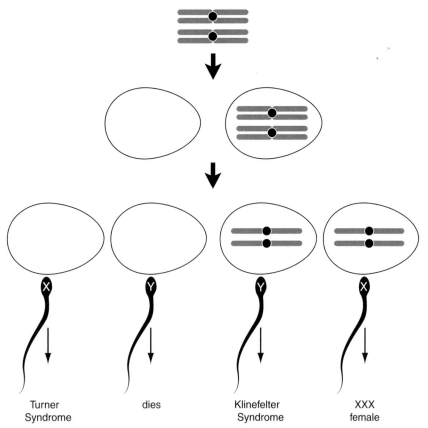

FIGURE 6.3    Meiotic nondisjunction in the mother resulting in Turner or Klinefelter syndrome.

and have some external feminization such as breasts and hips. Because the testes have atrophied, not a lot of testosterone flows through the male body. There is almost an even amount of estrogen and testosterone in the individual's system. It has been found that some Klinefelter males have social problems and some even have a subnormal intelligence level.

*(We are going to say this as clearly as we can. Not all Klinefelter males will have all of the characteristics just listed. Nor will whatever descriptors we use later for Turner syndrome, Down syndrome, or any other human disorder apply to ALL the people affected by that disorder. People are different: These disorders can manifest themselves quite differently from one person to the next. When a disorder is discussed we will try to give you a general description of the common features of that disorder. We know there will be exceptions, but it is the best we can do.—RSH and CAM)*

Refer to the story in Box 6.2 about Rob and Steve. By giving Steve testosterone he continued to develop more or less along the lines of a normal male, but as a consequence of low (uncorrected) levels of testosterone Rob began to experience feminization during puberty. This should make a lot of sense after the previous explanation of the development of secondary sexual characteristics and hormones. If it does not, please reread the section about hormones. This is crucial.

*Turner syndrome* is the consequence of another genetic mishap during meiosis (or mitosis, which will be discussed in the next chapter). These individuals are female with an XO genotype, so there is only one functioning X in these women. These individuals are always sterile with ovaries that appear as a rudimentary streak. Turner females may also be shorter than average and show immature development of the breasts and genitals. The frequency of this syndrome is 1/5000 births. The actual frequency of conceptions with Turner syndrome is much higher, perhaps a few percent, but 99.9% of XO conceptions abort spontaneously. Turner syndrome will be discussed in much greater detail in Chapter 7. (Please see Fig. 6.3 for meiotic nondisjunction that produces a Turner female.)

Two other types of genotypic abnormalities of the sex chromosomes, which fail to have fancy names, are *XYY males* and *XXX females*. Although XYY males do not exhibit any characteristic set of abnormal phenotypes, they are often taller than average males and some fraction of them may have behavioral problems. Although most of these males lead normal lives, it has been found that the frequency of XYY males is strongly increased in various kinds of prisons and mental institutions, especially among inmates greater than 6 feet in height. The frequency of XYY males is 1/1000 among live births, but may be as high as 20/1000 among inmates in some institutions. This matter will be discussed in far greater detail in Chapter 19. XXX females are not associated with any abnormal phenotype but there may be a decrease in fertility and, in some cases, a lower-than-average intelligence (as assayed by IQ).

## BOX 6.2 TWO STORIES OF KLINEFELTER SYNDROME

On August 28, 1970, in a small town in southern California, two young boys were born of separate families who had no social connection whatsoever. Robert was born into the family of two busy professionals and Steve was born to a single working mother. Both boys had normal childhoods. Eventually, Rob and Steve hit puberty along with the rest of their classmates and began to undergo bodily changes. However, unlike the rest of the boys in their age groups, Rob and Steve underwent a peculiar form of puberty. Rob's parents failed to realize that their son Rob was developing breasts, female-like hips, and possessed abnormally small testes. Steve also started to develop female physical traits but his mother sought a physician immediately.

After a careful examination of Steve, the doctor agreed that Steve was not developing like a normal male should. Steve was then put through a series of tests in search of an explanation for Steve's abnormal development. After his first blood test, the doctor found that Steve had a low level of testosterone flowing through his body; in fact, there was an almost equivalent amount of estrogen and testosterone in his body. A chromosome analysis (karyotype) showed that Steve possessed two X chromosomes and one Y chromosome, a disorder known as Klinefelter syndrome.

Steve began taking testosterone on a regular basis and the breast and hip changes seemed to abate. Instead, Steve developed facial and bodily hair, a strong upper body build, and a deep voice. Steve's classmates saw Steve as a normal male and there was no question of his masculinity. Nonetheless, Steve would always remain sterile and possess very small testes. Psychologically, Steve felt fairly stable and never felt that he was any less of a man. Steve's mother was extremely supportive of her son and treated him just as any mother would treat her teenage son.

Meanwhile, Rob was going through his pubescent years wondering why his body did not look like the rest of the males in his class. Rob did not have much body hair, was soft spoken, and did not possess much upper body strength. His classmates thought he was strange, and in high school gym class he always changed in the privacy of the bathroom.

As Rob got older he began engaging in socially unacceptable and rebellious behavior. His disorder was eventually diagnosed during his first stay at juvenile hall. Unfortunately, his emotional problems were so great at this point there was little that could be done to help his adjustment. One disorder, two very different outcomes.

Both Turner (XO) and Klinefelter (XXY) syndrome raise an interesting question: If all the Y does is to determine sex, then why are XO females so often lethal as early embryos and why are live-born females affected with Turner syndrome? Similarly, all normal females have two X chromosomes, so why is having a second X so deleterious in XXY males? For the answer, follow us into the next chapter.

# 7

# Sex Causes Problems:

## The Inactivation of the Second

## X Chromosome

*Cassi was a first-year student at a well-known medical school on the East Coast. Cassi loved medical school and was already at the top of her class. One day, she was sitting in her medical genetics class listening to a professor discuss genetic disorders and their effects on the individual. This topic excited Cassi, especially since she herself had a genetic disorder. The professor continued to go through a list of the disorders and eventually came to the disorder that Cassi knew best, her own.*

*"The Turner syndrome female carries only one X chromosome due to a loss of one of the X chromosomes during meiosis or early embryonic mitosis. She may be of shorter stature, possess rudimentary ovaries, thus making her sterile, have immature breast and genital development, and usually will be of below average intelligence."*

*An intense feeling of rage swept over Cassi. BELOW AVERAGE INTELLIGENCE, USUALLY? She had never heard that in her life! How could he say such a thing? She had met other women with Turner and not one of them displayed below-average intelligence. For the first time in Cassi's life, she felt that all of her awards, plaques, and honors were being cruelly taken away from her due to an X that she did not possess.*

*Cassi was very open with her disorder. She never felt that it was something that she had to overcome; it had never held her back from anything. She began to wonder what other people who knew of her disorder must think. "What if people think I got into medical school because of my condition?" "What if others think that the reason why I have done so well is because I am getting special attention or help?" Cassi felt that all of her hard work and discipline had been belittled by one simple statement that her genetics professor had so casually stated in class.*

*Cassie decided to speak to her professor the next day. "Dr. Wright, do you remember that brief comment you made about Turner syndrome females being of 'usually below average intelligence'?" Cassi asked as she took in a deep breath.*

*"Yes," he responded. She really did not have to say more. He knew that he was looking at one of the best students in the first-year class who was a Turner's female.*

*"I am a Turner's female, Dr. Wright, and I hardly believe that I am of below average intelligence. Yesterday, I felt that you were slapping all of the women with Turner syndrome in the face. Where did you get such untrue data?"*

*The next day during lecture Dr. Wright started lecture off with a brief epilogue to his last lecture on genetic disorders:*

*"Class, during the last lecture I spoke of genetic disorders and the effects on the individual. I would like to clarify something. When I explained that many of these individuals are sterile, have shorter or taller stature, immature somatic sex characteristics, or any of the other biological factors that are connected to their condition, those were facts. However, when I discussed psychological and mental capacity, I was providing you with isolated cases, studies I had read or my own personal experience. Please realize that we are still dealing with the individual and that there will be large variability in these characteristics. There is nothing written in stone about how these people will live their lives and what their mental capacity will be. Always remember, you are dealing with individuals. . . . Do not pass unnecessary judgment. You may hurt someone or make them feel unworthy."*

*Cassi smiled to herself and suddenly realized that she was one of the smartest people Professor Wright had ever met, and she had just educated him.—CAM*

### THE X CHROMOSOME: HOW UNIQUE IT IS!

The last chapter went on and on about the importance of possessing either two X's or an X and a Y chromosome. It also suggested that there is really only one known gene of importance on the Y (the TDF gene). Did it occur to you to ask why it is that males can survive with only one X, whereas a female possesses two X's? The X chromosome carries many important genes that are essential to life, so how are males capable of surviving with only one X? No autosomal monosomy is viable, so why is monosomy for the X chromosome a normal condition? Matters get worse when you think about Turner syndrome; why aren't most XO fetuses viable if all they are missing is a Y? The answer lies in the fact that the X is special in a number of ways. This chapter will give you a full explanation of just "how unique" the X chromosome is. We will then embark on a discussion of Turner syndrome, the disorder that affected Cassi.

## X INACTIVATION

As explained earlier, *monosomy* (possessing only one copy of a chromosome) is incompatible with life. In other words, a zygote carrying only one copy of one of the autosomes would not survive beyond the first few weeks of gestation. Similarly, a *trisomy* (possessing three copies of a single chromosome) is also incompatible with life, although there are individuals who do possess three copies of chromosome 21 (Down syndrome). Those who possess three copies of chromosomes 13 and 18 usually do not live beyond their first year of life; however, due to medical advancements, some do live but are usually extremely disabled. So, if all monosomies and most trisomies are lethal *in utero*, then how is it that males can survive with only a single X and females can survive with three X's?

The mystery of the X chromosome was solved by noting two key facts. First, the levels of enzymes encoded by genes on the X chromosomes are the same in the cells of both males and females. Studies conducted with cells cultured in the laboratory showed that levels of enzymes encoded by X chromosomal genes is the same in XY, XO, XX, XXY, XXX and XXXX cells. Second, as shown in Fig. 7.1, in a female who is heterozygous for two alleles of the X chromosomal gene called G6PD, each of her cells produces only one variant of this protein: either G6PD-A or G6PD-B. Thus, this woman is composed of two different populations of cells; she is a salt-and-pepper mosaic of both her parents X chromosome contribution. Such discoveries led to the conclusion that all but one X chromosome is inactivated in human cells. Hence, only ONE X chromosome is active in human cells, whereas the other is inactivated. In some of this female's cells, her

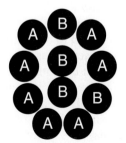

cells expressing the A or B
form of G-GPD in a
heterozygous woman

note: no cell expresses both

**FIGURE 7.1**    G6PD expression in individual cells of a heterozygous female.

father's X is expressed while her maternal X is inactive and in some of her cells, her mother's X is expressed while her paternal X is inactive.

You may now be wondering how does this work when an individual possesses three X's (XXX females are found occasionally, they appear quite normal), four X's (a few such individuals have been documented), or two X's and a Y (Klinefelter). The same rule applies: Only one X is active; the remaining X chromosomes are inactivated.

Thus, for most genes on the X chromosome, there is no difference in gene dosage between males and females (there are a few X chromosomal genes that escape inactivation in females, but they will be discussed later). This should fully answer the question as to why males can survive with only one X chromosome, whereas normal females possess two X's. This should also provide you with clear reasons why aneuploidy of the X chromosome is so much less harmful than the aneuploidy for autosomes. Put simply, an XXX embryo can survive because it inactivates two X chromosomes. Thus, like a normal cell (XY or XX), it has one functioning X chromosome. In contrast, the cells of an embryo carrying three copies of chromosome 21 have no way to simply inactivate the extra copy!

## TIMING, RANDOMNESS, AND SELECTION OF X INACTIVATION

X inactivation occurs in humans quite early in human development, when the embryo consists of approximately 32 total cells. The individual cells are indifferent as to which X they will inactivate: It will be either the paternal or the maternal X. Thus, this is a *random* event. However, in each individual cell, after either the maternal or the paternal X has been chosen as active, all mitotic descendants of that cell will be fixed. Some studies have shown that inactivation occurs earlier in extra-embryonic membranes (placenta, chorion, allantoic membrane), and, for some unknown reason, the paternal X is preferentially inactivated in those tissues. It is also important to note that the inactivated X is "re"-activated during female gametogenesis. During female meiosis it is necessary to have both X's active in order to carry out the meiotic process properly.

Because X inactivation occurs randomly and at an early stage in fetal development, it is possible for a tissue or an organ to be comprised entirely of either maternal or paternal X chromosomes. Thus, in the case of a woman who is heterozygous for a mutation that has caused color blindness in her male relatives, theory would say that she could become color blind if by chance all of the cells that will eventually form her eyes were to inactivate the same X homolog carrying the normal copy of the red–green color gene complex. However, this rarely happens. Why? By the time X inactivation takes place, there are many progenitor cells for the eye and we might only see problems if many cells independently inactivate the same X, which is highly unlikely. For X-linked diseases affecting tissues in which only a few progenitor, or stem, cells exist at the time of X inactivation, we might expect that sometimes we would see the disease, or some manifestations of the disease, turn up in females because of the inactivation of the normal X in most or all cells in that tissue (see Fig. 7.2).

In certain cases, there is cellular selection for nonrandom inactivation if a female is heterozygous for a mutation that causes cell death in the absence of a normal homolog. Mutations such as these are called *lethal mutations*. If the lethal-bearing X chromosome is inactivated, then the cell is fine. If the normal X chromosome is inactivated, then the cell will die. As shown in Fig. 7.3, cells of the embryo are under strong selection pressures to develop solely from cells where the normal X chromosome is activated. Such females will only express one of their X chromosomes in all of their cells. Please note that inactivation itself is quite random in these embryos, but the cells that inactivate the normal X die and thus do not contribute to the further development of the fetus.

## MECHANISMS OF X INACTIVATION

The inactivated X is a highly condensed blob of chromatin that sits against the edge of the nucleus in nondividing cells. This structure is called a *Barr body*. The cells in the human female possess a single Barr body, whereas an XY male possesses no Barr bodies. A male with Klinefelter (XXY), however, will possess a Barr body, whereas XXX cells will possess two Barr bodies.

To further explain the Barr body and how it manages to inactivate a chromosome, it is necessary to elaborate a bit more on the fine structure of chromosomes. The DNA and protein that make up chromosomes comprise a substance called *chromatin*. It is useful to differentiate chromatin into three types: *euchromatin, heterochromatin,* and *facultative heterochromatin.* Chromosomal regions that are composed of euchromatin possess active genes, whereas chromosomal regions that are composed of heterochromatin contain extremely simple sequences of DNA that are repeated thousands or millions of times in long string-like arrays. Because these heterochromatic chromosomal regions contain very few active genes, this region can be considered practically genetically inert. In fact, heterochromatic regions usually are not even capable of expressing genes that have been placed within their vicinity.

Heterochromatin is normally found surrounding the centromere, as well as near the tips of chromosomes (the tips are called *telomeres*). Scientists believe that the heterochromatic regions play some type of important structural role in chromosome function.

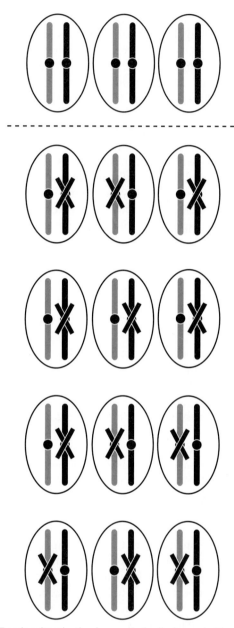

FIGURE 7.2    Random inactivation in a pool of cells descended from only three progenitor cells.

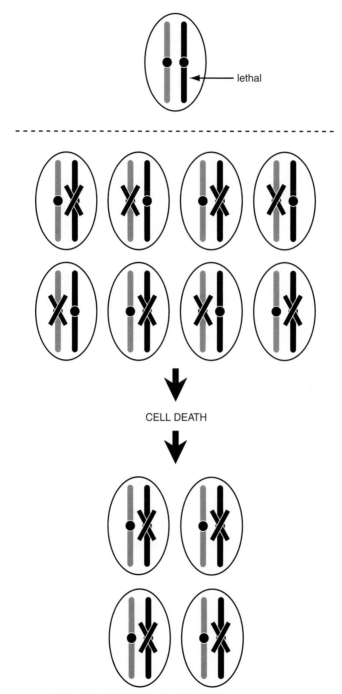

FIGURE 7.3 X inactivation in an embryo heterozygous for a recessive lethal mutation.

Some of the differences between euchromatin and heterochromatin and their roles in regulating gene function are still unknown. *(Scott claims that such questions are what's going to put his children through college someday.—CAM)* What is known is that heterochromatic regions replicate later than euchromatic regions and that heterochromatic regions stain darker with dyes than euchromatic regions. Also, almost all known genes lie within the euchromatic chromosomal region, whereas, as said earlier, heterochromatic chromosomal regions do not possess many active genes. However, it is worth noting that heterochromatic regions do contain genes that encode for ribosomal RNAs. Facultative heterochromatin has the unique ability to be active and appear euchromatic in some cells and to be inactive and appear heterochromatic in others. Thus, by making a region of a chromosome or an entire chromosome itself behave as facultative heterochromatin, the cell can shut that region of the chromosome or the entire chromosome off, thus preventing gene expression.

The X chromosome in human cells may be described as facultative heterochromatin: The active X chromosome behaves as a euchromatic chromosome, whereas the inactive X chromosome behaves as a heterochromatic chromosome. This is why the inactive X (the Barr body) appears darkly staining and highly condensed in the cell. In summary, the cell can achieve X inactivation by converting the entire X chromosome into heterochromatin.

The actual mechanism by which an X is inactivated remains a scientific mystery. The few things that are known about X inactivation are that inactivation requires and spreads out from a single site known as Xic. Also, as stated earlier, the inactivated X is "re"-activated in the germline and is then "de"-activated in early embryogenesis. DNA sequences are not in any way damaged or altered by activation. In any case, whatever the mechanism may be, daughter cells of any given cell inactivate the same X as the cell from which they originated, thus making this a mitotically stable process.

## THE INACTIVE X CHROMOSOME IS REACTIVATED IN THE FEMALE GERMLINE

Although one X does inactivate in the germ cell progenitors of early female embryos, it is eventually reactivated in oocytes prior to meiosis. This reactivation reflects a stringent requirement for two active X chromosomes in oogenesis. The absence of the second X chromosome causes rapid death (atresia) of oocytes during fetal development. The result is both sterility and small rudimentary ovaries. This phenotype is often observed in females with Turner syndrome and in XY females that lack the TDF gene.

XXY Klinefelter males also inactivate and reactivate the second X chromosome. The presence of the extra X causes death of the male germ cells during early puberty and subsequent atresia of the testicles. This testicular atrophy results in a great diminishment in the ability of many of these males to make testosterone. The resulting testosterone deficiency may explain many, if not most, of the characteristics of Klinefelter syndrome.

In summary, both females and males have only one active X in their somatic cells, but if the somatic cells only require one X, then why is Turner syndrome so dramatic?

## TURNER SYNDROME

Individuals bearing only a single X chromosome and no Y chromosome are said to have Turner syndrome. As noted earlier, XO females are sterile with ovaries that appear as a rudimentary streak. Turner's females may also be shorter than average and show immature development of breasts and genitals. More importantly, such females are the very rare survivors of a condition, being XO, that is almost always fatal in early fetal development. Indeed, 99.9% of all XO conceptions abort themselves spontaneously *in utero*. Thus, possessing only one X chromosome, minus a Y as well, is almost always lethal to the zygote, yet males can survive with only one X chromosome and we also know that females inactivate one of their X chromosomes in all of their cells. So why is that second X important? One would think if a normal female has only one X activated, what is the big deal with having only one X?

The answer lies in a conclusion that was made by scientists in Boston and London: There are essential genes that are present on both the X and the Y chromosomes, which are required in TWO doses. Moreover, such genes escape inactivation in cells with two or more X chromosomes. The existence of such genes provides a straightforward explanation for Turner syndrome. XO females possess only one copy of such genes, whereas both XX females and XY males possess two. These genes define a few homologous regions between the X and the Y chromosome, the so-called pseudoautosomal regions. Recall from the previous chapter that one such pseudoautosomal region performs the vital function of mediating X and Y chromosomal pairing and recombination during meiosis.

An example of this finding is the gene RPS4, which is present on both the X and the Y chromosome. RPS4 is responsible for encoding a protein that makes up part of the structure of the ribosome. The ribosome is a little factory that our cells utilize to assemble proteins. Reducing the copy number of RPS4 will thus reduce the number of ribosomes in half and so reduce the protein synthetic capacity of the cell by a similar amount. Moreover, because each ribosome requires a similar number of copies of most of its component proteins, one wouldn't want the level of RPS4 proteins to be much less than that of other ribosomal proteins encoded by autosomal genes. Thus, the easiest scenario to imagine is that there must be two copies of RPS4 in each cell. Having this gene present on both the X and the Y chromosome, and allowing it to escape inactivation in females accomplishes this goal. Indeed, the RPS4 gene escapes inactivation in females, as do all genes in pseudoautosomal regions.

XX females heterozygous for a deletion that includes the RPS4 gene display a Turner-like phenotype. This may not be too surprising. Studies that have been done with fruit flies have shown that the genes whose proteins contribute to the structure of the ribosome (so-called ribosomal proteins) are required in two doses for normal development. In flies bearing only a single copy of this gene, a so-called *Minute* phenotype is seen in which the flies develop much more slowly and display a large number of severe defects. One could easily imagine that such a defect in protein synthesis might also explain Turner syndrome, or at least the lethality of the vast majority of XO conceptions. Although such a model may well be correct, much more scientific research needs to be done to prove undoubtedly that possessing only a single copy of the RPS4 gene is responsible for Turner syndrome; there may be many other contributing factors involved, including additional genes from the pseudoautosomal region.

Given the very high lethality of Turner syndrome, why do such individuals ever make it to birth, much less beyond? The sterility of Turner females, like that of XY females missing the TDF gene, reflects the requirement for two functional X chromosomes in the female germline, but what about the other phenotypes? Where do they come from, and why are they so variable? Perhaps the answers lie in other genetic variations in the genomes of these individuals that slow development down long enough for XO cells to catch up. There may be another explanation; perhaps the reason for the survival of the rare Turner female and the vast phenotypic variability among such live-born females may be attributed to what scientists call *mosaicism.*

Some live-born cases of Turner syndrome are not composed solely of XO cells: They are composed of both XO and XX cells. The *mitotic* loss of a single X chromosome in one cell out of several cells present early in zygotic development may produce a combination of both XX and XO cells. Thus the resulting individual possesses both XX and XO cells. As long as an XX karyotype is present in those cells that *absolutely require* two X chromosomes, then the individual will survive. Those cells that do not require two X's will be able survive as either XX or XO cells. The more XO cells the individual possesses, the more severely affected the individual will be, whereas the more XX cells the individual possesses, the less affected the individual will be. This gives us a good basis for understanding the great variability in the phenotype of the Turner female.

The existence of genes such as RPS4 provides a useful way to understand Turner syndrome. As noted earlier, the testicular atresia seen in XXY males, the result of that second X chromosome, may explain most of the phenotypes of Klinefelter syndrome. Do these explanations suffice to explain all of the features of these disorders? Clearly not! We suspect that as our understanding of the X and Y chromosome structure and function grows, we will find a progressively better set of explanations for these disorders. We will also get a better explanation of the manner in which evolution has solved two of the problems created by our sex-determining system: How do you segregate nonhomologous chromosomes and how do you deal with the inequality of X chromosome dosage between males and females?

Chapter 8 will consider the development of another aspect of sexual development, namely sexual orientation. Before we leave this topic, however, we do want to note that the solutions to sex chromosome segregation and dosage compensation used in humans are not general. A look at fruit flies, worms, or wasps reveals fundamentally different mechanisms used to solve these problems (recall that it was the number of X chromosomes and NOT the presence or absence of a Y chromosome that determined sex in fruit flies). During the process of evolution, many living systems have apparently tried to tackle these problems and have created very different but equally effective systems. Why? Wouldn't it have been easier for all of life to do things the same way? We will say it again: Evolution is NOT Michelangelo and the Sistine chapel. It is a teenage kid with a broken car and no money. It did whatever worked. . . .

# 8

# GENDER IDENTIFICATION AND

# SEXUAL ORIENTATION

*There is a big difference between sexual orientation and gender identification. . . .*
*—A guest on a television talk show. The individual in question is a transsexual (male to female) who now considers herself to be a lesbian.*

In the last couple of chapters we have taken some care to draw out the manner in which the presence or absence of a gene called TDF determines the fate of indifferent gonads. We have then described the manner in which hormones produced by the ovaries or testicles control both primary and secondary sexual development, but we have been careful to limit ourselves to discussions of *biological* sex and to avoid the much trickier issue of *gender*. We scrupulously avoided terms such as "masculine" or "feminine" wherever possible, limiting ourselves to more strictly biological terms such as "male" or "female."

We, like most of our predecessors, chose this course of writing because the issue of gender identification is complex and difficult to address. Indeed, most human genetics texts simply avoid this issue entirely, considering it either the proper province for psychology tests or too "hot" an issue to handle, yet the issues are there. We know that there are such things as male sex roles and female sex roles, even if we have trouble defining them precisely. *(To paraphrase a Supreme Court Justice, "I may not be able to define masculinity, but I know it when I see it."—CAM Umm . . . but aren't appearances sometimes deceiving, Catherine?—RSH)* The ability of children to identify with and absorb these sex roles is referred to as *gender identification.*

## MECHANISMS OF GENDER IDENTIFICATION

Although most male infants come to think of themselves as boys and then men, and female infants to think of themselves as girls and then women, there are notable exceptions. These include transsexuals who believe that they are "a woman trapped in a male body" or vice versa. These individuals seem to have made a gender identification that is discordant with their biological sex. (Transsexuals, who are truly convinced that their self-identified gender role is in conflict with their biological sex, should not be confused with transvestites, men who dress in women's clothing. The origins and causes of transvestitism are quite unclear and confusing. Although some workers have suggested that some transvestites may be individuals who have internalized both gender roles and at various times express one or the other, the prevailing opinion is that transvestitism is less an example of failed gender identification and more an example of fetishistic behavior. The confusion on this point simply underscores our difficulty in understanding gender roles and gender-specific behaviors.)

Very little is known about the mechanisms that underlie gender identification, but they seem unlikely to be biological in nature. Recall that in the last chapter we talked briefly about children born with ambiguous genitalia. Depending on the exact condition of the infant, the treatment of those children often includes sexual correction, sometimes to a sex different than the chromosomal sex. These studies indicate that as long as the reassignment is done early, and the parents are comfortable with the outcome, the child will usually identify properly with the appropriate gender. Moreover, there are also very rare, but well-documented, cases of errors in sex identification in infants with ambiguous genitalia; for example, some parents brought home an infant they believed to be a boy who matured into a fully developed female during puberty. Again, these children usually identified with the sex role to which they had been raised and the changes at puberty were catastrophic. These studies suggest that sex role choice is not innate, but rather that children assume the sex role or gender identification that their parents teach them.

*(However, we would at the very least be remiss in not noting that there is some evidence to the contrary, i.e., that sex role identification may be either innate or determined quite early. For example, people often cite an example described by the noted authority on sexual differentiation, John Money, that relates a story of two identical twin males. Both children were born as normal males, but one twin suffered irreversible damage to the penis and scrotum*

*during circumcision. The surgeons were able to reassign the sex of the infant to female and, according to several sources, the individual matured into quite a healthy and happy young woman. It has become clear, however, that several inaccuracies have crept into this story over the years. Although the individual was raised as a girl and underwent estrogen supplementation during her/his early teens, she/he eventually chose to discontinue that course of treatment. The individual chose to reembrace his male sexuality and he now lives as a man. Because of such cases, a number of workers believe that some aspects of gender identification are biologically determined.—RSH)*

Our prejudice — and it is just that — is that much or most of what we call gender is learned, probably buried so deeply in our language and culture that we are not aware of the many ways in which we teach it to our children. *(My wife and I have gone to great efforts to avoid gender-specific toys as we raise our children. But the battle was hopeless. As I write this, my 5-year-old son's room is filled with toy airplanes, trucks, dinosaurs, and tanks, not to mention Power Rangers. My 9-year-old daughter's shelves are crammed with Barbie dolls, toy horses, and craft kits. My only hope is that as both children discover sports and computers, the dust will slowly gather on the boy- or girl-specific toys.—RSH)* Alternatively, one could also imagine that environmental cues picked up early in life act to stimulate or initiate the activity of one or another present biological pathway. Such a model might then explain the fact that the vast majority of people assume a gender identity consonant with their biological orientation. Errors, possibly genetic errors, in this pathway–choice mechanism might well explain transsexualism, although there is no strong evidence at this time to support or discredit this conclusion.

We will now turn our attention from the development of sexual or gender identities to the development of sexual orientation. To quote two major workers in this area, Simon LeVay and Dean Hamer, "Most men are sexually attracted to women, most women to men. To many people, this seems only the natural order of things — the appropriate manifestation of biological instinct, reinforced by education, religion, and law. Yet a significant minority of men and women — estimates range from 1 to 5% — are attracted exclusively to members of their own sex" (*Scientific American,* May 1994).

This statement by LeVay and Hamer raises some fascinating questions. First, just how is sexual attraction or orientation determined? Is it biological? Are there genes that direct males to be attracted to females and vice versa? Second, if sexual orientation is biologically programmed, then how are we to understand the etiology of cases where men choose men as lovers or women choose women? Could such people reflect genetic variation in genes for sexual orientation? If such genes and such variation do exist, then what are those genes and what do they do? These questions will be our focus for the remainder of this chapter.

The third question deals with the issue of the "natural order of things." We have no sense of the degree to which such behaviors are natural or unnatural. We leave such distinctions to others. To us, sexual orientation is but a variable phenotype within the human population, a phenotype whose etiology we wish to understand. However, we also understand that these issues are rather touchy for some people. We will try hard to be as sensitive and inoffensive as possible in the following discussion; we apologize in advance to those for whom we are not careful enough.

## GENETICS OF SEXUAL ORIENTATION

We define sexual orientation as the sex to which a given individual is sexually attracted. When, as is usually the case, a person is attracted to an individual of the opposite biological sex, that individual is referred to as *heterosexual.* In the case where people are attracted to others of the same sex, they are referred to as *homosexual.* In the last few years, we as a society have come to refer to homosexuals as "gay" men and "lesbian" women.

As noted earlier, sexual orientation, gender identification, and biological sex are usually concordant, but there are exceptions. It is especially important to disconnect the concept of sexual orientation from that of gender identification for two reasons. First, the vast majority of gay men and lesbian women have gender identifications consonant with their biological sex. Secondly, there is at least some evidence that, unlike sex role or gender identification, sexual orientation and partner choice may very well have a genetic basis.

Prior to the start of this decade, there were two lines of evidence to suggest that male homosexuality might be genetic. The first line of evidence came from studies of *heritability.* Heritability will be discussed in much more detail in Chapter 17, but for now we will simply define it as a tool for asking just how much of the variability for some trait within a given population is genetic and how much is environmental. One way to estimate heritability is to compare how often identifical twins are identical (*concordant*) with respect to some trait versus the concordance in fraternal twins. Presumably, identical twins have both identical genotypes and identical environments, whereas fraternal twins have different genotypes but identical environments. Thus if identical twins are much more frequently concordant for a trait than fraternal twins, the heritability is said to be high (i.e., much of the phenotypic difference in a trait that is observed in the population is said to have a genetic cause).

There are many problems with basing arguments about much of anything on heritability, primarily because any estimate of heritability is only good for one population at the time that estimate is made. Still, as long as one is careful not to make such comparisons and not to make too many assumptions, heritability can be a reasonably good indicator of just how much of some difference in the population *might* be genetic. In the case of homosexuality, such estimates are suggestive of an important role of genes in determining the phenotype. For identical male twins of whom one twin is gay, in more than 50% of cases the other twin is gay as well. Fraternal twins of gay males are substantially less likely to be gay (24%), as are other male siblings (13%). Similarly, for lesbians, in more than 50% of the cases the other twin is gay as well. Fraternal twins of lesbians are substantially less likely to be gay (16%), as are other female siblings (13%).

These data need to be looked at carefully. The first thing these data tell us is that the determination of sexual orientation cannot be wholly genetic. If it were, the concordance of identical twins would be 1.0, as it is for traits such as color blindness, Down Syndrome, or cystic fibrosis. Clearly, genotype alone cannot account for those 50% of cases in which the twins were discordant. However, genetically identical individuals are more likely to be concordant than genetically different individuals if some aspect of the trait is genetic. Moreover, the second thing you should have noticed is that brothers of gays or sisters of lesbians tend to be gay or lesbian far more frequently than expected (the frequency of homosexuality among American males is generally thought to be about 1–5%).

How are we to explain these data? Well, one explanation is that there may be genotypes that predispose individuals toward one orientation to others, but that these genotypes interact with the environment. The allelic differences, if indeed they exist at all, cannot be fully *penetrant.* They must predispose rather than direct.

In support of a genetic basis for male homosexuality, there are many pedigrees in which male homosexuality appears to segregate in a predictable and sex-linked fashion through a given kindred. The pedigree shown in Fig. 8.1 is an example. Face it, if you didn't know the phenotype under consideration, you would have glanced at the pedigree, and thought "sex-

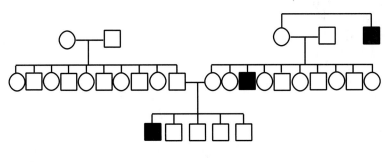

■ INDICATES A HOMOSEXUAL MALE

FIGURE 8.1    A pedigrees for male homosexuality (redrawn from Fig. 3 of Hamer *et al.*).

linked recessive" to yourself, and moved on. It was pedigrees such as this that caused Dean Hamer and colleagues at the National Institutes of Health (NIH) to begin a careful study of the genetics of male homosexuality in the early 1990s.

The initial subjects in Hamer's study were 76 self-identified gay men and their relatives over the age of 18, as well as 38 pairs of homosexual brothers and their relatives. The researchers recruited subjects through an AIDS clinic, through local gay organizations in Washington, DC, and through advertisements in gay-oriented magazines and newsletters. Before we go any further, we need to think about this study group. This population consists of gay men open enough about their sexuality to agree to both participate in this study and involve their extended families. In other words, all of these men were fully "out." *(But aren't there lots of gay men and lesbians who are not this open, not this out? Certainly there are such people, but they aren't in Hamer's study group. Does that bias the results? Maybe. Maybe not. But it's a really good question, hold onto it!—RSH)*

Okay, so they defined their population, but how do they assess the phenotype, especially among relatives? *(This is a nontrivial issue. I give this chapter as a lecture in one of my courses at the University of California. I am now used to students coming to my office afterward to ask questions such as "Well, once when I was at camp, I was with another guy. Does that mean I'm gay?" or "I have a boyfriend and all, but something happened with myself and another girl last year." Sometimes these are difficult categories to pigeonhole people into. But to do genetics, we must define a phenotype clearly and unambiguously.—RSH)* Well, Hamer and colleagues used two methods to ascertain the phenotype: self-assessment (just ask the people themselves and then write their answer down on a piece of paper) and a set of psychological tests known as the Kinsey scales (see Fig. 8.2).

Amazingly, both the self-assessment of the relatives and the assessment by the original members of the study group (the *probands*) were remarkably accurate. To quote the Hamer paper, "All (69/69) of the relatives identified as definitely homosexual verified the initial assessment, as did most (27/30) of the relatives considered to be nonhomosexual; the only possible discrepancies were one individual who considered himself to be asexual and two subjects who declined to answer all the interview questions." *(Wow, can you imagine, someone gets a call from some person in Washington where the caller wants to ask a bunch of real personal questions about just what this person does in bed and with whom. And some people decline to answer the questions! Go figure. . . .—RSH)* Thus, again quoting Hamer, "describing individuals as either homosexual or nonhomosexual, while undoubtedly over simplistic, appears to represent a reliable categorization of the population under study." Those graphs suggest that there are no bisexuals in the Hamer population. Does Hamer's data say that such

THE KINSEY SCALES

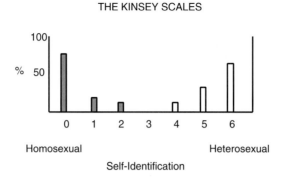

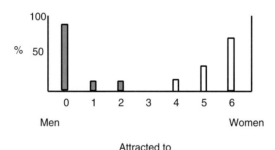

FIGURE 8.2   The Kinsey scale analysis of Hamer's population. Filled bars denote self-identified homosexual men, while open bars denote self-identified heterosexual men.

individuals don't exist? No, Hamer's data say only that such people don't exist in his study group. Again, this is a very highly selected and precisely defined study group.

Hamer's analysis yielded some fascinating conclusions. They confirmed previous studies by noting that the brother of a gay man had a 14% chance of being gay as compared with a 2% chance for males in the general population. Hamer and colleagues also noticed something even more interesting among more distant relatives of these gay men. Maternal uncles and sons of maternal aunts had a higher chance of being gay (7–8%) than expected for the general population, but no such effect was observed for paternal uncles or sons of paternal aunts. This is highly suggestive of an X-linked determinant for sexual orientation. (Remember, paternal uncles or cousins on the father's side cannot share an X with the homosexual male in question, but maternal uncles or cousins can! (See the schematic pedigree diagram in Fig. 8.1.) Thus, Hamer only saw a high frequency of concordance among relatives who could share an X chromosome, evidence again for sex linkage of a gene or genes with poor penetrance.

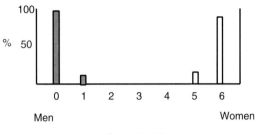

Fantasize About

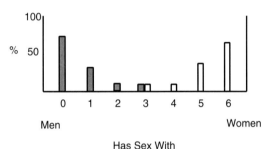

Has Sex With

FIGURE 8.2 (*Continued*)

To study this effect further, Hamer's group further refined their study group to 38 families in which at least two sons were gay. They excluded any family in which the father was gay (ruling out any case of father-to-son transmission) or that had more than one lesbian relative. It was hoped that by excluding other causes of homosexuality, this population might display further increases in the rates of homosexuality among maternal uncles and maternally related male cousins. It sort of worked. Maternal uncles and sons of maternal aunts had a higher chance of being gay (10–13%), and again no such effect was observed in paternal uncles or sons of paternal aunts. However, the ratios were still lower than those expected for a simple Mendelian trait (50% for a maternal uncle and 25% for a son of a maternal aunt).

To sort this out, Hamer and colleagues fell back on what I call the first rule of genetics: If a gene exists, you can map it; and, if you can map a gene, then it exists. They now focused on only 40 pairs of gay brothers. Realize how important this was. No matter what other environmental conditions need to be met, or other genes need to be present to develop homosexuality, these conditions must all be there in these males. If there really is an important gene on the X chromosome that determines homosexuality, and if the mother was heterozygous for that gene (a likely possibility), then these brothers should share a specific region of one of the mother's X chromosomes, the region bearing the allele predisposing them to homosexuality.

Consider thinking about it this way: If a woman is heterozygous for the color blindness allele (cb), and two of her sons are affected, it is because both inherited the cb allele from their mom. Because recombination is frequent on the human X, approximately five exchanges per bivalent on the long arm of the X alone, one doesn't expect the brothers to share the same alleles for various genes at other sites, but they should share the cb allele and other closely linked alleles as well.

As shown in Fig. 8.3, Hamer and colleagues analyzed the inheritance by these pairs of brothers of a large number of genetic markers distributed at various points along the length of the X chromosome. We will return to a full discussion of genetic markers in Chapter 12. Suffice it to say for now that they are simply innocuous allelic differences between homologous regions on homologous chromosomes that can be very easily assayed and that let us distinguish between the two chromosomes! For most of the X chromosome, the genetic markers or alleles at each site were randomly distributed between the two brothers. They were as likely to have two

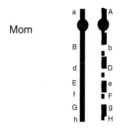

Mom

Two Homosexual Sons

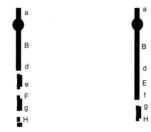

FIGURE 8.3    Genetic markers and sibling pair analysis.

different alleles as they were to both share one of the two given maternal alleles.

However, for one region, Xq28, near the tip of the long arm of the X,
the two homosexual brothers shared the same alleles in 33 out of 40 cases.
This finding is very highly significant and provides strong evidence for an
important gene in this region. Although this is a very strong result, you do
need to note that there were seven pairs of brothers who carried different
alleles within the Xq28 region. Thus, even in this highly refined population,
the Xq28 region cannot account for all cases of homosexuality. Nonetheless,
the basic result is still indicative of some *correlation* between the genotype
at Xq28 and sexual orientation phenotype in a large fraction of these sibling
pairs. In a more recent study, Hamer and colleagues have repeated this
mapping and extended their studies to include heterosexual brothers of
the two gay brothers initially studied. Not surprisingly, these heterosexual
brothers carried the alleles in the Xq28 region that were shared by their gay
brothers much less often (22%) than would be expected by chance (50%).

We should point out that Hamer's data also suggest that whatever genes
might be at Xq28 that affect sexual orientation in men, there is no evidence
for that gene or any other genes affecting sexual orientation in women.
Although the heritability of lesbianism is as high as it is for male homosexuality, very little is known about a genetic basis for lesbianism, or indeed if
one exists at all.

In summary, Hamer's data suggest, and only suggest, that there may be
a gene, or genes, in region Xq28 of the X chromosome that affect sexual
orientation. However, not everyone in the scientific community agrees with
that suggestion. Many workers worry about the small sample size of the
study group. Other workers are trying to repeat the results using different
populations. The final verdict is anything but "in." Finally, even if Hamer
and his group are correct in assigning a role to some region of the X
chromosome in determining sexual orientation in their population, it is not
at all clear how generalizable this result is to the general population.

So why did we bring this up? What if Hamer and friends are wrong?
Are you holding a chapter of useless, or even wrong, information in your
hands? Have all of us wasted our time, not to mention Academic Press's
paper? We think not. Whether or not Hamer is right about this gene and
this place on the X, we are confident that the approach we have described
will be the type of approach that will definitely answer these questions.
Their paper is a paradigm for studies to come.

Beyond those caveats, suppose Hamer and friends are right. Just what
kind of things might such a sexual orientation gene specify? How might it
work? In studies even *more* controversial than the genetics just described,
Simon LeVay has presented evidence for a structural difference between
a small region in the brains of gay and heterosexual men thought to be
involved in controlling sexual orientation. As interesting as these results

are, there are serious concerns about the correctness of LeVay's interpretation. This does not mean that LeVay is wrong, just that the scientific community is far from fully persuaded on this matter. *[For those interested in a critical assessment of both LeVay's work and that of Hamer and others, we recommend a pair of articles by LeVay and Hamer (the first article) and by William Byne (the critique) in the May 1994 issue of Scientific American.— RSH and CAM]*

## SUMMARY

With this chapter, we close both this section and our formal discussion of sexual differentiation in humans. It is hoped that you have thought at least a little about the mechanism by which this process is controlled. Perhaps you have spent a moment contrasting the roles of the AR and TDF genes or even thought about those genes in relation to genes that might control such things as sexual orientation. It is hoped that you have sat back and asked yourself, "Why not let testosterone or estrogen control sex role identification or sexual orientation?" or "Why might evolution have selected against total genetic determinism in these processes, allowing instead for these processes to have strong environmental components?" *(If you have had these questions, or something even close, please call Scott right away. He has a lab bench waiting and several graduate students who are eager to make a geneticist out of you.—CAM)* Maybe you are even wondering if Catherine and I were just a bit too quick to dismiss a role of genes in sex role identification. Perhaps, perhaps not.

It is our hope that as you continue on in this book, and as we discuss other issues relevant to human genetics, that these questions will stay with you. We very much want you thinking about questions such as "Well, what population did they study and how do you define the phenotype?" and "How do we parse out the effect of genotype from environment here?" It is also hoped that you don't forget about either sex linkage or X inactivation, as these concepts will come up over and over again. We now turn our attention to the rest of the genome and to the larger issues of human biology and disease. Sex was just the beginning. . . .

*(A concluding personal note: Shortly after Hamer's paper came out, I was interviewed by a local paper. The paper asked for comments on the work and I did the best I could. The afternoon after the article came out, I received an anguished call from a man whose son had just come home from college and announced that he was gay. The poor man was distraught and confused. Referrals to P-FLAG, a support group for families of gays and lesbians, were not what he wanted. What he did want was for me to assure him that Hamer was right. He wanted to be told in certain terms that his son was born that way, genetically preprogrammed to be gay. He wanted*

*me to assure him that his son was born with an X chromosome, one derived from his mother, that made this announcement inevitable. I couldn't tell him that. All I could do was suggest that he had loved his son before the announcement and that his son was still the person he loved. I explained the Hamer paper as best I could several times, caveats and all. I was never able to offer the concrete assurances he wanted or felt he needed. He kept saying that if he could be sure that his son's homosexuality was genetic, then it would be okay. I would like you to think about that.—RSH)*

# When Meiosis
# or Mendelian
# Inheritance Fails

Back in Section I we spent a lot of time discussing the behavior of chromosomes and the mechanisms of heredity. This section will discuss the consequences of the failure to transmit genes in the proper fashion. Chapter 9 will discuss the consequences of failed chromosome segregation and attempt to identify some of the causes that may underlie nondisjunction. Chapter 10 will discuss the changes in genes that can happen as chromosomes pass through the genome, both genetic changes such as mutation and reversible changes known as "imprinting."

This chapter will introduce some of the most basis tools of human genetics, such as chromosome analysis (karyotyping) and DNA analysis. It will also *introduce* the concept of prenatal testing and diagnosis.

# 9

## FAILED CHROMOSOME
## SEGREGATION AND
## THE ETIOLOGY OF
## DOWN SYNDROME

*His name was Earl and I met him when we were both in high school. We were both freshmen and in the same physical education class. The similarities stopped there. Earl had Down syndrome, a disorder caused by an imprecise segregation of chromosomes into the egg from which he arose. Like virtually all kids with Down syndrome, Earl was severely mentally retarded and had been that way since birth. His intellect had stopped somewhere around that of a five year old, but his body never got the message. Because of his limitations and because of the facial features that are characteristic of Down syndrome, Earl became the butt of an awful lot of high school humor. Kids couldn't resist making fun of the way he walked, ran, or talked.*

*For four years I spent one hour a day in class with Earl. For various reasons, we became friends. Earl never did figure out why people made fun of him, but he knew that they did. Once one of our high school sports heroes tripped him in the hallway during break. The humiliation he felt seemed to hurt worse than the bloody lip.*

*During my junior year of high school, the March of Dimes held a public lecture on the basis of birth defects. For reasons that have long faded into a mist of high school memories, I made my father drive me to that lecture at a nearby (nearby by California standards) college. There I learned for the first time about genes and chromosomes, but mostly about Earl and about me. I developed a passion or, in Catherine's words, "an all-consuming obsession" with understanding how heredity works and how genes make us what we are.*

*This book, especially this chapter, is a child of that obsession.—RSH*

We spent a lot of time earlier in this book talking about meiosis. We talked about the importance of precisely segregating homologous chromosomes so that each sperm and each egg ends up with 23 chromosomes, one of each pair. We made the point that doing meiosis properly is crucial for normal reproduction. The failure of two chromosomes to segregate properly is called *nondisjunction*. Nondisjunction could occur either because two homologs failed to pair and/or recombine or because of a failure of the cell to properly move the segregating chromosomes on the meiotic spindle (those track-like structures made of microtubules on which the chromosomes pull themselves to the poles). Indeed, considerable evidence exists that much of human nondisjunction may be due to failures of the processes that move chromosomes to opposite poles at meiosis I. These failures or errors can include defects in the spindle integrity, in the motors on the chromosomes that move them to opposite poles, or in whatever proteins hold the sister chromatids tightly together during the first meiotic division.

Regardless of its physical mode of origin, the result of nondisjunction is the production of gametes that are *aneuploid* (i.e., gametes that carry the wrong number of chromosomes). When such a gamete is involved in a fertilization event, the resulting zygote is also aneuploid. Section II discussed sex chromosome aneuploidy, such as Turner syndrome, Klinefelter syndrome, and XYY males. Like Turner syndrome, most autosomal aneuploids are virtually always incompatible with life. We are not aware of a baby ever being delivered alive that carried one or three entire copies of chromosome 1, chromosome 2, or chromosome 3. . . . Well, you get the point. It's not that such zygotes don't arise, they do. As discussed later, meiosis is really pretty sloppy in humans, and autosomal nondisjunction occurs at a reasonably high frequency. We just don't think much about it because most nondisjunction events in humans result in the formation of inviable zygotes that abort spontaneously quite soon after conception. Humans just don't tolerate *aneuploidy* very well at all. Most aneuploid zygotes are just not viable and result in early spontaneous abortion. This is especially true for *monosomies*, zygotes with only a single copy of a given chromosome. With the exception of the X or Y chromosome, *monosomy* (just one copy of a given chromosome) is simply not compatible with life and leads to spontaneous abortion.

However, a few types of trisomic zygotes are capable of survival. These are trisomy 21 (Down syndrome), trisomy 18, and trisomy 13. We begin with a discussion of Down syndrome.

## DOWN SYNDROME OR TRISOMY FOR CHROMOSOME 21

Trisomy 21 is by far the most common and best known genetic defect. It is also the single most common cause of mental retardation among individuals outside of institutions. Babies with this disorder grow slowly and have poor muscle tone. They have characteristic facial anomalies, most especially slanting, or epicanthic, eyes and small, frequently low-set noses. They also have rather short fingers and short broad hands. They have a wide skull that is somewhat flatter than usual at the back, and the irises of the eyes often have obvious spots. In many cases, the mouth appears to remain partially open due to a protruding tongue.

Perhaps the most commonly known aspect of Down syndrome is mental retardation. IQs normally range from 25 to 50 (the average IQ in individuals who do not have Down syndrome is 100), but some children do show higher levels of mental functioning, including near-normal IQs and the ability to read and write at the 6th- to 12th-grade levels. There is serious controversy, and some increasing degree of optimism, regarding just how much children with Down syndrome can be expected to achieve. Clearly some children with Down syndrome greatly exceed our expectations and grow up to be happy and reasonably self-reliant adults, but many are severely limited. Growing evidence shows that certain types of early educational intervention, especially computer-assisted teaching, may be of real help to children with Down syndrome. Moreover, in these times many adults with Down syndrome may be expected to live either semi-independently or independently and often are able to enter the work force. (In some cases such individuals seem to do better in so-called "sheltered workshops," but other individuals are able to find work in various aspects of the public and private sectors.)

Half of the children born with Down syndrome are born with severe heart malformations. These and other life-threatening conditions are so severe that some of these children die before age 5. However, for those children who survive the fifth year of life, the average life expectancy is 50 years. Even so, these individuals are at high risk for leukemia and for a degenerative brain disorder similar to Alzheimer disease. Adult males are sterile, but adult females are fertile; and yes, from the few scattered reports available, it appears that half of their children are born with Down syndrome. *(On one hand this result makes good sense—half of the eggs produced by such a female should carry two copies of chromosome 21. But given that some 80% of Down syndrome fetuses spontaneously abort, it's hard to see why the final result should be a 1:1 ratio.—RSH)*

Children with Down syndrome possess on extra copy of a region on the long arm of chromosome 21. How do we know that? Well, most Down syndrome children possess an extra chromosome in each of their cells. We know this because we can take blood from anyone and examine their chromosomes by a process called *karyotyping*. In this process we can capture some of their white blood cells in the part of the cell cycle called mitosis when their chromosomes are well separated and condensed. An example of such a mitotic figure lying on a glass microscope slide is shown in Fig. 9.1. Chromosomes can often be distinguished from each other by length, by centromere position, and by a staining process called *banding* (again, see Fig. 4.2). Banding is a modern-day process that seems like alchemy in which chromosomes that have been affixed to the surface of a

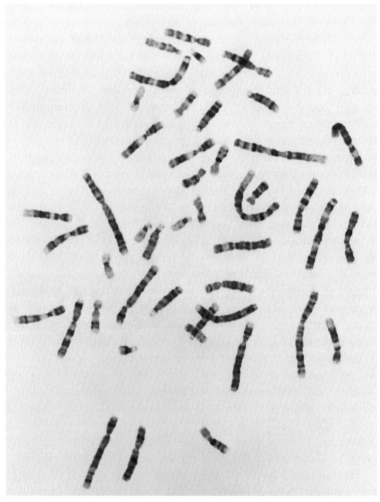

FIGURE 9.1    Mitotic chromosomes (a spread).

glass microscope slide are stained with dyes. These dyes bind to some regions along the length of each chromosome but not to others. By virture of these techniques, all 23 different human chromosomes can be differentiated (Fig. 9.2). Both members of a pair of autosomes (known as homologous chromosomes) will of course have the same length, centromere position, and banding pattern (a set of dark and light bands that make the chromosome look like the bar codes used by automated check-out devices in stores). Children with Down syndrome usually carry an additional copy of chromosome 21.

In about 5% of Down syndrome cases, a baby is found that carries only a part of chromosome 21. Clearly an extra copy of even part of one of the smallest human chromosomes may have disastrous results. Why should triplication of a chromosomal region be bad? The answer lies in the complex interactions between various genes in our genomes and of the proteins they encode. Many genes produce proteins that act to regulate other genes, and thus the amount of protein they produce must be tightly regulated, as either too much or too little can be disastrous. In other cases, many different proteins must combine together to form large complex structures in the cell. In those cases the exact concentration of each protein component may well be critical. We still aren't sure exactly which triplicated genes are the real culprits in producing the various components of Down syndrome, but very good scientists are working hard to find out.

If we could answer that question, could we *cure* this disease? We just don't know. Our guess, and it is only a guess, is *maybe*. On one hand, even if we knew exactly what was wrong for any given component of the syndrome, we are unlikely to be able to fix the whole array of problems. We base this pessimistic prognostication on the evidence that some of the most profound problems arising from the triplication of chromosome 21 arise during development before birth, in structures such as the lungs and heart. Once that damage is done we may simply have to rely on traditional medical and surgical processes for help. However, some of the Down syndrome problems, such as leukemia and Alzheimer diseases, develop after birth and may become subject to prevention or improved medical intervention if enough is understood about the genetic basis of this disorder. On the other hand, if the most conspicuous component of this disorder, mental retardation, is truly due to the triplication of a single gene, then it *may* be possible someday, with early enough prenatal diagnosis (see later), to correct or at least ameliorate that problem. It remains to be seen whether the study of triplicated chromosome 21 genes will give us any capability to intervene in postbirth developmental processes that might impact IQ and other capabilities.

Consider this: What if we could identify potential parents who are likely to be at risk, identify such fetuses very early, and try to fix it then? Great

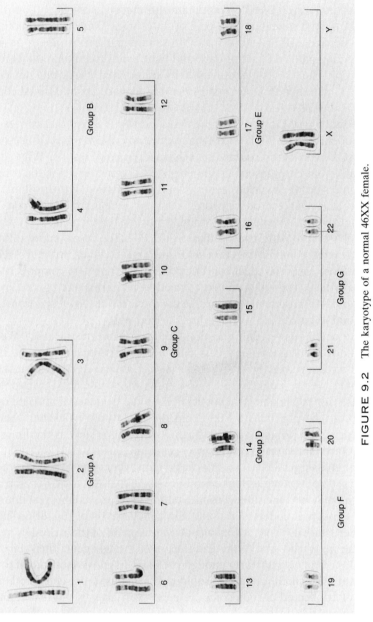

FIGURE 9.2    The karyotype of a normal 46XX female.

question! So how might we do that? Well, right now our crystal ball is really hazy, especially since we do not even know what most of the critical genes or proteins are, but maybe someday we will have enough information to be able to help. What is clear is that if we are ever going to accomplish such a goal, we are going to have to understand the causes of the nondisjunction that results in trisomy 21. We will also need to be able to identify those potential parents at the greatest risk and to develop much better techniques for prenatal diagnosis. Only then will we be able to think about prenatal therapies! Each of these issues will be considered.

## MOST CASES OF DOWN SYNDROME ARE DUE TO NONDISJUNCTION IN THE MOTHER

Virtually all cases of trisomy 21 are due to nondisjunction at the first meiotic division in the mother. We know this because we have developed several methods for determining which chromosome came from which parent. Two of these methods, banding polymorphisms and the analysis of DNA markers, are in common use. The nature and use of banding polymorphisms are described in Box 9.1, whereas DNA analysis is described

---

### BOX 9.1 BANDING POLYMORPHISMS

Each chromosome has distinctive banding patterns. However, sometimes banding patterns for the same chromosome can slightly differ among individuals. Because these differences occur at the same sites on homologous chromosomes they may be thought of as allelic polymorphisms. In the schematic diagram shown in Fig. 9.4, we consider a couple in which each copy of chromosome 21 can be distinguished by polymorphisms in the banding of the centromeric region. The consequences of nondisjunction at meiosis I (MI) versus meiosis II (MII) in the mother are shown in Figs. 9.5 and 9.6.

Clearly, the consequences of nondisjunction at meiosis I in the mother are quite different than for meiosis II. In the case of MI nondisjunction, both maternal polymorphisms are present in the cells of the child with trisomy, whereas in the case of nondisjunction at MII, the child with trisomy possesses two copies of one of the maternal chromosomes. Moreover, and please draw this out for yourself, the chromosomal constitution of the child with trisomy is also quite different than would be observed had the nondisjunction occurred at either division in the father.

in real detail in Section IV. Using these techniques we can show that the extra copy of chromosome 21 almost always comes from the mother. Why? Well, the answer to that question is currently in rather hot dispute, but the best guess is that in male meiotic cells the failure of two autosomes to properly pair and segregate results in the cessation of meiosis and, indeed, in cell death.

In other words, there appears to be a checkpoint in male meiosis that asks whether all of the chromosomes are properly paired and ready to segregate from their partners. If the answer to that question is "no," then the meiotic cell may be doomed and the potentially aneuploid sperm are never produced. Such checkpoints apparently do not exist in most female meiotic cells (*oocytes*). In oocytes the cell seems committed to completing meiosis despite whatever failures may occur. Thus, although the checkpoint system in sperm is not foolproof and some aneuploid sperm do get through, it does work efficiently enough to result in a substantially reduced frequency of this kind of nondisjunction in sperm as compared to eggs.

Why is there such a curious sexual dichotomy here? Well, think about it this way: every time a male ejaculates he releases some 200 million sperm. If one or a few of those sperm were to be aborted early in the process of sperm production because of a meiotic failure, the probability of the remainder of that group of sperm (the other 199,999,999 sperm) finding an egg would probably not be greatly diminished, especially when the cost is compared to that of an aneuploid sperm finding the egg, an event that usually results in death of the embryo.

Because a human female usually only makes one egg a month, the investment in producing that egg is quite high. If an error occurs, there is no normal sibling gamete to take the aneuploid egg's place. Thus, perhaps the cost of continuing is no greater than the cost of stopping. Regardless of whether these ideas are correct, and they may well be, it is clear that aneuploid sperm are made far less often than aneuploid eggs.

## THE MATERNAL AGE EFFECT

Not only is Down syndrome normally a consequence of nondisjunction in the mother, but the frequency of Down syndrome births increases dramatically with advancing maternal age (see Fig. 9.3). The risk is about 1/1,500 for women under age 25. By age 35 the risk increases to somewhere between 1/300 and 1/100. By 40 it may be as high as 1/100 to 1/50. By the mid-40s the risk is in the range of 1/25, an increase of almost a hundredfold. (*If you are a college-aged woman reading this, you have probably started to do some math. "Hmmm, I'm in my early 20s now. Two more years to finish college, but four for graduate/law/medical/business school, a few years to get established, a few more to find a guy who isn't crazy . . . ." Yes, it is*

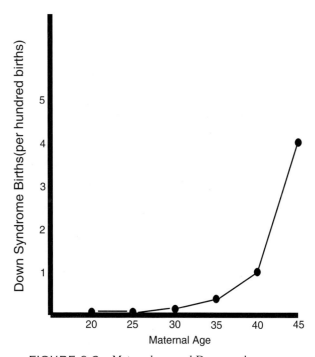

FIGURE 9.3    Maternal age and Down syndrome.

*something to think about.—RSH)* We should note that age-dependent in-creases for the frequency of nondisjunction are not observed in men, per-haps partly for the reasons described in the previous section.

The basis for the maternal effect is unknown, but it is likely to be a consequence of the long delay between prophase and the first meiotic

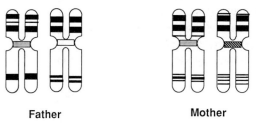

**Father**          **Mother**

FIGURE 9.4    A schematic diagram of chromosome banding polymorphisms for chromo-some 21 in two parents.

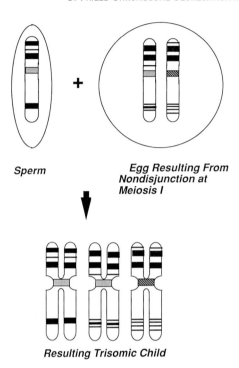

Sperm                Egg Resulting From
                     Nondisjunction at
                     Meiosis I

Resulting Trisomic Child

FIGURE 9.5    Consequences of nondisjunction in the mother at meiosis I. (Note that polymorphisms remain heterozygous in the disomic egg.)

division in human oocytes. Recall that human oocytes begin meiosis long before the female is born. They arrest in the late meiotic prophase, after pairing and recombination have taken place, some months before birth, where they stay in a state of suspended animation until the girl/ woman hits puberty some 10–13 years later. Then several oocytes are given permission to restart meiosis each month, but usually only one of these is actually allowed to complete the first meiotic division! Realize then that the egg ovulated by a 45-year-old woman has been stalled part way through meiosis for 45 years. One can imagine that quite a lot could go wrong in that period of time, and apparently often does. However, just what actually does go wrong remains a bit of a mystery. (Indeed, the absence of an age effect on nondisjunction in men may also reflect the fact that, unlike female meiosis, male meiosis is a continuous process, with no built-in halts.)

Some people argue that the egg's ability to build a normal spindle deteriorates over the years, whereas others propose defects in the mechanisms that hold sister chromatids together or in the resolution of some types of

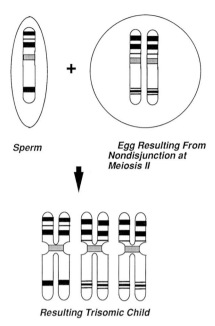

Sperm    Egg Resulting From
Nondisjunction at
Meiosis II

Resulting Trisomic Child

FIGURE 9.6    Consequences of nondisjunction in the mother at meiosis II. (Note that the egg now carries two copies of the same maternal chromosome.)

recombination events. I have my own model to explain age-dependent nondisjunction in human females. Obviously, I like that idea more than the other existing models, but you just need to know that many people have good ideas and that no one yet has the answer.

Still, only about a quarter of kids with Down syndrome are born to mothers over 25. Think about that for a minute. Yes, the risk is much higher for older moms, but then far more women are bearing their children in their 20s than in their 40s.

Why is it important to understand the cause (etiology) of nondisjunction, especially in older mothers? The answer lies in our desire for better tools for prenatal diagnosis. In the best of all possible worlds we would like to be able to give every woman the best estimate of her risk prior to conception. In addition, we need to develop effective and preferably nonintrusive mechanisms for early prenatal detection of Down syndrome fetuses.

The following section describes briefly the available types of prenatal diagnosis for Down syndrome. More detailed discussions of each of these techniques will be found in the chapter devoted to prenatal diagnosis.

## PRENATAL DIAGNOSIS OF DOWN SYNDROME

There are two lines of diagnosis for Down syndrome: maternal age and a maternal blood test for a fetal protein known as $\alpha$-fetoprotein (AFP). Given that the risk that a woman over 35 carries a child with Down syndrome ($\sim$1/100) well exceeds the risk of the various diagnostic procedures, most mothers in this age group are advised to seek testing. $\alpha$-fetoprotein is a fetal protein that can cross the placenta and is found in the mother's blood supply. High levels of this protein in the woman's blood can indicate that the nervous system of the fetus has failed to develop properly. However, a low level of $\alpha$-fetoprotein in the mother's blood may also indicate the presence of a fetus that is trisomic for chromosomes 21, 13, or 18.

Because the number of false positives for this AFP test can be quite high (see Chapter 21), a better test has recently evolved. This test, called the "triple screen," measures two other chemicals in the maternal blood stream, human chorionic gonadotropin (HCG) and estriol (E3), as well as AFP. The combined levels of these three components of the mother's blood predict the presence of a Down syndrome fetus in 60–70% of the cases and show a much lower rate of false positives. The triple screen is now in wide use as a screening tool for Down syndrome, as well as several other fetal anomalies, in women 35 and over.

Two other risk factors include family history and chromosome anomalies in the parents. If the parents in question have had a previous aneuploid fetus or liveborn child, this increases the risk of a trisomic child in this and subsequent pregnancies. Similarly, if one parent is known to carry an altered form of chromosome 21 (e.g., a translocation in which pieces of the normal 21 have been rearranged to lie on another chromosome), then the risk of an aneuploid conception goes up greatly. Concern about possible aneuploidy in the fetus is also raised in cases where a couple has had an unusually high number of miscarriages. The occurrence of those miscarriages raises the possibility that one of the parents might carry a genetic aberration or chromosome rearrangement (see earlier) that increases the probability of producing a trisomic fetus.

In those cases where risk factors exist, or where a "positive" result is obtained using a screening test such as the triple screen, it must still be followed by other tests such as *amniocentesis*. Basically, a needle is inserted into the uterus around the 13th to 16th week of pregnancy, and a small amount of the fluid surrounding the fetus is removed. The withdrawn fluid contains a substantial number of fetal cells that can be used for chromosome or DNA analysis. A second less common test is called *chorionic villus sampling* (CVS). This test can be performed from 8 to 12 weeks after conception and requires an actually biopsy of the the tissue that will form the placenta. This test, as well as others like it, are described in more detail in Chapter 21.

*(It is sound editorial practice to stop this discussion here, without explaining why prenatal diagnosis, especially early prenatal diagnosis, is desirable. To go further, to drop the other shoe, surely means offending someone, or perhaps almost everyone. Nonetheless, we feel strongly that the question needs to be asked: If we can't cure Down syndrome, and we currently can't, then why are early prenatal diagnostic tests so incredibly important? The answer, at least in our view, lies simply in choice. Giving a woman or a couple a fetal diagnosis of Down syndrome as early in her pregnancy as possible allows her/them to choose either to terminate the pregnancy or to prepare to raise a baby with very special needs. We offer no judgment here on either of these two choices, but rather simply note that correct information, offered early, provides the soundest basis on which a woman/couple can make her/their choices.*

*Clearly, the eventual goal, once we are able to consider a prenatal therapy, is to be able to improve the alternatives that go with the word "choice." Consider the question raised earlier, "Could we develop an early detection system, figure out what gene products are in excess, and then find some way to inhibit them?" Well, could we? "Could we think about a cure?" On one hand, it sounds a bit like science fiction, more deserving of a place in an Isaac Asimov novel or a short story by Harlan Ellison than in a textbook on human genetics. I can see my colleagues in the genetics biz just shaking their heads that I should even raise such a silly possibility. But then, in case you haven't noticed, this isn't your ordinary college text and this is no ordinary possibility. I am old enough now to expect that the things we now think are impossible will be tomorrow's day-to-day certainties. Besides, whether most of us admit it or not, it is possibilities like this, kept firmly in the back of minds and discussed late at night in bars at scientific conferences, that keep my colleagues and me in our labs until the wee small hours of the night most days of the week. The answer is "yes, someday we just might, maybe." If we ever do develop such a therapy, very early prenatal diagnosis will be essential.*

*As for so many diseases in which the underlying genetic cause has been identified, whether it's nondisjunction or mutation of a gene, we are currently in a kind of limbo in which we can diagnose things that we cannot cure. However, we can see far enough over the research horizon to realize that identication of these genetic causes IS eventually going to give us ways to fix lots of things and to make things lots better even when we can't bring about a true cure. So for those of you whose first reaction is that we should stop going after the genes that get used in prenatal diagnosis for purposes of making choices about the continuation of a pregnancy, we want you to stop and look into our crystal ball with us. We want to show you that the long-term goal of these studies is to get beyond our current limitations to a point where the word "choice" includes the alternative of helping the baby.—RSH and CAM)*

## OTHER HUMAN TRISOMIES

Excluding rare exceptions, only two other human trisomies have been reported among live births. These are *trisomy 13* and *trisomy 18*. Trisomy 18 is observed at a very low frequency (1 out of 8,000 live births). The incidence at conception is much higher, but most of these embryos abort spontaneously. These usually have a characteristic set of malformations along with significant neurological deficiencies. There are usually severe cardiac problems, and survival beyond 6 months is very rare. Trisomy 13 is even less frequent, occurring approximately in 1 in 25,000 live births. The phenotype includes severe nervous system malformation, cleft lip and palate, and a host of other defects. Again, these babies live very short lives. Like Down syndrome the incidence of these trisomies increases dramatically with advancing maternal age.

Trisomies for other chromosomes are not viable, but they do indeed occur. Indeed, as determined by karyotyping fetuses that have aborted spontaneously, other trisomies and even *triploidy* (trisomy for the whole genome) are relatively common at conception, and their frequencies increase with advancing maternal age. However, all of these aneuploid conceptions lead to spontaneous abortions. Indeed, errors of chromosome segregation during meiosis must be considered a major killer of humans from conception onward.

## REPRISE

And so we have described the mechanistic origin of the story that was Earl's life, in which I was for some time a member of the chorus, that began with an error of the meiotic process, an error whose seeds may have been laid before his mother was even born. I keep wondering though, is it really fair to define Earl in terms of the meiotic error that produced him? Indeed those of us who took the time and effort to know him, who have been touched by his life, know that he is much more than that, so much more. . . .

# 10

# EXTREME MUTATION:

## TRIPLET REPEAT SYNDROMES

*The funeral hadn't been easy for Donna. She had sobbed into her fiance's arms throughout the service. Her grandmother, whom she loved more than words could tell, had died just a month before Donna's wedding. Not that they hadn't seen her grandmother's death coming. Grandma Ellen had died of Huntington disease. Death had been slow, taking its careful measure of time. Donna had cared for her grandmother through much of the last few years. She had missed her senior prom because her mom was sick and someone had to be with grandma. But Donna was happy to be helping Grandma Ellen; after all, Grandma Ellen was really the woman who raised her when she was little, since mom was on the road all the time. Donna knew that she was more like her grandmother than anyone else. Donna was an almost perfect copy of the picture of Grandma Ellen on her wedding day that adorned her mantle. They laughed at the same things, especially at Donna's mother. Neither of them ever could take that woman very seriously. When grandma got sick and Donna's mother could no longer care for her alone, it just seemed natural for Donna to become grandma's nurse and her babysitter. Donna was with her when death finally played out its game of chess. Everyone knew that Donna was just like her grandma. Donna's mother had once said that the two of them were clones. Once when Donna beat the daylights out of the boy who had teased her mercilessly in fifth grade, it was Grandma Ellen who prevented a severe punishment by telling Donna's mother, "The girl's got my genes, honey. She can't help being ornery. She's got my genes." Donna's tears focused on grandma's genes. Grandma died of Huntington disease. Huntington disease is caused by an autosomal dominant mutation. So the question, put all too simply, is which of Grandma Ellen's genes does Donna have?—RSH*

Back in Section I, we discussed Mendel's first law, the law of the purity and constancy of a gene. We made the point that genes themselves were not changed by passage through a given individual regardless of the phenotype or experiences of such individuals. In other words, we made the point that genes were constant "immutable" entities. But in reality genes can change in a stable and heritable fashion and we refer to this process as *mutation.*

Mutation is the genetic process that actually alters the base sequence of the DNA. It can include changes from one base pair to another (for example, A–T to G–C), deletion of one or more base pairs, or insertion of one or more bases. To some degree mutation can be considered to be a spontaneous process. DNA polymerase, the enzyme that executes DNA replication, is an unbelievably accurate enzyme. Still, it inserts the wrong base at a frequency of about one error in every 10,000,000,000 bases replicated. That number may seem small, but remember that every time the human genome is replicated, DNA polymerase must copy approximately 6,000,000,000 base pairs. Thus, on average, slightly less than one new base pair mutation will occur every time the human DNA complement is replicated. If that number seems small to you remember that a human being will go through some 1,000,000,000,000,000 complete replications of his/her DNA (cell division) in his/her lifetime. Thus, a very large fraction of the cells in our bodies might be expected to carry one or more base change mutations.

Some, but certainly not all, such mutations in a given gene can result in changes that either alter the protein that gets produced from that gene or prevent the expression of that gene. However, a large fraction of mutations may not have any discernible effect on the gene in which, or in whose vicinity, they occur. These "silent" mutations might include base pair changes in sequences between genes or in introns. They might also result in base pair changes within the coding regions that either do not alter the amino acid to be incorporated (recall that many amino acids can be specified by two or more codons) or result in the incorporation of a new amino acid that does not disturb protein function.

Realize that the vast majority of these new mutations will occur in somatic cells and that even those deleterious mutations that do occur will likely have little or no effect. In most cases, the resulting impairment in gene function will be "covered" or "masked" by the unmutated copy on the normal homolog. Moreover, even if such mutations were to result in the death or impairment of a single cell and its somatic descendants, it is unlikely that the loss of a single cell or cluster of cells would be terribly deleterious to the organism. (We will however consider a rather dramatic exception to this generalization when we discuss the genetics of cancer in Section IV.)

Perhaps of more interest to us is the frequency of mutations in the germline. What fraction of human gametes might be expected to carry a new mutation that will impair or prevent the proper function of a given gene? Based on assaying the frequency of those new mutations in known genes that have phenotypic consequences, scientists have concluded that each gene in the human genome will be mutated (that is to say functionally altered, not just changed silently) only once in every 100,000 gametes. By this metric, Mendel does not seem to have been so far off — mutation is a very rare process indeed. But remember, the actual frequency of DNA changes in a germline are much higher because of the types of "silent mutations" mentioned above. The observed mutation rate measures only those changes that dramatically alter gene function. Accordingly, whenever you think about mutation rate, it will be important to stop and ask yourself whether you are looking at all heritable changes happening to the DNA or whether you are looking at changes that are detectable as a phenotypic change in a living person.

However, it is worth remembering that some agents in the environment can greatly increase the mutation rate. We will return to this issue in the chapter on the genetics of cancer. Moreover, there are a few types of DNA sequences that undergo spontaneous mutations at far higher frequencies than those considered earlier. Specifically, this type of high-frequency alteration involves the rapid amplification or contraction of certain clusters, tandemly repeated simple sequences such as dinucleotides (e.g., CACACA-CACACACACA), which will be discussed as a form of genetic marker later in Chapter 13, or trinucleotides (e.g., CGGCGGCGGCGGCGGCG-GCGGCGG), which can occur in the coding sequence of genes. This is to say that under some conditions, runs of a trinucleotide repeat (the kind of simple sequence repeat we are concerned about in this chapter) can undergo expansion and contraction at high frequencies. In most such cases, the frequency of such mutations at these sites appears to increase with the number of copies of the trinucleotide repeat.

It turns out that such rapid triplet repeat expansions play an important role in the etiology of a number of common genetic diseases. These disorders are referred to as triplet repeat disorders and represent a rather egregious violation of Mendel's first law. One of the best known of these disorders is Huntington disease.

## HUNTINGTON DISEASE

As is obvious from Fig. 10.1, Huntington disease (HD) is an autosomal dominant disorder with an incidence of approximately 5 cases per 100,000 people. The mutation that causes this disorder maps to band 16.3 on the short arm of chromosome 4. HD is also known as Woody Guthrie disease,

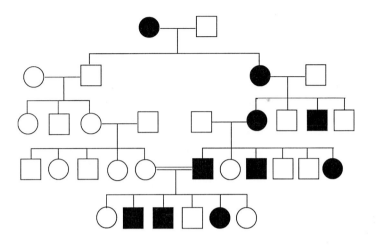

| ■ | OR | ● | INDICATES AFFECTED PERSON |

○━━━□    = CONSANGUINEOUS MARRIAGE

FIGURE 10.1    A pedigree for Huntington disease.

after one of its most famous victims. *(If you don't know who Woody Guthrie was, turn off the hard-rock CDs and take a course in the origins of the folk movement in America.—RSH)* It usually becomes evident in the fifth decade of life, but an earlier onset is often seen when the HD allele is passed on from the father. Symptoms include the progressive development of uncontrolled or jerky movements that can include cognitive and psychiatric effects as well.

Because of its late age at onset, the discovery of a genetic marker that could be used accurately to identify people carrying HD mutations was a bit of a mixed blessing. Consider the case of Donna, whose situation was just described. Donna's mother Carol (age 38) is fine, but Donna's 57-year-old maternal grandmother (Ellen) just died of Huntington disease. (This pedigree is drawn out in the first three generations of Fig. 10.1.) Because Donna is thinking about starting a family, she approaches her physician to request a test for the HD mutation. Donna has made the decision that if she does carry the HD mutation—she has a one in four chance of carrying it—she does not want to pass on the problems that she witnessed while caring for her Grandmother to her child. She would rather opt not to have children, and perhaps adopt, or choose to have each fetus tested for the disease prenatally, with the intent of terminating pregnancies that would result in an HD-bearing child.

These are understandable concerns that lead her to want the test. So fine, a test is available, and Donna wants it, so what's the problem? Carol, Donna's mother, is the problem. Because if Donna turns out to carry the HD mutation, then Carol must carry it as well. What if Carol doesn't want to know? One can easily imagine a hundred reasons why Carol might not want to know. The disease is incurable, Carol has had her kids, and a positive test means living the next couple of decades with the certainty of succumbing to her own mother's fate. Does Donna have the right to have this test done if it might reveal information about her mother's health status? Usually, the rules of medical testing tell us that patients do not have to have a test done if they do not want it, and that rules of confidentiality allow a patient to determine whether their relatives will know the results of any medical tests they have done. In this case, if Donna has the test done, not only might this tell Carol something she doesn't want to know, but it also would tell Donna something that Carol might not have been willing to tell other family members, even if she wanted to know herself.

We have no easy answers to this dilemma, nor do the medical geneticists. There is a high-tech gambit, known as *embryo selection,* that might let Donna be assured that her own children will be free of the HD mutation. Basically, this technology requires that Donna and her husband conceive their children by a process called *in vitro fertilization* (fertilization in a test tube). Once the zygotes have started to develop, doctors can test them for the presence of the HD mutation. Only embryos that do not carry the HD mutation are implanted into Donna's uterus and allowed to develop into babies. Using this technology would allow Donna to be sure that she does NOT pass the HD mutation onto her children without necessarily providing any information about Donna or her mother's genotypes in terms of the HD gene mutation. The doctors doing the test will know, but they will keep their silence if so requested. But realize that in this process the doctors may have to discard any HD mutation-bearing embryos they identify. This latter step raises troubling ethical questions for some people in our society. We will talk about this and other options in the chapter on genetic counseling at the end of the book.

*(Indeed, we will tell you a lot more about the razzle-dazzle frontiers of human reproductive biology, but the answers won't get clearer. We would be really happy if you took some time to put yourself in Donna's and Carol's shoes and think about their dilemma. Whose right to know or "not to know" is greater? Talk to your roommates, friends, or even your parents about it. The answers you come up with may tell you a lot about yourself.—RSH and CAM)*

So how do working medical geneticists solve this dilemma? Carefully, very carefully. Each situation is approached carefully, with much opportunity for discussion and counseling. There have been research studies whose sole purpose has been to try to sort out these types of human issues. It is actually reassuring to realize that along with all of the money being spent on finding genes, research funds are also being spent to try to solve the ethical quandries that go with and to determine how to best work with people who are trying to sort out such profound dilemmas.

Dr. Nancy Wexler, who played a leadership role in the international consortium that cloned that HD gene, has applied the name Tiresias complex to the dilemma of making the choice regarding whether to be tested for something for which there is no cure. The name comes from the blind seer Tiresias who, in *Oedipus the King* by Sophocles, said, "It is but sorrow to be wise when wisdom profits not." In describing the Tiresias complex, Dr. Wexler asked, "Do you want to know how and when you are going to die, especially if you have no power to change the outcome? Should such knowledge be made freely available? How does a person choose to learn this momentous information? How does one cope with the answer?" (Wexler, N. S., "The Tiresias Complex: Huntington's Disease as a Paradigm of Testing for Late-onset Disorders." *FASEB J,* **6,** 2820–2825, 1992). (We note that Wexler herself is at risk for HD and has chosen not to be tested.)

If the ethical issues associated with this disease are complex, the kind of mutational change found in the HD gene, and other genes like it (see below), is also odd enough to define a new class of genetic lesions, the triplet repeat syndromes. As shown in Fig. 10.2, the mutation underlying Huntington disease is due to the amplification of the triplet codon repeat CAG (which encodes glutamine, as shown back in Table 3.1). This CAG repeat is located in a gene cleverly called IT15 (for "interesting transcription unit 15") before it was ever realized that this was THE interesting transcription unit that an international alliance of Huntington researchers had pursued for at least a decade. Normally, this gene includes a stretch of 10–36 copies of the glutamine codon arranged as one long tandemly repeated array. In mutant HD alleles, the copy number has expanded to anywhere from 37 to 86 copies of the repeat (and maybe even as high as 100 copies). For Huntington disease and the other triplet repeat disorders described later, it seems clear enough that the severity of this disorder and, to some extent, the age at onset are related to the number of repeats. Basically, the larger the repeat number, the earlier the onset.

In the case of Huntington disease the repeated triplet is in the coding region of the gene. Thus the mutant allele encodes a protein (huntingtin) with a greatly increased number of tandem glutamines within its amino acid sequence. For reasons we do not understand, the presence of this protein is damaging (think of it as poisonous, even though that may not

Normal HD Allele

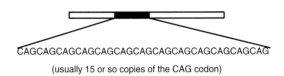

CAGCAGCAGCAGCAGCAGCAGCAGCAGCAGCAGCAG
(usually 15 or so copies of the CAG codon)

Disease-Causing HD Allele

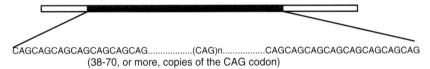

CAGCAGCAGCAGCAGCAGCAG................(CAG)n.................CAGCAGCAGCAGCAGCAGCAGCAG
(38-70, or more, copies of the CAG codon)

FIGURE 10.2    Triplet repeat expansion at the HD locus.

be the actual chemical or biological mechanisms) to a very specific subset of the cells of the nervous system. Thus it is not the absence of a normal protein that creates the disease state, but rather "poisonous" effects of the protein produced by the mutant allele. (Consider this a prime example of why a given mutation might be dominant.) One of the most perplexing things about this protein is that the mutant form of the protein results in the death of a very specific set of cells in the brain, even though the supposedly "poisonous" protein is actually present in many other cells that do not die. We do not yet understand this or why it usually takes so very long in life for this mutation to exert its deleterious effects. However, cases of early onset HD are known, which are correlated with very large numbers of CAG repeats.

The mechanism by which new mutations arise in the HD gene remains unclear. As shown in Fig. 10.3, most workers think that when the DNA polymerase attempts to copy the tandem repeat of CAG codons in the normal gene it sometimes "slips" or "falls back" and then recopies the same set of codons or "falls forward" and then continues copying, thus reducing the number of copies in the replicated strand. If such errors are left unrepaired, new mutations that are expansions or contractions are generated.

Regardless of the mechanism by which instability occurs, it is clearly present and it gets worse as the number of copies of the repeat rises. Using a technique called *polymerase chain reaction* (PCR), which is introduced in the next section, Norm Arnheim and colleagues were able to measure

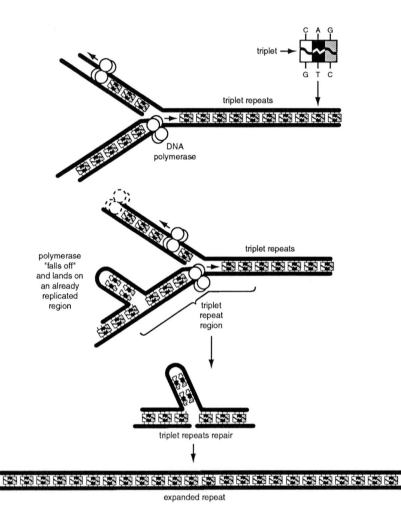

FIGURE 10.3 A possible molecular mechanism for triplet–repeat expansion.

the mutation rate in terms of the expansion or contraction of the repeat number for this gene by analyzing the genes in a single sperm! Normal- or average-sized HD alleles (15–18 repeats) showed three contraction events (reduction in the number of repeats) among 475 sperm. Even at that low level the mutation rate is an astonishing 0.6%. However, when they looked at a man bearing a normal allele with 30 triplet repeats, the mutation rate in terms of expansions and contractions went up 11%, i.e., 11% of all of the sperm carrying this allele carried a variant copy of the HD gene. (Remember what was said earlier, the standard mutation rate is on the order of 1 in 100,000. The mutation rate here is 10,000 times higher than that!) An allele with 36 repeats showed a mutation rate of 53%, and 8% of all sperm bearing this allele carried expansions large enough to cause the disease. Disease-causing alleles, with 38 to 51 repeats, showed expansions or contractions in 92–99% of the sperm carrying these alleles.

Simply put, as the number of repeats increases, so does the frequency of expansions. (Curiously, the frequency of contractions also increases up to 36 triplets, but falls off as the copy number of the allele increases above 36.) So one can imagine how these mutations arise: One small increase makes a second more likely than another and so on. Realize that although the mutation rate from a normal allele (15 to 18 repeats) to an intermediate and unstable allele (~30 repeats) is quite low; however, once the repeat number does get above 30 the mutation rate to a disease-causing allele is quite high.

Thus, the larger the copy number the greater the chance of further changes, up or down, in that copy number. Moreover, the number of copies among various HD alleles is predictive of both the age at onset and the rapidity of disease progression. *(This appears to be simple, the more gluta-mines at that position in the protein, the more poisonous this protein be-comes!—RSH)*

These observations may partially explain the phenomenon called *antici-pation* seen in some HD pedigrees. Basically, anticipation means that the disorder seems to get more severe with succeeding generations. Recall that HD is inherited as a simple autosomal dominant. Still there is an oddity in the inheritance pattern: Most cases of juvenile onset are the offspring of carrier fathers. One can understand this result by realizing that the HD alleles carried by the fathers are themselves very unstable and predisposed to further increases in triplet repeat copy number. Although some research-ers used to believe that anticipation only reflected little more than bias of ascertainment (and indeed, for some diseases, when enough investigation is carried out this does seem to be true), there is now a very good reason to believe that anticipation, in fact, exists and reflects the increase in copy number over several generations.

## TWO OTHER WELL-KNOWN TRIPLET-REPEAT EXPANSION MUTATIONS

### SPINAL BULBAR MUSCULAR ATROPHY [TRIPLET REPEAT EXPANSION IN THE ANDROGEN RECEPTOR (AR) GENE]

As noted earlier, X-linked spinal bulbar muscular atrophy (or Kennedy disease) also results from the amplification of a CAG triplet in the coding region of the gene that encodes the testosterone receptor (AR). Affected males begin to show symptoms in their 30s and 40s that include muscle weakness and neurological defects. Some, but not all, also show signs of mild testosterone insensitivity, including infertility and testicular atrophy.

Normal alleles of this gene have an average of 20 CAG repeats, whereas disease-causing alleles have more than 40 or 50 copies of this repeat. Realize again that loss-of-function alleles of this gene result in AIS females (i.e., complete somatic feminization). Instead of causing dramatic sex reversal, amplification of the trinucleotide repeat causes atrophy of the spinal–skeletal musculature. Testosterone receptors are clustered in such muscles. The repeat amplification in spinal bulbar atrophy creates a protein that can still function in normal sexual development, but apparently is inadequate with respect to the long-term maintenance of these muscles. Thus a mutation that creates an altered form of this protein creates a very different phenotype than the loss-of-function mutation. *(We might also add that there may be somatically arising missense mutations in this same gene that are associated with the development of prostate cancer. Different mutations confer different phenotypes in different cell types.—RSH)*

Again the expansion and contraction rate for normal alleles of this gene (~0.7%) is high relative to mutation rates at other genes, but it is much lower than the mutation rate (27%) observed for the top of the size range for the mutant or expanded CAG repeat alleles. Once again, the expansions seem to outnumber contraction mutations by 3 to 1, and the greater the number of CAG repeats, the earlier the age at onset.

### MYOTONIC DYSTROPHY

This is the most common form of muscular dystrophy in adults (a frequency of 1 per 8,000 individuals) and it is inherited as an autosomal dominant trait. The mutation that causes this disorder maps to a single locus at band 13 on the long arm of chromosome 19. The mutational event is the expansion of a CTG triplet repeat located within the transcribed region of a gene whose product is required at the nerve–muscle junction. However, unlike the cases discussed earlier, the amplified repeat in myotonic dystrophy mutations does not lie within the coding region, but rather in the 3'-untranslated region of the message encoding a protein called *DM protein kinase* (DMPK). (Go back and look at Fig. 3.8 to remind yourself that the structure of the mRNA includes some parts of the message that do not contribute to the coding sequence that ends up in the protein.)

Most people carry five copies of this repeat (CTGCTGCTGCTGCTG). In individuals with myotonic dystrophy, the number of copies of this repeat can expand to 100 or even 1000 or more in severely affected individuals. The greater the copy number of the repeat, the earlier the onset and the more severe the symptoms. As is the case for the triplet repeat mutations described earlier, expanded alleles of this gene are extremely unstable. The instability at the myotonic dystrophy gene does, however, differ from that observed for the two cases listed earlier in one important way: Large expansions occur predominantly when the mutation is transmitted through the female germline. (Expansions and contractions are, however, observed to occur in the male germline as well, and in some cases these contraction mutations appear to be responsible for the *reversion* of a mutant to a normal allele.)

## OTHER TRIPLET-REPEAT DISORDERS AND OTHER DISEASES

Several other triplet repeat mutations are also known in human beings. The amplification of a CAG repeat in the coding region of three other genes is individually responsible for the dominant mutations that cause *spinocerebellar ataxia type* I (SCA1), *dentatorial pallidoluysian atrophy* (DRLPA), and *Machado–Joseph disease.* For SCA1 and DRLPA there is an excellent correlation between the number of repeats and the age at onset: the higher the triplet repeat number and the earlier the age at onset, the more severe the disease. Also, virtually all expansions appear to occur in the paternal germline.

## TRIPLE-REPEAT MUTATIONS AND FRAGILE SITES

Several other genes displaying triplet repeat expansion were first identified because these mutations are associated with a very obvious change in chromosomal morphology, namely a site of breakage on the chromosome when cells from patients carrying these mutations are grown in certain culture conditions. These chromosomes are called "broken" if either there is a clearly separated fragment or the two parts of the chromosome appear to be connected by only a thin thread. These sites of breakage, diagrammed in Fig. 10.4, are referred to as *fragile sites.* The best known of these fragile sites is called FMR1 (discussed at length in the next chapter), which is associated with the amplification of a CGG repeat. FMR1 triplet-repeat expansions are the underlying genetic lesion for the so-called fragile X syndrome.

The fragile X syndrome is a common form of inherited mental retardation, affecting 1 out of every 4,500 liveborn males. [Fragile X syndrome is also seen in females, but at a much lower frequency (less than one in 9,000), and females are much less severely affected.] The disorder is characterized by severe mental retardation, large ears, and large testicles. As shown in Fig. 10.5, when cells from males with this disorder are grown in certain types of culture media, one frequently observes breaks at a specific site near the tip of the long arm of the X chromosome (Xq27), hence the name of the syndrome: fragile X. Both the fragile site and the fragile X syndrome are due to a triplet-repeat expansion mutation in a gene called FMR1 that maps to band Xq27 on the X chromosome. This gene is required for normal development of the brain and the testicles and probably of other tissues as well.

## THE GENETIC ANALYSIS OF THE FRAGILE
## X SYNDROME

Let's take a look at the fragile X family in Fig. 10.5. A careful examination of this figure reveals several genetic oddities. First notice that all males exhibiting fragile X syndrome in this pedigree can trace the Xq27 region of their X chromosome back to a single unaffected grandfather or great grandfather, whom we will refer to as the founder. Moreover, his daughters, all of whom are carrier females, are fine. But more than one of these females produced affected sons and daughters. Therefore, even though all of the individuals affected by fragile X can trace their X chromosomal lineage back to the founder male, the disease is not present in any of his sons or daughters, nor is he affected.

Moreover, in an X-linked recessive disorder, you expect that about half of the sons of a carrier female will be affected. If we look at the pedigree in Fig. 10.5, we see that less than half (approximately 40%) of the males are affected in families where we know the mother must be a carrier. This departure from the expected value of 50% reflects the fact that sons can inherit the mutant FMR1 allele from their moms yet be unaffected. (Affected and unaffected individuals bearing the same Xq27 alleles are denoted in Fig. 10.5 by filled symbols and open symbols with a dot in the middle, respectively.)

Perhaps more surprisingly, all of the daughters who inherit the FMR1 mutation from their dads are normal, while 50% of the girls who inherit the FMR1 mutation from their mothers are affected. Daughters who inherit the FMR1 mutation from their mother tend to exhibit a milder degree of

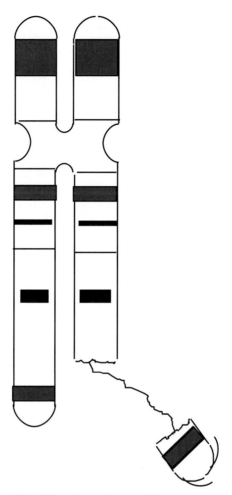

FIGURE 10.4   A fragile X chromosome.

mental retardation than their affected brothers. The best explanation for this phenomenon is that half the cells in their nervous system will express only the normal FMR1 allele they inherited from their dad. These cells will be normal. The other half of the cells will express the mutant FMR1 allele and display a fragile site. Depending on which cells most crucially require FMR1 expression and which X chromosome is inactivated in those cells, a female can exhibit mild to moderate mental retardation.

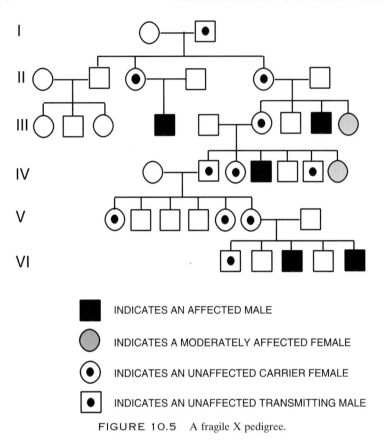

■    INDICATES AN AFFECTED MALE

◕    INDICATES A MODERATELY AFFECTED FEMALE

⊙    INDICATES AN UNAFFECTED CARRIER FEMALE

[•]   INDICATES AN UNAFFECTED TRANSMITTING MALE

FIGURE 10.5    A fragile X pedigree.

## THE MOLECULAR GENETICS OF THE FRAGILE
## X SYNDROME

As noted earlier, the gene FMR1 is expressed and required in the brain, the testes, and a few other tissues, and in the normal gene there is a trinucleotide repeat. In this case the amplified triplet is CGG. As shown in Fig. 10.6, the normal FMR1 gene has some 30 copies of this repeat (ranging from 6 to 49) in a region that is transcribed but is upstream (5′) of the coding region of the FMR1 mRNA. If we examine the FMR1 gene in males who can pass along the disease but who are not affected, we see that they are genetically different. These transmitting, or carrier, males have from 50 to 200 copies of this trinucleotide repeat in their FMR1 gene. Apparently, this limited amount of expansion does not affect the function of the gene.

So can we identify a difference between these carrier males and the males who actually end up affected? Yes! Affected males and females carry

Normal FMR1 Allele

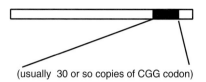

(usually 30 or so copies of CGG codon)

Premutation Allele

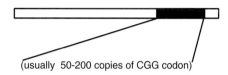

(usually 50-200 copies of CGG codon)

Disease-Causing 'Fully-Expanded' Allele

(usually 200-1300 copies of CGG codon)

FIGURE 10.6   Triplet–repeat expansion in the FMR1 gene.


200 to 1,300 copies of this repeat. This means that some somatic cells in a single individual can have 200 copies, some 210, some 220, some 500, and so on, up to more than 1,000. Thus, affected males really are mosaics. Their cells differ enormously in the number of copies of the triplet repeat. (Notably, their sperm carry an unchanged, unexpanded carrier allele most likely as a result of selection in the germline for low-repeat-number-bearing cells.)

Thus, there are not two, but three alleles of FMR1: a normal allele (6 to 49 copies of the repeat), a carrier or *premutation* allele (50 to 200 copies), and a hugely expanded disease-causing allele (200 to more than 1,000 copies). Males carrying either a normal allele (6 to 49 copies of the triplet repeat) or a carrier allele (50 to 200 copies) are phenotypically normal. Carrier females possessing a normal and a premutation bearing allele are also normal.

However, although a carrier female possesses only the normal and pre-mutation alleles, she can produce affected children. This is explained by

## BOX 10.1 THE LAIRD MODEL FOR THE FRAGILE X SYNDROME

Sometimes when researchers are having trouble solving a problem, what is needed is for someone to come in from the outside with a completely different way of thinking about the problem. A large piece of the genetic mysteries surrounding the fragile X syndrome was worked out by a *Drosophila* geneticist at the University of Washington named Charles Laird. *(Charles was one of Scott's teachers and is one of his friends, so Scott was able to follow the genesis of this idea as it happened.—CAM)* Indeed, it happened something like this: Charles called me up a few years ago when I was still a professor at that medical school in New York. He told me that he had just heard a seminar about fragile X and knew that there just had to be a simple solution that the human geneticists were missing. He then told me that he intended to go to lunch and figure out the inheritance of fragile X. I wished him a good lunch and with a slight smile of disbelief went back to doing whatever I did at that medical school in New York. As smart as I knew Charles to be, and he is very smart indeed, I did not expect him to solve a problem that had been vexing other geneticists for two decades, especially over a plate of bad chicken salad at the University of Washington cafeteria. As it turns out Charles called me the next morning with an answer. More importantly, his answer was a significant contribution to solving the problem. *(Go figure.—RSH)* As you think about his model, please remember that he figured all of this out long before any of the molecular biology had been done.

Charles reasoned that the original founding male had to pass down to his daughters an X chromosome bearing a *premutation*. He argued that the premutation was a DNA lesion in some gene, FMR1, that by itself was harmless and could be carried and transmitted faithfully by males without producing a phenotype. Thus, males carrying the premutation-bearing X can never produce affected sons or daughters. However, daughters carrying the premutated X can have both affected and unaffected kids who carry the same X chromosome. Charles' hypothesis explained this process by arguing that when the premutation-bearing X passed through a female germline, and *only* a female germline, it had a one-half chance of being *activated* in such a way that it could produce a phenotype in the next generation.

Charles supposed that the basis of this imprinting lay in the mechanism of X inactivation. Specifically, he imagined that the primary effect of premutation was to *locally* block the reactivation of an inactive X in the female germline. (Recall that both X's must be made active in oocytes, so the X inactivated during early embryogenesis must be reactivated in the female germline prior to meiosis.)

The beauty of Charles' model is that it explains how this female can produce both normal and affected sons bearing the premutation. If the oocyte is one in which the X chromosome without the premutation was originally inactivated, then there will be no block to reactivation. The premutation will be irrelevant and a normal son or daughter will result. But if the original inactivation event involved the premutation bearing X, then the FMR1 premutation will locally block reactivation and a partially inactivated X chromosome is passed on to the offspring producing the disorder.

There are now reasons to think that expansion of a premutation may be much more complicated than just inactivation and failed reactivation, but there is no question that Laird got the basic components (a premutation acted upon by imprinting in the female germline) dead right! The molecular details will follow in due time.

You might not have predicted that a background in *Drosophila* genetics was what was needed to make a breakthrough on human hereditary mental retardation. However, sometimes that much difference in viewpoint is exactly what is needed to bring about a paradigm shift, a new perspective based on a completely different set of underlying assumptions, a different body of background information, or the application of different methodologies or thought processes.

It can actually be quite an instructive thing to go look at the life stories of Nobel Prize winners and ask, "What subject did they study in school?" and then ask, "What subject were they researching that got them the prize." Some of the combinations are truly surprising! For instance, Max Delbruck, one of three 1969 recipients of the Nobel Prize in Physiology or Medicine, received his Ph.D. in physics before switching to biology to elucidate the processes by which bacterial viruses replicate themselves inside their host cells. In Delbruck's Nobel lecture, "A Physicist's Renewed Look at Biology — Twenty Years Later," he showed his unusual way of looking at things through the perspectives of both the physical (nonliving) and biological (living) worlds.

the observation that premutation alleles have a very high probability of expansion in the germlines of carrier females: the larger the copy number of the premutation bearing allele, the greater the likelihood of expansion in the female germline. The events that promote this instability still remain under investigation. (One model proposed by Charles Laird is described in Box 10.1.)

Fully expanded alleles (200 to 1,300 copies) cause the phenotypes associated with fragile X syndrome. Expansion to this level of triplet repeats is correlated with a chemical modification (methylation) of triplet repeats themselves and of other nearby regions of DNA. As discussed more fully in the next chapter, methylation likely serves to inactivate the FMR1 gene.

A similar tandem repeat expansion mutation, also associated with mental retardation and located just 600,000 bp distal on the X chromosome to the FMR1 mutation, results in a second fragile site mutation in the FMR2 gene (fragile X E or FRAXE). Other similar fragile site mutations have also been characterized elsewhere in the genome.


## SUMMARY

Basically, the human genome is a pretty stable place. Errors in replication seem to occur rarely and, when they do occur, they are rapidly corrected. However, certain triplet repeats appear to represent an Achilles heel for the replication and repair systems. These sites mutate at an increased frequency after the first expansion mutations occur. Some of these expansions, the ones that exhibit primarily paternal instability, appear to reflect errors in male meiosis. Others, such as the expansion arrays that underline the fragile X syndrome and myotonic dystrophy, appear to reflect events occurring in the female germline. Triplet-repeat expansions are not the only curious elements taking place in our germlines. There are others. But to tell you about the processes we need to discuss the most egregious violation of Mendel's laws, something called *imprinting*. If mutation and nondisjunction are "violations," imprinting is a capital offense. But that story warrants a new chapter . . . .

# 11

## IMPRINTING OR EPIGENETIC CHANGES IN GENES AND CHROMOSOMES

As if things like triplet repeat syndromes were not troublesome enough, the field of human genetics is replete with even more curious violations of Mendelian inheritance. Consider, for example, Mendel's second law, the one referred to as the "the law of the gene." Basically, this law says that each trait is specified by a gene and that each of us carries two alleles of that gene, one from each parent. For Mendelian inheritance to work the paternal copy has to be equal, at least in terms of potential function, to the maternally derived copy. Thus, for a recessive disorder, an individual heterozygous for a recessive mutation should be phenotypically normal, regardless of which parent donated the normal allele. For a dominant disorder, a heterozygous individual should be affected no matter which parent contributed the affected allele, and that almost always works. However, there are a few, *very few,* examples of a phenomenon called *imprinting,* in which the activity of a gene is determined by whether it came into the zygote via the egg or the sperm.

The process of imprinting basically involves presetting an active or inactive status of a gene, chromosome region, or entire chromosome as it passes through one of the two germlines. Thus a given gene, or genetic version, might be preset to be inactive by passing through the male germline and/or preset to be active while passing through the female germline or vice versa.

*(As best as we can tell, the phenomenon of imprinting is restricted to a few genes. Among those genes, whether a gene is activated or inactivated as it passes through the paternal or maternal germline depends on which gene it is. Thus the decision to activate or inactivate is really the product of the combination of which gene it is and which germline it is passing through.—RSH)*

Realize that such changes due to imprinting cannot be permanent and must be erased and reset in each generation as a given chromosome passes from mother to son to granddaughter and so on. Therein lies the difference between mutation and imprinting: Mutation actually changes DNA sequences, resulting in stable and heritable mutations, whereas imprinting changes the activity or potential activity of a gene or chromosome by some modification of the DNA that can be, and almost always is, erased in the germline of the next generation. Because whatever changes underlie the process of imprinting are not stable and do not act as Mendel predicted, we refer to these processes as *epigenetic.* *(Just think of epigenetic as meaning that the actual DNA sequence does not get changed. If it started out with the sequence AAGCTTG, it continues to have the sequence AAGCTTG throughout these epigenetic processes, yet somehow the gene gets turned "on" or "off" or "up" or "down" depending on the epigenetic imprinting that takes place. If that doesn't make any sense yet, it will if you just keep reading!—RSH)*

We will begin by discussing imprinting as a chromosomal phenomenon and then as a phenomenon affecting smaller genetic regions. As we go on, please remember that imprinting is rare. Fortunately, most genes do exactly what Gregor Mendel said they should do.

## IMPRINTING AT THE CHROMOSOMAL LEVEL

Those of you who can remember back to Chapter 7 *(let's face it, there have probably been at least a few parties since then—CAM)* may recall that X inactivation is not random in extra-embryonic membranes of mice and probably not in extra-embryonic membranes of humans either. Instead, in the cells that form the chorion and the allantois, it is always the paternal X chromosome that is inactivated. This is true despite the fact that the cells that form the fetal tissues undergo random inactivation.

This nonrandomness is carried a bit further in some of our vertebrate cousins. For example, in marsupials *(that's scientist jargon for kangaroos and platypuses—CAM)* it is always the father's X that is inactivated in every cell of the animal. Again, note that when that same X is reactivated and passed on from mother to daughter, it now escapes inactivation (see Fig. 11.1). In a more Amazonian example, there is an overgrown gopher called *Microtus oregoni* in which all nongermline female cells simply destroy and discard the father's X chromosome *(I keep wondering what Freud would say about this process, but that's another book—RSH).* The mealy bugs go a bit further; females inactivate the entire paternal genome! Notice that the word here is inactivate, not toss out. The paternal genome is still

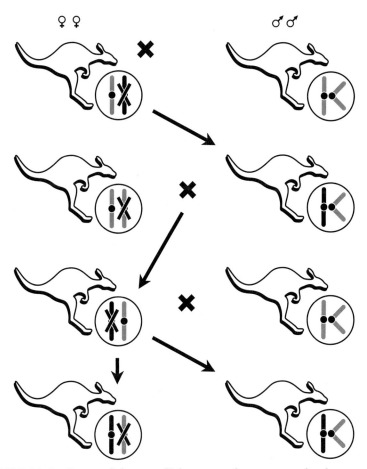

FIGURE 11.1     Passage of a kangaroo X chromosome from great-grandmother to grandfather to mother to daughter and son. Follow the darker X chromosome (the bent chromosome denotes the Y chromosome).

there and still available to be passed onto subsequent generations, but it remains inactive during the daily life of the female mealy bug.

There is a common thread here. Genes and chromosomes appear to be marked or imprinted by passage through the male or female germline. This imprinting is, however, not permanent because it can be erased and reset at each generation. Clearly, the vast majority of genes do not show really obvious signs of imprinting or we wouldn't be able to do Mendelian genetics. So how important is imprinting? Consider the following two examples.

## IMPRINTING AT THE LEVEL OF THE MALE
## AND FEMALE PRONUCLEUS

One can fertilize mouse embryos in a petri dish and manipulate the sperm and egg nuclei (called the pronucleus or pronuclei) prior to their fusion. For example, the female pronucleus of one zygote can be destroyed and replaced by transplanting it into the female pronucleus from another zygote (see Fig. 11.2). The result is a perfectly good embryo, which can

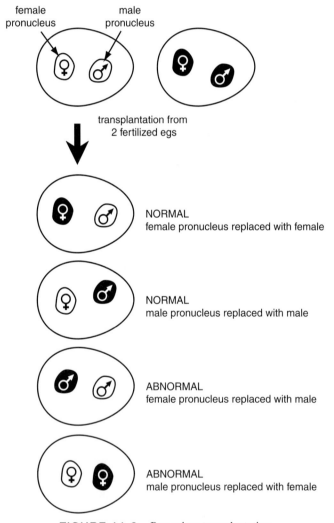

FIGURE 11.2    Pronuclear transplantation.

produce a very expensive mouse when implanted into a female who believes that she is pregnant. It is also possible to destroy the male pronucleus and replace it with some other male pronucleus.

*What can't be done successfully is to fuse one male pronucleus with another male pronucleus or a female pronucleus with a female pronucleus.* If one attempts to create an embryo with two female pronuclei, embryonic development starts out quite normally. Indeed, the fetal tissues develop quite nicely, but the extra-embryonic membranes develop improperly. If the experiment is reversed, namely fusing two male pronuclei, the extraembryonic membranes develop fine, but the embryonic structures develop abnormally. *The only reasonable conclusion from these experiments is that both male and female pronuclei are essential for normal development.* Apparently, some genes are differentially programmed or even inactivated during passage through gametogenesis in the two sexes. An example of this is provided by Prader–Willi and Angelman syndromes. These two disorders, while quite different in terms of their pathology, are due to mutations or deletions in the same small region of chromosome 15.

## IMPRINTING IN SMALL GENETIC INTERVALS

Both Prader–Willi and Angelman syndrome are usually associated with newly arising deficiencies that include band q11 on chromosome 15. When that deficiency is inherited from the father, the offspring exhibits Prader–Willi syndrome (obesity, small hands and feet, hypogonadism, and mental retardation). When the deficiency is inherited from the mother, the offspring exhibits Angelman (the "happy puppet") syndrome. Occasionally, one finds patients with Prader–Willi that do not display a deletion. Astoundingly, these individuals can be shown to possess two maternal copies of chromosome 15 and no paternal copies!

*(Such individuals are examples of what we call uniparental disomy: the result of meiotic or mitotic nondisjunction events resulting in an individual bearing two identical copies of a chromosome derived from a single parent. As shown in Fig. 11.3, one could imagine MII nondisjunction in the mother followed by mitotic loss of the paternally derived homolog in the resulting zygote. Or you might imagine an aneuploid egg with two copies of chromosome 15 being fertilized by a sperm that is also aneuploid in a complementary manner, i.e., missing chromosome 15.—RSH)*

Thus it is not the lack of two copies of this region that causes Prader–Willi, it is the lack of a paternal copy. Clearly, the paternal copy is essential for some function that the maternal copy cannot provide.

Similarly, one also finds patients with Angelman syndrome that do not display any deletions in the critical region of chromosome 15. These individuals can be shown to possess two paternal copies of chromosome

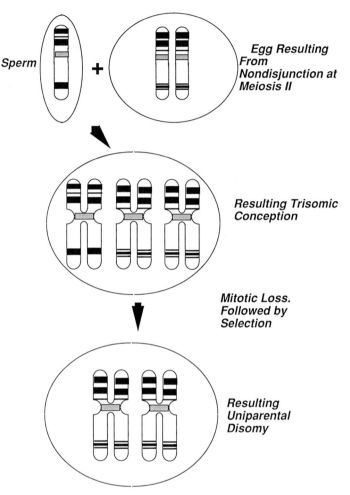

**Sperm**

**+**

**Egg Resulting From Nondisjunction at Meiosis II**

**Resulting Trisomic Conception**

**Mitotic Loss. Followed by Selection**

**Resulting Uniparental Disomy**

FIGURE 11.3    The generation of uniparental disomy.

15 and no maternal copies! Again, it is not the lack of two copies of this region that causes Angelman; it is the lack of a maternal copy. Clearly, the maternal copy is essential for some function that is quite different from that for which the paternal copy is required, a function that the paternal copy cannot provide. The bottom line here is that the same genetic region is programmed to perform two different and essential functions, depending on which of the two germlines it has passed through. Obviously, daughters can erase and reprogram the genes they obtained from their fathers and sons can reprogram what they received from their mothers.

## HOW IMPORTANT IS IMPRINTING?

If the truth must be told, we're not at all surprised by imprinting. The different histories of a male and female pronucleus are reflected in the proteins bound to their DNA and in their degree of compaction. This difference must have provided a fertile substrate on which differential systems of gene regulation could be developed. Indeed, we would be shocked if this phenomenon didn't exist.

## IMPRINTING AND HUMAN BEHAVIOR (MAYBE?)

Recently, psychological studies of women with Turner syndrome have raised the possibility that a region on the X chromosome that controls behavior and/or personality might also be subject to imprinting. According to work by David Skuse and his collaborators, Turner syndrome women who received their single X chromosome from their mother had more problems in social adjustment than did Turner syndrome females who received their sole X from their father. According to these workers, Turner syndrome females who inherited their X from their father displayed superior skills in those areas that mediate social interaction. These authors infer from these data that there is a region on the X chromosome that plays an important role in establishing the patterns of "social cognition," that is imprinted in such a fashion that is not expressed from the maternally derived X chromosome.

There are problems with this interpretation, primarily because the distinction isn't really "all or nothing." It simply isn't the case that all Turner syndrome females with a maternally derived X chromosome behave one way and all Turner syndrome females with a paternally derived X chromosome behave another; rather, it is a matter of degree. (For example, 72% of the Turner syndrome girls in this study with a maternally derived X exhibited difficulties in social interaction as compared to only 29% of the paternally derived cases.) Nonetheless, it is an intriguing finding that raises some curious possibilities.

*(My prejudice, and it is just that, is that if this result holds up, it will not be because there is an "imprintable" region on the X chromosome that controls social interactions. Rather, I suspect that, if this result is correct, it means that there is an X chromosome gene(s) involved in building the nervous system and that passage through the male or female germline may play a subtle role in determining that gene's pattern of expression in the next generation. I could then imagine that the altered social skills are a downstream consequence of improper expression of that gene.—RSH)*

As you can imagine, this result quickly got the attention of the scientific and popular press. One possible interpretation of these data is that men, who get their X from their mother, might be more vulnerable to disorders of social condition (like autism) than are women, who receive one X chromosome from both parents. The popular press took things even further in somewhat recklessly suggesting that the studies of Skuse and his colleagues provided a way of understanding why men and women sometimes seem to differ in their patterns of social interaction and aggression. However, it seems to us that extending results from Turner syndrome females to the general population might need to be done with some great degree of caution.

Regardless of the more global implications (or lack thereof) of this result, the work of Skuse provides a dramatic suggestion that imprinting may have quite significant implications for human development. It thus behooves us to figure out just how it might work.

## HOW DOES IMPRINTING WORK?

So far we have dealt with imprinting on about the same level that brother Gregor Mendel looked at his pea "genes." We have also been looking at organisms and phenotypes and modes of inheritance. Now lets try moving down to the molecular level and look at how we think imprinting could take place. Keep in mind here that we do not yet know all the possible causes of imprinting, and even the parts we will talk about here are only partly understood at this point, so bear with us. What we are offering is just an example that will hopefully help you understand what is, frankly, a strange subject.

### CHEMICAL TAGS

Okay, since we brought it up in the previous chapter, let's talk about methylation, the process of adding a "methyl" group onto an existing molecule. Hmmm, that definition doesn't seem all that helpful. Let's try again. We are NOT going to worry about the chemistry and biochemistry of this situation anymore than we worried about what chemical bonds hold the different parts of a DNA molecule together. All you really need to do is to think of a "methyl group" as a small chemical tag that can be tacked on to bases in the DNA and change how that base "looks" to the enzymes in the cell that "read" the DNA sequence. It is interesting to note that some of the cells enzymes seem to be able to "see" through the disguise created by adding a methyl group while other enzymes cannot. So plain old cytosine, the C in the DNA sequence, looks just like methylated cytosine to some proteins, but it can look quite different to other proteins. The most popular models for how the methyl groups are acting does not involve "seeing" through disguises; rather, it calls for the presence of methyl groups to change the local structure of the DNA, like how tightly it is coiled or how loose and "open" it is.

If you compare the pattern of where the methyl tags are located in the same cell type in males and females, activated vs. inactivated copies of the X chromosome, or eggs vs. sperm you will find lots of differences in which bases are methylated. You may recall that many differences in cell functions are the result of differences in the proteins present in the cell, which result from differences in the RNA transcripts being made, which, in turn, are the result of whether something in the promoter region of a gene is turning that gene on or off or up or down. Now we don't want you to think that regulation of gene expression is a simple matter of whether there are methyl tags stuck on some of the bases because the real story is much more complex. One of the many complex factors affecting gene expression is the presence or absence of methyl tags on certain bases of the promoter region of a gene. Hence, one of the factors affecting the proteins that are present and the functions carried out is methylation.

The fact that gene expression and pattern of methyl tags both vary like this does not prove that one thing is causing the other. However, there are lines of experimental evidence to support the idea that methylation affects gene expression and that changes in the pattern of which bases are methylated lead to changes in gene expression. So let's just presume this useful model: The addition of methyl groups to DNA in promoter regions changes how the promoter DNA "looks" (whether through local disguise or through global changes in the DNA structure) to proteins that bind to the promoter DNA as part of the process of regulating the production of the RNA that will lead to synthesis of the protein gene product.

The pattern of which bases are methylated is different in eggs and sperm. For some genes and chromosomal regions the "maternal" or "paternal" pattern of methylation persists through development and can still be detected in the child or adult. Earlier, we examined the case of Prader-Willi and Angelman syndrome from a genetic perspective, where sex of the parent contributing the expressed gene copies determined which of the phenotypes would be present. Now let's take a look at what is being found on the molecular level. Researchers have proposed that there is a region of about 100,000 base pairs containing several genes and showing differences in methylation on the maternal and paternal copies of the genes within the region to which the Prader-Willi and Angleman phenotypes map. More than one gene seems to have its expression regulated by methylation of specific parental origin.

Much more research will be needed to determine just how the methyl tags are having their effect, which gene(s) are the critical gene(s) in the region, or to tell what other factors might be complicating the picture. Until then, we do know that the methylation is affecting gene expression, that the pattern of methylation in imprinted genes is derived from the pattern of methylation in egg or sperm, that the pattern of methylation changes (i.e., methyl tags get added, or Cs that were previously methylated end up free of methyl tags) under some circumstances, and that this "epigenetic" methylation mechanism is complex and technically very difficult to study.

If we put this information together with the phenotypic descriptions of imprinting earlier in the chapter, we end up with a molecular model that works to explain what we know so far about imprinting and that will help us design further experiments aimed at understanding the role of specific genes (and the methylation and expression) in imprinted phenotypes. Basically, a given region of DNA can be modified by methylation as it passes through a given germline (male or female) and is loaded into a gamete. This modification serves to direct various aspects of gene expression in that region of the DNA of the developing zygotes. The crucial point is that this modification can be removed or erased in the germline of that newly developing fetus and then reimposed or imprinted according to the sex of that zygote. Go back and look at the example of X chromosome imprinting in the kangaroo in Fig. 11.1. We can understand this by imagining that some region that controls X chromosome expression is imprinted in the male germline in a fashion that directs its inactivation in the subsequent generation. We need only then to imagine that such imprints are removed or "erased" in female germlines to explain this curious pattern of inheritance.

As we noted earlier, imprinting doesn't really surprise us. The differences in the biology of the germlines and the existence of erasable DNA modification systems created an ideal substrate for the evolution of a curious set of ways of regulating genes. It may seem a bit messier than we would like, but as we have said before, "evolution is not Michelangelo and the Sistine chapel. It is a teenage kid with a broken car and no money. It just does whatever works!"

# HUMAN GENES

---

This section discusses the methods used to physically isolate and characterize human genes. It also discusses some common human genetic diseases from what is often referred to as the "molecular point of view." This is to say how such genes were in fact cloned and isolated, how their protein products were characterized, and what all that high-tech molecular biology has told us about the disease itself. In no case will we be able to tell you that such knowledge has led to a cure for the disease in question. In that sense we will talk to you about "maybe tomorrow" and the virtue of patience. These are valuable lessons.

# 12

# INTRODUCTION TO
# GENE CLONING

This chapter describes some of the basic techniques used to isolate and characterize specific human genes. We are going to present a rather simplified approach to isolating the human TDF gene. The story we will spin is not an exact tale of how this gene was isolated. Such a rendering lies somewhat beyond the scope of this chapter. Rather this is a tale of how this gene "might have been isolated," a tale designed to teach you the basics. The next chapter will describe more modern, although not necessarily more complicated, versions of these tools and the means by which gene isolation in humans is currently accomplished. The succeeding chapters will describe the actual methods used to isolate several human genes. Up until now, we've tried really hard to avoid detailed considerations of how things are actually done in the lab, i.e., "techniques." Unfortunately, we can't do that any more. The concepts here are inextricably wedded, without the possibility of divorce, to the techniques that created them.

*(Hello, my name is Julia Richards and I will be adding my two cents worth occasionally throughout the rest of the book. Like Scott, I am a geneticist; indeed Scott and I were in some graduate school classes together a long time ago—no Scott, I am not going to tell them how long ago! Scott studies flies, I study humans. Sometimes the similarities between what he does and what I do can be pretty surprising, but the differences also can be quite profound. Scott has been avoiding discussions of techniques because of his "phobia" of long discussions about scientific "tools," or of "tool-building." Basically, Scott thinks that if you want to learn about tools, you should go to a hardware store. But sometimes you need to know about the tools because the types of "tools" you use can affect the type of information*

*you can get back. One of the things that makes the biggest difference between Scott's choice of scientific tools and mine comes from the difference in how much DNA there is in a fly cell vs. the amount of DNA in a human cell. A lot of the other differences are ethical and legal.—JER)*

## WHAT IS CLONING?

Unfortunately, for those of us who labor at writing textbooks, the term cloning has two meanings in contemporary genetics. The first, perhaps more common, meaning refers to the production of genetically identical organisms. Yes, we know that some folks in Scotland took the diploid nucleus from a mammary gland cell (hence the new sheep's name) of an adult female sheep and introduced it into a sheep egg. By some clever trickery the egg was tricked into acting as if it had been fertilized and it developed to produce a live-born kid named Dolly. Realize that Dolly is now a precise genetic copy of the sheep from which the mammary cell nucleus was taken, a true clone.

Reporters and legislators reacted to Dolly's birth as if the sky were falling. "Clearly," they argued, "this cloning created some horrid new frontier in technology." One wonders how such ostensibly bright folk could have missed the fact that Dolly and her genetic progenitor are little more than identical twins (okay, so they were born a few years apart!) and, at least to the best of our knowledge, identical twins are not viewed as a serious threat to our society. Perhaps what drove the outcry and the furor was that a few scientists were able to do in the laboratory with extreme difficulty what nature does capriciously. More likely, people worried that if one could make one clone then perhaps one could make 50 or 100 Dollys by this method.

Most folks probably don't find this too scary a possibility for sheep; indeed most sheep look pretty much alike to most people anyway. However, we can imagine the fear of making multiple copies of certain humans, especially those humans, and we can name several living and dead, of whom one copy seems to have been more than enough. As we go to press it looks unlikely that this cloning technology is going to work in humans. It is already pretty clear that one can't do similar things in the mouse, and certain crucial aspects of early human development parallel more closely those of mice than sheep! The smart money seems to ride on the idea that, other than identical twins, there won't be much human cloning going on in the near future. Certainly, the newspapers have been full of announcements by one group that plans to try, but the jury is still out and our crystal ball is hazy at best.

When we use the term cloning in this book, we mean something very different. The term cloning means the ability to isolate and make a "gazillion" copies of a specific piece of human DNA. In other words, taking a single fragment of human DNA, usually a gene or part of a gene, and making a pure preparation of many identical copies of that gene (hence the term cloning). Our basic trick here is that we take a piece of human DNA and use a biological or chemical process to replicate that DNA molecule, and only that molecule, as many times as we desire.

*(Scott, is gazillion the technical term? I know that when I worked with clones in graduate school I would routinely produce about 10,000,000,000 copies of the important clones I was studying. Why would I want so many copies? Because many of the things we do with DNA are biochemical tests that are not very sensitive and use a lot of DNA in each test.—JER)*

## WHY WOULD ANYONE CLONE HUMAN DNA AND WHAT WOULD THEY DO WITH IT?

The human genome is very large, and there are many questions about individual genes that cannot be asked if a whole genome worth of DNA is present. Why? Because the biochemical technologies available to us are not sensitive and specific enough to detect individual pieces of information buried in such a large background of information. If the piece of DNA to be studied is isolated (i.e., separated away from the rest of the DNA in the target genome) and present in pure form in a test tube (or inside a bacterial cell that can serve as a kind of "life-support" system for the cloned DNA outside of the human cell), it is possible to

- determine the sequence (the linear order of As, Cs, Gs, and Ts) along the piece of DNA,
- use the base sequence of the gene to determine the structure (in terms of the sequence of amino acids) of the protein produced by that gene,
- use the DNA sequence information as the basis for looking for mutations in the gene,
- use the DNA sequence to create diagnostic tests for genetic diseases,
- produce the protein that is the gene product and then study the protein's function, and
- use the piece of DNA or the protein produced from it to treat genetic diseases.

All of these are good things to be able to do. So how can we accomplish them? To explain the process of gene cloning we need to go back to the beginning. How do we first isolate a specific fragment of DNA?

## ISOLATING AND CLONING DNA FRAGMENTS FROM LIVING ORGANISMS

Finding a particular gene is not an easy process. There are so many different genes in the human genome that finding a particular gene almost seems like finding a needle in a haystack. So how do geneticists isolate, amplify, and purify a gene? More specifically, how was the TDF gene isolated and cloned, and what did we learn from that process? The cloning and molecular anaylsis of TDF is a terrific story, but to tell that story we need to discuss the stuff that science fiction movies are made of. We need to tell you about cloning, about recombinant DNA, and about (gasp!) genetic engineering.

To begin any type of gene isolation, we must be able to clone (isolate and make identical copies of) defined DNA fragments in an "easy to work with" organism, such as a bacterium or a yeast cell. The cells in which a specific DNA fragment will be cloned are called *host* cells. *(Just think of the host cell as a microscopic life-support unit with an assembly line capable of copying the piece of DNA you are studying.—JER)*

The essence of cloning is simple enough. We begin by isolating what we call *target DNA*. As discussed in Box 12.1, target DNA is DNA from the organism bearing the gene that we propose to clone. So if we wanted to clone the TDF gene, we would need to have a male volunteer contribute a blood sample containing our target DNA. *(Why might it be a very bad idea to attempt to clone TDF from the convenient sample of human female DNA already languishing in your freezer? Wouldn't that be easier? Once you've answered this question, you should know that one or two "horror stories" like this are staples of the lab lore of virtually every lab I know!—RSH)*

We also need to obtain DNA corresponding to our cloning vector. A *cloning vector* is a DNA fragment capable of replicating in *E. coli*. As

---

### BOX 12.1 YOU ONLY HIT WHAT YOU AIM AT: TARGET DNA

Target DNA is the piece of DNA that you want to clone. Sometimes the target sample is a pure preparation of the DNA fragment you are trying to clone. More often, the target DNA is contained in a much more complex sample that includes more than just the piece you are after. The sample you start with often contains your target mixed in with the rest of the complete genome of the organism you are working with. Your cloning experiment is not a search for a needle in a haystack, it's a search for one particular straw in the haystack, because the piece you want may be very hard to tell apart from the rest of the haystack even once you have it in hand.

discussed in Box 12.2, most cloning vectors were derived from bacterial viruses or from small circular minichromosomes, called plasmids, that are carried by many bacteria. Cloning vectors can ferry your target DNA into a bacterium and maintain them there. *(Indeed, one of the earliest cloning*

---

### BOX 12.2 THE CLONING VECTOR: A VERY USEFUL MOLECULAR TOOL

The cloning vector is an extra piece of machinery that we add to the cell's assembly line. It is a DNA molecule that possesses the capability of replicating itself in a host organism. It is useful to us because it can also replicate a gene or a piece of DNA that we have attached to it. Some cloned DNA exists in the cell as a linear (long straight) piece like DNA on a human chromosome, but much of the cloning we will talk about uses a plasmid as a cloning vector. As shown in Fig. 12.2, plasmids are small, circular DNA molecules (vectors) that possess the ability to replicate well and autonomously in bacterial cells. Plasmids do not integrate into the chromosomes; rather, they will remain separate in the form of small and easily isolated circular DNA molecules which are present at high copy number. In many of our examples we will talk about cloning with plasmid vectors, partly because plasmid vectors are so common and easy to use and partly because their use is easy to illustrate.

The key parts of a plasmid cloning vector are a bacterial origin of replication (the place on the DNA at which the copying process starts and without which copying does not happen) and a gene for antibiotic resistance (a nifty genetic toy that kills any host cells keep and copy the plasmid whether it wants to or not). Any bacterial cell harboring this gene for antibiotic resistance gains the ability to resist some nasty bacteria-killing drug. Why would you need a selectable marker like an antibiotic resistance gene? Bacteria are VERY efficient and they have a strong tendency to get rid of anything they don't absolutely need and they have a tendency to not do things they don't have to. If you don't put some kind of selective pressure on the bacterial cell, you can put your plasmid into the bacterial cell but the cells that have not kept it around will take over.

Bacterial viruses can also be used as cloning vectors as can chromosomes from yeast, a simple single-celled eukaryote. We will discuss these vectors and their uses later in this section.

*vector was a virus known as Charon 4. This vector was named after the mythical boatman of ancient Greek mythology who ferried dead souls across the river Styx and into hell. Do you suppose that the name was a metaphor for the direction in which many of us thought this new technology might be leading mankind? Kinda scary, huh?—RSH Actually, those who know Fred Blattner, the creator of the Charon vectors, realize that he actually holds an optimistic view of the positive things that molecular biology and cloning could accomplish.—JER)*

We then cut our target DNA and our cloning vector with a form of biological "scissors" called a *restriction enzyme* (see Box 12.3). The neat thing about such restriction enzymes is that they leave the cut DNA with "sticky ends" (see Fig. 12.1). We can then link the sticky ends of the cut target DNA to the cut sticky ends of the cloning vector (see Fig. 12.2). In this case, the sticky ends of a fragment of target DNA comes together with the same sticky ends from the plasmid by good ol' complementary Watson–Crick base pairing. *(Sort of like two sides of a zipper coming together.—JER)* We then use a biological "stapler" called *ligase* to fasten the cut products to each other (see Box 12.4 and Fig. 12.2). We call this process *DNA ligation.* If we do this right, each piece of cut target DNA is now safely locked or *"recombined"* into a closed

---

### BOX 12.3 RESTRICTION ENZYMES: SCISSORS THAT CAN CUT DNA

Restriction enzymes are a key feature in the cloning process. They are commercially available proteins that cut DNA into pieces at certain defined sequences, such as GAATTC. They cut at specific sites on the cloning vector and within the DNA sample from which we would like to clone our gene. Cutting a particular DNA molecule with a specific type of restriction enzyme results in the creation of a discrete set of smaller fragments of a particular size. Thus cutting many copies of the DNA molecule will produce many copies of each of those smaller fragment sizes.

Most of these enzymes cut the double-stranded DNA molecule in such a way that the cut on the "Watson strand" is several base pairs away from the cut on the "Crick strand." The result is that each fragment has a small single-stranded tail at each end. See Fig. 12.1 for an example of how the restriction enzyme *Eco*R1 cuts the DNA molecule.

before cutting

... A A C G C A G|A A T T C A A G G A G A C T A ...
... T T G C G T C T T A A|G T T C C T C T G A T ...

... A A C G C A G          A A T T C A A G G A G A C T A ...
... T A G C G T C T T A A          G T T C C T C T G A T ...

after cutting

FIGURE 12.1    Cutting of a DNA molecule by a restriction enzyme.

circular DNA molecule containing the target DNA and the cloning vector (see Fig. 12.2).

Next we use a process known as transformation to put the "recombined" or recombinant DNA into the host cell (see Boxes 12.4 and 12.5 and

---

BOX 12.4 LIGASE: THE ENZYME THAT RESEALS THE STICKY ENDS

Another very handy enzyme (also available from various biotechnology companies) is called ligase. If two different pieces of DNA have each been cut with a particular enzyme like EcoRI, the single-stranded pieces at each end of the DNA fragment act as "sticky" ends. The two single-stranded tails of DNA stick to each other by putting the single-stranded part from the first fragment together with the single-stranded part of the other fragment to recreate a double-stranded segment of DNA. This association then gets chemically sealed in place by ligase (see Fig. 12.2). The important thing about this process is that after ligation (the process of using ligase to seal two pieces of DNA together) the region where the sealing event took place is now chemically indistinguishable from parts of the DNA that did not get cut and glued back together. Why does this matter? If a sticky end from a piece of human DNA comes together with a sticky end from a piece of vector DNA and then gets ligated, the host cell cannot tell where the vector leaves off and the inserted piece of human DNA begins.

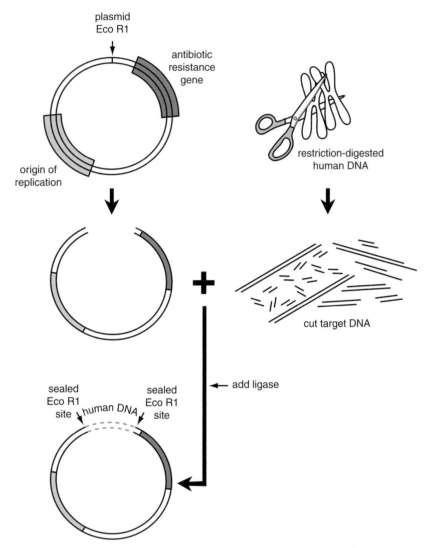

FIGURE 12.2   Inserting a DNA fragment into a plasmid.

Fig. 12.3). We now let the host cell make huge numbers of copies of it for us. Once the host cell has made copies, we can then isolate the cloned copies away from everything else in the host cell. Voila! A large number of pure copies of the thing we want to study. *(OK, so in real life it isn't always that simple, but cloning experiments are basically all variations on that same series of steps.—JER)* Thus, to clone a piece of DNA four things are necessary: *target DNA,* a *cloning vector,* a way to fasten target and

BOX 12.5 THE HOST CELL: A MICROSCOPIC FACTORY

The host cell is like a microscopic life-support unit or a miniature factory with an assembly line capable of copying the piece of DNA you are studying. Host cells for cloning are single-celled organisms, often bacteria such as *Escherichia coli,* but more complex organisms such as yeast cells have also been used as hosts for clones. Beginning with very early cloning experiments, researchers made a point of producing host cells that were genetically crippled so that they cannot survive outside of the laboratory environment.

An important reason for using single cells such as bacteria is that you can spread the cells out on the surface of a petri dish and a clone of cells will grow as a little colony at the point occupied by the single cell. If you spread the original cells out far enough apart, then you can isolate an individual colony away from the other colonies after they are grown. If all of the cells in the colony are descended from the one cell that landed at that place on the petri dish surface, then they should be genetic clones of each other and have the same genetic content.

Thus, if you cut up a large target into many different pieces of DNA, ligate them together with vector, and transform the whole mess into bacterial host cells, you will have a mess of cells that each contain a different piece of the human genome. But, if you spread those cells out far enough to let them form separate clonal colonies, then you can pick separate colonies off the plate and have a pure preparation of a different piece of the human genome separately cloned in each of the separate colonies.

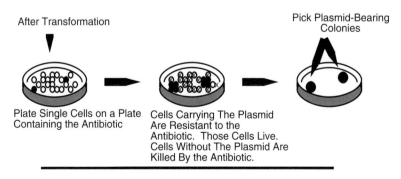

After Transformation

Pick Plasmid-Bearing Colonies

Plate Single Cells on a Plate Containing the Antibiotic

Cells Carrying The Plasmid Are Resistant to the Antibiotic. Those Cells Live. Cells Without The Plasmid Are Killed By the Antibiotic.

O denotes a cell without a plasmid ⊘ denotes a dying cell

● denotes a cell with a plasmid ◉ denotes a dividing cell

● denotes a colony of antibiotic-resistant (plasmid-containing) cells

FIGURE 12.3   The process of transformation.

vector together, and a *host cell* in which to make copies of the "recombined" target DNA–vector combination.

The basic strategy for making clones is thus described in this simple flow chart.

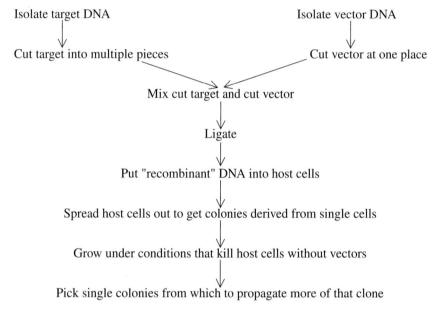

Isolate target DNA                                                    Isolate vector DNA

Cut target into multiple pieces                          Cut vector at one place

Mix cut target and cut vector

Ligate

Put "recombinant" DNA into host cells

Spread host cells out to get colonies derived from single cells

Grow under conditions that kill host cells without vectors

Pick single colonies from which to propagate more of that clone

By applying this technique, one can clone millions of different sequences from the orignal target DNA. Indeed, if you make a large enough set of clones, called a *library,* you can be fairly sure that each possible fragment from your target DNA is represented at least once. It's a neat trick, but so far it hasn't taught you anything, despite the fact that you have now done a rather large amount of work (Box 12.6).

*(Scott, do they realize that you just told them how to make recombinant DNA? Yes, folks, that famous term that you run across in newspaper articles is the very thing just made in your experiment. DNA from two different sources were cut with restriction enzymes and then pieces from each original DNA fragment were "recombined" to form a new DNA fragment, a recombinant DNA fragment containing some material from each or the two original sources.—JER)*

## A MOLECULAR LOOK AT SOME
## SIMPLE VIRUSES

Indeed, just cloning random things can be pretty boring. Yep, its DNA. So what? The important things about cloning are the tools that

BOX 12.6 TRANSFORMATION: PUTTING DNA INTO A HOST CELL

Taken together the two key vector elements allow you to transform a culture of bacteria with a given plasmid. The term transformation really just means that you "transform" or change the characteristics of the bacterial cell by putting DNA into it that gives the cell the ability to express those new characteristics. First, you just soak the bacteria in a solution carrying lots of copies of your plasmid. Under the right condition some of the bacterial cells will actually take some copies of the plasmid into the cell (Fig. 12.3). Once inside the cell, the plasmid can use its bacterial origin of replication to begin making lots of copies of itself. It will also begin expressing that antibiotic resistance gene that lets the cell survive the selective pressure of the antibiotic. Thus by adding an antibiotic to the culture you can kill any cell that did not incorporate the plasmid. When you are done, the only cells left in your petri dish or your culture flask have one or more copies of the plasmid.

allow you to isolate clones of specific DNA fragments. Such specific cloned fragments can provide you a way to get at information you could never get, or at least not easily get, without the clone. To help show an example of this, let's do an experiment that uses cloning to help answer a question about some viruses that we think may be related to each other.

Suppose that you isolated five different, but morphologically related, viruses from five mice in a mouse colony where many of the animals were affected with a flu-like illness. Now, imagine that you (yes, you, the scientist) wanted to determine how related these viruses are by comparing certain aspects of their DNA sequence. "DNA sequence? Where did we get DNA sequence?" you ask with a confused tone in your voice. Don't worry, you actually already know how to do this. Remember that restriction enzymes only cut at specific DNA sequences. Thus, if you look at the sites on a piece of DNA where a given enzyme cuts, you are getting information on the positions of that specific target sequence. For example, GAATTC is the sequence cut by *Eco*R1, our favorite restriction enzyme. So, if *Eco*R1 cuts a 3-kb piece of DNA into two 1.5-kb pieces, you know that the sequence GAATTC is found just about in the middle of that 3-kb piece.

So if one takes two pieces of 3-kb long DNA and compares the sites at which a given restriction enzymes cleaves them, one gets some insight into their DNA sequences. As shown below, two identical pieces of DNA (A and A′) should produce exactly the same pattern of restriction enzyme cleavage. But two different pieces of DNA (e.g., A and B) might produce very different sets of cleavage products.

(X denotes a cutting site for EcoR1)

<table>
<tr><td>────────────────────────────</td><td>A</td></tr>
<tr><td>X              X</td><td></td></tr>
</table>

─────  ───────────  ───────────────     Restriction fragments
following cleavage

<table>
<tr><td>────────────────────────────</td><td>A′</td></tr>
<tr><td>X              X</td><td></td></tr>
</table>

─────  ───────────  ───────────────     Restriction fragments
following cleavage

<table>
<tr><td>────────────────────────────</td><td>B</td></tr>
<tr><td>X         X         X</td><td></td></tr>
</table>

──────  ────────  ────────  ────────     Restriction fragments
following cleavage

How can you tell what size fragments are produced? The answer lies in a technique known as *agarose gel electrophoresis* (see Box 12.7 and Fig. 12.4). This technique allows you to separate DNA fragments according to size, a fashion that is easily visualized. Once you understand this technique you can easily apply it to the analysis of our five viral samples.

We begin by comparing the restriction patterns of the five DNA samples by comparing the fragment sizes as determined by agarose gel electrophoresis. Visually, this is a little like deciding whether two products in a store are the same by comparing the patterns of the two bar codes used to label the products. In this case, the DNA is isolated from each virus, cut with the EcoRI, and then the sizes of the resulting DNA fragments are evaluated by running them side-by-side on a gel. After running the gel for enough time that the fragments are well separated, you can compare the positions of your experimental fragments with the positions of fragments of known size.

Notice in Fig. 12.4 that all five viral DNA samples produce four bands of different sizes following restriction digestion. Each virus has fragments A, C, and D; however, only four of the viruses have the 3.5-kb frag-

BOX 12.7 GEL ELECTROPHORESIS SEPARATES DNA
ON THE BASIS OF LENGTH

A gel is a rectangular slab of a gelatin called agarose, through which DNA will move when propelled by an electric field (see Fig. 12.4). It makes perfect sense that in this gel, the smaller pieces move faster than bigger pieces because the former can wriggle more easily through the tiny pores in the agarose. Thus, after a given amount of time we would find that the smaller fragments had moved a greater distance from the starting point (called the "well") than had the larger fragments. However, fragments of the same size will move identically, making a visible line (called a "band") on the gel.

When a DNA molecule is digested with a restriction enzyme and analyzed by gel electrophoresis (or "looked at on a gel" as we usually say), the restriction digestion will lead to a pattern of several fragments (corresponding to visible bands) of different sizes (see Fig. 12.4). The pattern of bands will be characteristic of that particular DNA molecule, and if you prepare additional copies of that molecule, cut them with the same enzyme, and analyze them on a gel, you will see the same fragment sizes you saw the first time you did the experiment. If you want to know what size a fragment is, then you compare it to fragments of known size run on the same gel in an adjoining lane carrying a size standard as shown in Fig. 12.4. (Take our word for it, you always want to include a lane with fragments of known size to which you compare your unknown fragments. As you can see in Fig. 12.4, our fragment A has run just as far through the gel as the known fragment that is 5 kb, or 5000 bp, in length. From this we would conclude that the fragments are of similar length and that fragment A is also about 5 kb in length. Fragment C has migrated to a position on the gel that is midway between the location of the 1-kb and 2-kb fragments, leading us to think that fragment C is approximately 1.5 kb in size. If we wanted to estimate the sizes of larger fragments, we would use a different set of longer fragments in our size standard lane.

ment labeled B. Instead of the 3.5-kb fragment B, sample 3 has a different fragment that is smaller than 3 kb. (Be sure you understand how to tell the sizes from the gel. Can you tell what the sizes of fragment B would be?)

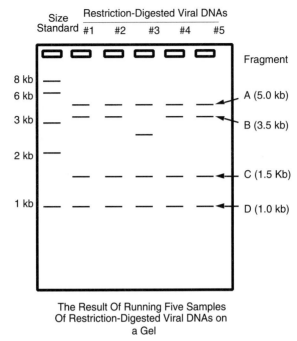

The Result Of Running Five Samples
Of Restriction-Digested Viral DNAs on
a Gel

**FIGURE 12.4**    Gel electrophoresis of restriction-cut DNA from the five virus isolates.

Now comes the first question:

Does band A really correspond to the same band (really the same sequence of DNA) in each of the five samples?

To provide an answer to that question you will want to know whether the DNA sequence of the five fragments is the same. *(No, you don't have to determine the actual sequence of the two DNA fragments to answer that question. Questions like this get asked all the time without sequencing anything.—JER)* To figure this one out you will rely on a trick called *DNA–DNA hybridization* (see Box 12.8). Basically you are going to ask the question:

Can the bases on the Watson (or the Crick) strand of isolate 1's copy copy of fragment A carry out base pairing with the Crick (or Watson) strand of fragment A from the other viruses you are studying?

Or, put more simply:

Do those 5-kilobase restriction fragments obtained by restriction-digesting the other viral DNAs have the same DNA sequence as fragment A in 1?

---

### BOX 12.8 DNA–DNA HYBRIDIZATION

In Chapter 3 we talked about double-stranded DNA and the fact that the two strands of the DNA are paired to each other based on their complementary sequences. If you take a piece of double-stranded DNA and separate it into its two separate single strands, those strands can come back together again and reform the double-stranded structure. An even more interesting observation is that if you take two pieces of double-stranded DNA that are identical to each other (piece 1 and piece 2) and separate them into single-stranded DNA, the Watson strand from piece 1 can pair with the Crick strand of piece 2 (see Figs. 12.6 and 12.7) and vice versa. Depending on the conditions under which we do the experiment, we can even get sequences that are not a perfect match (that is not complementary at every single base pair) to join into double-stranded DNA.

If we are careful to use experimental conditions that require an essentially perfect match, we can use this hybridization process to ask whether two sequences are the same or not. If we start with two different samples of DNA and make them single stranded, we can do tests to tell whether any pairing between the two samples took place (as opposed to strands from the same sample finding each other again). If pairing between the two different samples occurs, then they are the same (or very similar). If no pairing occurs, then they are different.

---

There is a crucial insight here that you need to understand before we go further. If two fragments (call them A1 and A2) have identical, or even very similar, DNA sequences, then the Watson strand of A can base pair with the Crick strand of B and vice versa!

What do you need to be able to carry out such an experiment? You need two things. First, you will need to obtain a reasonable quantity of pure fragment A DNA from one of the samples such as in this example, isolate 1. Second, you need a "copy" of the gel shown in Fig. 12.4 in which the DNA is mounted on something far less fragile and more manageable than a gel. To obtain this you are going to do something called a Southern blot (described below). So let's go get those two reagents.

## CLONING FRAGMENT A

When you look at the gel in Fig. 12.4, you can see that fragment A has been separated away from the rest of the virus. Let's start with a gel just like the gel in Fig. 12.4. After running the gel, the restriction-digested virus DNA fragments from virus isolate 1 are separated according to size. Remember that the gel material is soft. *(Think of jello jigglers!— JER)* So you can simply cut out a block of the gel that contains fragment A and then purify the fragment A DNA from the gel slice. We can now take that purified fragment A DNA and put it into solution with a plasmid cloning vector that was also cut with *Eco*R1. Realize that both the vector and fragment A have the same sticky ends. The fragment A DNA molecules can then Watson–Crick base pair with the plasmid ends and the resulting molecule can be resealed with ligase, just as we have done above.

*("Don't the fragment A molecules ever link up (anneal) to other fragment A molecules? Don't plasmid molecules ever anneal to other plasmid molecules?" you ask plaintively. Yes, they do. But we usually have ways of getting rid of such misbegotten ligation products. Just trust us, this isn't usually a big problem.—RSH and JER)*

We can now transform our ligation mixture into *E. coli* and select transformant colonies by antibiotic resistance. Once we pick the resulting plasmid-bearing colonies we can grow up as much bacteria carrying the cloned fragment A as we want. We can easily isolate the plasmids borne by those bacteria and obtain a pure preparation of fragment A-bearing plasmid. We can then isolate fragment A from that fragment A-bearing plasmid by restriction digestion and we can do this in quart quantities if we so choose.

In this case, our main issue is yield. The amount of the cloned fragment A that we can prepare by isolating and restriction digesting the virus DNA from a virus and then running that DNA out on a gel is quite small compared to the amount we can get with a clone. In other cases the issue may be ease of preparation of the DNA. Many of the cloning vectors have been designed to be very easy to work with. Often the target DNAs we are studying, such as viruses or human cells, may provide us with many technical difficulties that make it hard to get DNA that is pure or DNA that is not degraded (broken down into little pieces). However, most frequently we want to clone something so that we can get it away from the rest of the target DNA.

Okay, you have the cloned Fragment A DNA. You use chemicals to both denature it (separate it into single strands) and to "label" it with a chemical tag; either a radioisotope or a fluorescent molecule. We won't worry about the nature of such tags, suffice it to say that these tags allow you to follow the fragment A molecules for the rest of the experiment. For the rest of this chapter, the label will be a radioisotope. This labeled and denatured collection of fragment A molecules is referred to as a *probe*.

The next step is to use our cloned fragment A probe from isolate 1 to inquire about the relationship of that fragment to the 5-kb fragments in the other virus DNA isolates. We are going to do this by asking if the cloned fragment A pieces composing the probe can Watson–Crick base pair with the other 5-kb fragments. Realize that such pairing will only be possible if those fragments share a substantial amount of shared DNA sequence (i.e., if they are very similar or the same). This technique is called *DNA–DNA hybridization*. We'll show you how to do that in a minute. First, we need to move the DNA sitting in that gel onto a more stable and useful platform.

## THE SOUTHERN TRANSFER: A PRACTICAL "COPY" OF A GEL

To isolate DNA from each of the five viruses in a way that lets you carry out the DNA–DNA hybridization test, you need to transfer the DNA out of the gel onto a sturdier and less mushy substrate. Having cut the viral DNA with *Eco*RI and run it on a gel just like the one in Fig. 12.4, you then put the gel into a basic solution that *denatures* the DNA (which is an acid, after all) molecule into single strands (Watson is parted from Crick). Next, you take the gel and transfer the single-stranded (denatured) DNA within it onto a piece of filter paper in such away that the DNA on the filter paper is a direct, same-size image of where the DNA was located within the gel (Fig. 12.5A). The crucial point here is that the transfer of DNA fragments from gel to paper is precise (like transferring ink from a stamp block onto paper), the fragments of varying sizes remain in separated bands just as they were on the gel. This process is called Southern blotting (after Ed Southern, the scientist who invented it). You are finally ready to proceed.

## DNA–DNA HYBRIDIZATION

DNA–DNA hybridization takes place when you soak the filter paper in a solution that contains the fragment A probe. As shown in Fig. 12.5B, the labeled probe DNA and the DNA on the filter can then reanneal by Watson–Crick base pairing (go back to Chapter 3 if you want a refresher on this topic). It is essential that you now remember that *hybridization will*

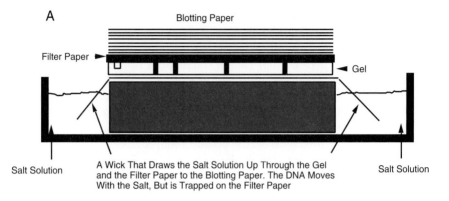

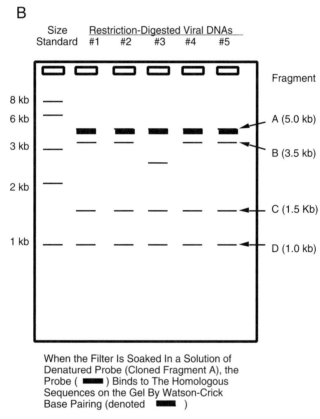

FIGURE 12.5   A Southern blot. (A) How a Southern blot is done. (B) Hybridizing probe (fragment A) to the filter paper after transfer.

*occur only at those sites on the filter bearing DNA that is homologous to fragment A.* (Remember, the probe cannot bind to the filter itself, it binds only to complementary DNA strands affixed to the filter during the process of Southern blotting.) These particular sites can be seen by exposing the filter to X-ray film. The radioactive label on the probe molecules will expose the X-ray film wherever the probe is bound. The resulting picture is called an autoradiogram (see Fig. 12.6). Note that in this case a single black line (band) appears on the film at a position that corresponds to the position of the 5-kb fragments on the filter and before that on the gel. This tells you that, whatever may be different in the rest of the viral DNA samples, at least some of the sequence that composes fragment A is shared between these viral samples.

### OKAY, BUT WHAT ABOUT FRAGMENT B?
### IT WAS DIFFERENT IN ISOLATE 3

Excellent point! The questions are as follows:

1. Are the 3.5-kb fragment B molecules seen in isolates 1, 2, 4, and 5 really the same sequence? And if they are,

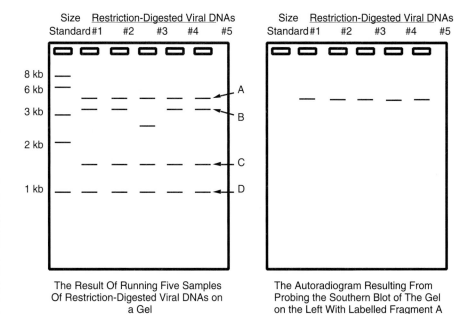

The Result Of Running Five Samples Of Restriction-Digested Viral DNAs on a Gel

The Autoradiogram Resulting From Probing the Southern Blot of The Gel on the Left With Labelled Fragment A

FIGURE 12.6    The gel and corresponding autoradiogram when probed with fragment A.

2. Is the 2.5-kb sequence seen in isolate 3 related to the sequence defined by these fragment B molecules?

The first step is to clone fragment B in the same fashion described above and to make a labeled probe. Now, we can repeat the experiment that we performed above using fragment B as a probe. (Basically, we digested the viral DNAs with *Eco*R1, ran them on a gel, made a Southern transfer filter, and hybridized that filter to the fragment B probe.) What do we end up seeing? In Fig. 12.7, you can see the autoradiogram containing the image of the places on the filter where hybridization took place.

The first thing we can conclude is that we have an answer to our question. On at least a superficial level, the 3.5-kb fragments in samples 2, 4, and 5 appear to be related in sequence to (and probably are the same as) the 3.5-kb fragment B that we cloned from sample 1.

Now look at sample 3. It is missing fragment B, but the "new" fragment in that lane also hybridizes to the probe *(the band "lights up" is how we sometimes express this—JER)*. When we go back to our lab notebooks and look at what we know about the mice from which we isolated the viruses, we find out that strains 1, 2, 4, and 5 cause the illness present in the mouse

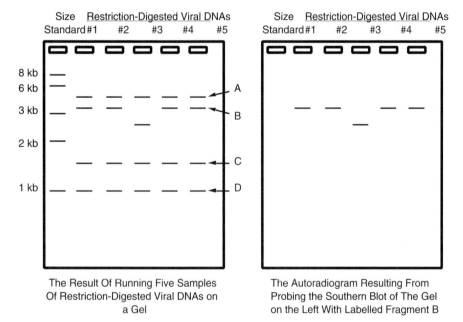

The Result Of Running Five Samples Of Restriction-Digested Viral DNAs on a Gel

The Autoradiogram Resulting From Probing the Southern Blot of The Gel on the Left With Labelled Fragment B

FIGURE 12.7    The gel and corresponding autoradiogram when probed with fragment B.

colony but that strain 3 does not appear to cause disease, even if we pass it on to additional mice.

Since the novel fragment hybridizes to fragment B, it is not unreasonable to propose that the mutation that "knocked-out" the ability of the virus in sample 3 to cause disease is associated with a deletion of 1 kb of DNA from somewhere in fragment B. One could easily imagine that fragment B contained a gene that is required for the virus to be able to make some protein essential for causing disease and that the deletion of 1 kb within that gene has effectively killed that gene's ability to make its product. We have now hypothesized that the new fragment is essentially fragment B with a piece missing out of it and that the viral strain with the missing piece does not cause the disease.

Does that prove that the 1-kb deletion in fragment B is the physical basis of the mutation that knocked out this virus' ability to cause the mice to lie on the couch and watch "Oprah" for days at a time, Kleenex and nasal spray in hand? No, it doesn't. There could be one or another change in the DNA somewhere else in the virus that is responsible for the phenotype. You have several choices for different kinds of experiments to help prove that the deletion is the cause of the functional difference. You might determine the complete DNA sequence of both viruses, the normal and the mutant, and show that the only sequence difference is the missing 1 kb. For most organisms, and most questions of this kind, the sizes of the pieces of DNA to be compared (in this case two whole virus genomes) are much too large for such an experiment to be possible. Basically, you need to do one of two experiments to prove that the "cold-causing" gene resides in the missing 1-kb segment: (1) specifically delete the same 1 kb from a normal virus and show that this synthetic deletion is also unable to cause the flu, or (2) repair the deficiency on the mutant virus and show that the flu-causing ability comes back. These are tricky experiments, but they are doable. Actually, working on a virus that causes disease in mammals is a lot more complex than this for a variety of technical, safety, and ethical reasons. We presented this very hypothetical example simply to introduce you to the techniques involved.

We have begun our introduction to gene cloning by using a virus whose entire genome is small enough that restriction-digests are composed of only a very small number of bands. But the human genome is gigantic. Restriction digesting of human DNA results in the generation of millions of fragments of varying size. Thus, running human DNA out onto a gel results in a smear of DNA fragments and discrete bands are not usually visible. That is okay. Southern blotting, at least when we use a probe that corresponds to a sequence present only once in each haploid genome (or twice in each diploid cell), is so sensitive that individual fragments can still be detected by autoradiography against the background smear that is the rest of the

DNA in the target genome. It is this trick that we will use to clone the TDF gene in the following section.


## CLONING THE TDF GENE

Now that we have a good basic understanding of gene cloning and how it works, we are going to journey into "The Search for Male-Specific Clones" (it sounds like a movie). If we are to search for specific genes only found in the human male, take a wild guess what chromosome we would utilize in our search. Of course, the Y chromosome. The Y chromosome has the unique property that it is usually only found in the DNA of males.


### ISOLATING CLONES OF Y CHROMOSOMAL DNA

Let's assume that human chromosomal DNA has been cut into bite-size pieces by restriction enzyme digestion and then ligated into awaiting plasmids. These plasmids are then transformed into E. coli so that each transformed cell acquired only a single insert-bearing plasmid. These cells are then diluted and plated (spread out) on many petri dishes containing nutrients on which each of the well-separated single bacteria will grow into a separate colony in which all of the bacteria are descended from (and theoretically identical to) the first bacterial cell that landed at that spot in the petri dish during the "plating" procedure. Each colony can be "picked" up off the petri plate and used as a pure source from which large numbers of descendants of the first bacterial cell can be grown and used as a source of DNA containing the recombinant plasmid that carries the "insert" fragment of interest. The result is that it is possible to combine all of the colonies to form what is called a library of clones, but it is also possible to isolate individual colonies and to thus separate the "insert" DNA from that colony away from the rest of the genome. If our library is large enough, then every sequence in the human genome should be present at least once in one of the colonies.

Each of these clones can now be tested for male specificity (i.e., for a Y chromosomal location) by doing a Southern blot experiment. Male and female total genomic DNA can be restriction-digested separately and run on adjacent lanes of our old friend the agarose gel. The DNA in the gel can then be blotted onto pieces of filter paper (this unbelievably expensive paper is called nitrocellulose) where each filter is hybridized to a different clone. Male-specific clones will, on autoradiography, display a band on the male DNA lane but not on the female DNA lane (see Fig. 12.8).

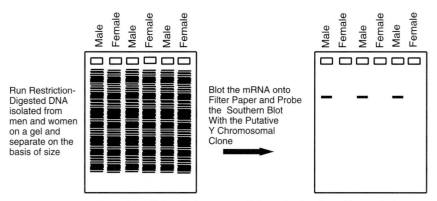

Run Restriction-Digested DNA isolated from men and women on a gel and separate on the basis of size

Blot the mRNA onto Filter Paper and Probe the Southern Blot With the Putative Y Chromosomal Clone

FIGURE 12.8    Identifying Y chromosomal clones by Southern blot analysis.

*(When I started graduate school back in—oh, never mind what year— we made clone libraries essentially in the fashion that I have just described (with differences that don't really matter here) any time we wanted to isolate clones from an organism or individual. Now there are so many more options available. Frankly, it is now possible to simply take a catalog off a shelf and order clone libraries from enterprising companies that have them stockpiled. Some of these libraries are amazing, including libraries made not from whole human DNA but instead from whole chromosomes that have been sorted away from other chromosomes. Thus, it is possible to call up one of these companies and say, "Send me a library of clones from the Y chromosome," and within a few days the library would arrive in your lab. Or you might decide that starting out with the entire Y chromosome was really leaving you searching too wide a field. You might just decide to spread some chromosomes from male cells on a slide under a microscope and "microdissect" (manually cut out and remove from the slide) a piece of the Y chromosome. This would be tedious because this process would have to be repeated about 40 to 80 times before enough material was obtained to use in making a library of clones representing the interesting part of the Y chromosome. I am sure I will have a chance to tell you about more of the currently available tricks, including some that let you answer your questions without even doing experiments at a lab bench, as we continue to look at TDF cloning.—JER)*

Now we have a set of clones derived from Y chromosomal DNA sequences. To map these clones to specific positions on the Y chromosome, we will use a technique called *fluorescence in situ hybridization* (FISH). This technique allows you to hybridize your cloned DNA sequence, or probe, to metaphase chromosomes that have been spread out and affixed to the surface of a glass microscope slide. *(Yes, hybridize. It's the same process used on Southern blot filters: Label the probe with something that can be visualized, make it single stranded, and then let pieces of the probe sequence find complementary bits of sequence with which to base pair on the single-stranded target. The target in this case is chromosomes spread out on a slide instead of DNA that has been transferred from a gel onto a filter.—JER)* Once the chromosomes are bound to the slide they are then treated in such a way as to denature the DNA within the chromosomes. The probe is usually labeled with a fluorescent tag. We are now asking the single-stranded probe to find and base pair with the identical sequence on the denatured Y chromosome.

Once the probe DNA has bound to a complementary site on the chromosome, the "tagged probe" now emits a fluorescent signal that can be visualized easily by standard microscopic methods. *(Scott tells me that these "standard fluorescence microscopes" can put you back $50,000–$150,000!— CAM. Let's see . . . what do I want, a new microscope or a bright red Ferrari; a new microscope or a really fast bright red Ferrari . . . ?—RSH)* Once the fluorescent tag is seen under this expensive microscope, you will be able to determine the exact position of your cloned sequence on this chromosome. Because our gene of interest is the TDF gene, you would choose to study only those clones that hybridize to the appropriate region on the short arm of the Y chromosome. Realize that clones hybridizing to the long arm of the Y are of no interest to us. Why? Because by the time we arrived at trying to clone the TDF gene we knew that it mapped to a spot on the short arm of the Y chromosome. Although FISH is a common way to position clones within the human genome, there are newer and faster ways coming along. One of these, called radiation hybrids, is described in Box 12.9.

You are now down to a population of clones derived from the correct chromosome arm. But you want more than that. You want the TDF gene itself! How do you sort through those clones for the ones that are very near or contain the TDF gene? You would start by identifying those clones that were absent from the Y chromosomes of XY females carrying a deletion of the TDF gene (or which were present in XX-TDF males). Thus, this is a compare and contrast type of situation.

## CHROMOSOME WALKING

The techniques just discussed only get you into the right region of the chromosome. However, even the best tests can only get you into a region

BOX 12.9 OTHER MEANS FOR POSITIONING A GIVEN
CLONE WITHIN THE GENOME

There are many ways to map a clone within the human genome, besides FISH. One way that you ought to at least hear about is to use something called somatic cell hybrids. These hybrid cells have been formed by creating a rodent (usually mouse or hamster) cell line that contains some pieces of the human genome that gets replicated and carried along with the rodent DNA. There are a lot of different kinds of hybrid cell lines available. In some cases a cell line (lots of cells descended from one original cell) contains several human chromosomes. In other cases the hybrid cell line may contain a single human chromosome. One important recent advance in the mapping of human genes was the development of cell lines called radiation hybrids. To create these cell lines, human DNA was broken up into very small pieces with radiation before being put into the rodent cells. A large number of these cell lines have been created, and each one has been characterized extensively to determine which pieces of the human genome stayed in the rodent cell line.

If you want to know where a clone came from, you can test a set of radiation hybrid cell lines to determine which of the cell lines have sequences from your clone and which ones do not. A computer program compares your pattern of pluses (my clone is present) and minuses (my clone DNA is not present) to the pluses and minuses for many other known pieces of DNA. You score a "hit" when you find another piece of DNA that has the same pattern of pluses and minuses as your clone. A statistical test is done to determine the probability that your clone comes from some place in the genome very close to the piece of DNA that gave the same pattern as your clone.

Currently, at the Stanford Human Genome Center, where maps derived from use of radiation hybrid information are being built, a panel of 83 radiation hybrids has been used to place more than 10,000 bits of DNA sequence called STSs (sequence tagged sites) on a very dense and detailed map. The screening system for placement of your particular piece of DNA onto this map is so easy that it is being done regularly in labs throughout the world and often lets you map your DNA piece not just to the right chromosome, chromo-somal arm, or band in the banding pattern; it lets you map things to a much greater level of resolution than you can "see" with FISH.

of 50,000 to 100,000 bp (also known as 50 to 100 *kilobase pairs* or kb). *(Scott, I should be so lucky most of the time as to only have to search through 100 kb of DNA! It is quite common to play recombinant blindman's bluff in an area of a million base pairs or more.—JER)* Thus, you must have a way to quickly find clones in this region or, in other words, the clone that you have has to aid you in cloning other adjacent DNA sequences in this given region. To do so there is a technique called *chromosome walking.* The gist of walking is that it allows you to use the sequences at the ends of the piece of DNA that you start with as probes for cloning adjacent regions of DNA.

Please look at the 100 kb piece of DNA diagramed below. Cleavage sites for three different restriction enzymes (labeled R, S, and B) are indicated on this piece of DNA.

```
                    Probe                          TDF gene
                    ------                         ----------
----------------------------------------------------------------------------
|       |   |    |          |   ||   |        |    |    |    |
R       S   B    S          R   RB   S        B    SR   S    B
```

The location of the TDF gene and your initial cloned sequence (denoted as the "probe") are indicated. Notice that the probe is some 30 kb away from the TDF gene. Now, let me redraw this map, indicating each enzyme site separately.

```
                    Probe                          TDF gene
                    ------                         ----------
----------------------------------------------------------------------------
|                           |   |             |
R                           R   R             R
```

```
                    Probe                          TDF gene
                    ------                         ----------
----------------------------------------------------------------------------
    |       |                   |        |        |
    S       S                   S        S        S
```

```
                    Probe                          TDF gene
                    ------                         ----------
----------------------------------------------------------------------------
        |                       |        |             |
        B                       B        B             B
```

Let me now give each of those fragments a number:

```
                              Probe                    TDF gene
                              ------                   ----------
------------------------------------------------------------------------
  |                         |   |                   |
  R                         R   R                   R

         1                    2       3                4

                              Probe                    TDF gene
                              ------                   ----------
------------------------------------------------------------------------
      |          |                 |         |       |
      S          S                 S         S       S

    5          6                 7         8     9    10

                              Probe                    TDF gene
                              ------                   ----------
------------------------------------------------------------------------
          |                     |         |                    |
          B                     B         B                    B

    11                        12        13           14
```

   Observe that fragments 1–4 will be present only in the bite-size cuts (library) made with DNA digested with enzyme R; fragments 5–10 in the library made from DNA digested with enzyme S; and fragments 11–14 only in the library made with DNA cut by enzyme B. So the key question that comes to mind is: "How do I get from the clone I have to the gene I want?" Good question!

   Your initial clone corresponds to fragment 2. If you probe the second DNA library (the one made with the S-cut DNA) with this probe, you can clone fragment 7. If you go back to the enzyme R-cut DNA library and use fragment 7 as a probe, you can clone fragment 3. Fragment 3 can then be used to reprobe the S-cut library and you can pull out fragment 8. Unfortunately, that really does not help much. Fragment 8 is completely encompassed by fragment 3. This can be solved by using either fragment 3 or 8 to probe the B-cut library and pull out fragments 13 and 14. Fragment 14 ends in the gene of interest, but could also be used to pull out fragments 10 and 4 from the appropriate libraries.

Realize that the preceding narrative assumes a unidirectional walk. In fact, when you start you usually don't know which way you need to be walking. Most walks go in both directions, assuming that one of the two directions has to be correct. (After all, a base sequence is really a single-dimensional object that allows you to move in only two directions.) Usually, you can figure out which end of your walk is going in the right direction at some point part way through your walk.

The point is that you move along a chromosome from a defined starting point by isolating fragments of DNA that overlap each other. There are also some tools that make it much easier. For example, libraries are made by physically shearing the DNA in a manner that creates random breaks. The resulting pieces can then be cloned by some clever chemistry. In this case the same piece of DNA, corresponding to your cloned fragment, might be present multiple times in the library on a different sized fragment.

```
                              Probe
                              ------

-----------------------------------------------------
|          |    |    |         |    ||   |
R          S    B    S         R    RB   S

        ---------------------------------------------------
        |    |    |         |    ||   |         |
        S    B    S         R    RB   S         B

                ---------------------------------------------------
                |    ||   |         |    ||    ||   ||
                R    RB   S         B    SR    S  ** B
```

Walking is made more efficient by the existence of libraries in which bacterial viruses (*phages*) are used as vectors. These *phage libraries* can carry much larger inserts than most plasmids. Indeed, a new system using artificial chromosomes in baker's yeast results in libraries in which the average insert is anywhere from 100 to 2,000 kb in length. Such libraries are referred to as *yeast artificial chromosome* (YAC) libraries and are now commonly used tools in human molecular genetics. Although these libraries have bigger inserts, the basic approaches and tools are often the same. If you have a YAC clone and you want to walk into a neighboring region of DNA, isolate a probe from near the end of the YAC insert and use that probe to identify other YACS that also contain the probe sequence.

*(Actually, in 1998, you MIGHT have to do that experiment to be able to walk between YACs, but for many regions of the human genome you would not use lab bench experiments to accomplish your walk. Instead, you would go to your computer and get on the Internet, where you can CLONE BY PHONE! On the Net there are databases that contain enormous amounts*

*of information about the human genome. One of those databases contains information about thousands of YAC clones and their location on the human genome map. Much of the information that placed YAC clones on that map came from experiments fundamentally similar to the walking experiments just described. Now that the work to walk between YACs has been done, all YOU have to do is dial in, identify the YAC you started with—YACs in the database all have unique identifier names called addresses—and look up the addresses of the YACs that overlap the YAC you started with. At the end of the process, you get to call up a commercial company that keeps stockpiles of the cells descended from each of the YACs in the database. Place your order for YACs with the addresses you just looked up in the database, and the YAC clones show up in your lab within days without your having pipetted ANYTHING! Actually, with a variety of databases around the world holding rapidly increasing amounts of information about the human genome, we find that there are more and more times when we really DO clone by phone before going on to another experiment. Of course, just having your clone-by-phone clones isn't enough. You still face the same problems that Scott describes next.—JER)*

The really crucial issue is how do you know when you reached the TDF gene or maybe even walked right by it? The DNA alphabet is not going to spell out: STOP RIGHT HERE: IT'S THE TDF GENE!

## FINDING THE TDF GENE WITHIN YOUR WALK

Realize that as you walk along the chromosome there are no neon signs telling you that you have found the gene of interest. Several features of the DNA of normal and mutation-bearing chromosomes can assist you in this effort.

First, if you are very lucky, you will find XY females that arose due to mutations in the TDF gene. If you are even luckier, mutants will be due to an aberration that creates a major abnormality in the DNA sequence within the gene instead of a single base pair change. These kinds of aberrations include deletions, insertions of foreign DNA, or translocations whose breakpoints lie within your gene. Such aberrations will alter the restriction digest pattern of the gene when compared to a normal allele. (Recall the information given earlier on cloning and the demonstration that deletions and insertions created changes in the size of restriction fragments recognized on a Southern blot for a given probe). Figure 12.9 displays such a change on a Southern blot that is associated with a small deletion in the TDF gene. Such changes in the restriction pattern can in fact create road signs announcing that one has walked to and is now walking through the gene of interest. (Caution: There are differences between individuals that have nothing to do with the disease that can change this kind of pattern on a Southern blot. This difference in pattern, called a *restriction fragment*

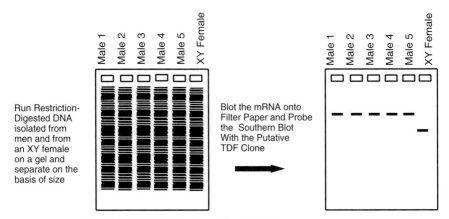

**FIGURE 12.9**    Southern blot of TDF gene rearrangement.

*length polymorphism* (RFLP), will be described further in Chapter 13. In principle, when we use the kind of Southern blot hybridization experiment described here to look for certain kinds of changes that might represent the mutation we are looking for, we usually have to do many additional experiments to tell whether it is a meaningless RFLP or whether it is actually a functional change in the gene of interest.)

Second, it's a good bet that gene sequences are important enough to be conserved during evolution. So real genes, unlike the spacer DNA sequences that separate them, should be conserved. In this case, the real TDF sequence should be found in the DNA of most male mammals. One can test this by performing a Southern blot using DNA from males and females of various organisms (called a zoo blot) and probing with parts of your walk. Any part that gives a male-specific signal is a candidate for the TDF sequence.

Third, if a given region of DNA contains your gene, then it should be expressed (transcribed) in the appropriate sex and tissue and at the right time during development. Transcription can be assayed by *in situ* hybridization to mRNA carried by cells in a given tissue. The cells or tissues are mounted on a glass slide and hybridized to a labeled probe as if one were doing *in situ* hybridization to chromosomes. Cells expressing the gene light up like neon signs, whereas cells not expressing the transcript are not labeled. You can tell hybridization-to-message from hybridization-to-genes because the message is so much more abundant and the signal corresponding to hybridization to message is usually distributed throughout the cytoplasm.

*(I wish this always worked; sometimes it does. Sometimes, however, the level of transcription is too low for your gene to be detected by this method, which makes life a lot harder.—JER)*

One of the criteria for identifying the TDF gene was that it was transcribed in the indifferent gonad at the appropriate time during development. (It's important to remember here what we've told you before: To determine maleness the TDF gene needs to act at only one tissue at one very early time in development. Indeed, the TDF gene is expressed only in the Sertoli cells of the indifferent gonads at a very early time.)

At a finer level of scale, remember that techniques exist to sequence any given DNA. Note that if a region of DNA does contain the gene of interest (as opposed to being a piece of junk or spacer DNA that separates genes), then the sequence of that region should contain a region of base pairs capable of encoding a protein. This is called an *open reading frame* (ORF). ORFs are long runs of bases that can specify amino acids and are not interrupted by stop sequences.

Computer programs are available that can analyze your sequence and look for many additional characteristics associated with sequences that occur within genes as opposed to sequences found between genes. Unfortunately, there are no points in the sequence itself that wave red flags at us to say, "Over here, I am a gene, come study me." However, there are a variety of computer approaches to figuring out where genes might be within the sequence, including checking to see whether your sequence matches the sequence of any of the cloned copies of mRNAs produced in human cells. Not all human genes are represented in the databases yet so this kind of computer search will still miss some of the genes located in the area you sequenced. There are also some very difficult lab-bench approaches that can identify points in your sequence that have high potential to be a gene. The bottom line, however, is that we still often end up having to do experiments similar to those described in this chapter, and we mostly have to slog our way through huge amounts of information — whether we produced it at the bench or someone else put it out there on the internet — in order to identify the gene we want.

*(As the Human Genome Project progresses, we will get more and more out of computers, but we cannot yet foresee a point in time when we will really just sit around like ancient philosophers, contemplating profound genetic questions while we wait for computers to tell us the answers.—JER)*

Finally, mutant and wild-type alleles can be sequenced. Finding the exact molecular position of several mutations that alter gene function provides good evidence as to the whereabouts of the gene. This worked well for the TDF gene. A small number of XY females still possess the DNA believed to correspond to the TDF gene. These individuals were shown to carry mutations in the coding region of the putative TDF gene that prevented the normal function of that gene. This provided very strong evidence that the proposed TDF gene (also called the SRY gene in some books) did in fact encode the testis-determining factor.

DNA sequencing can be a slow process. If any part of the sequence is already known, DNA sequencing can go much faster. For example, if the sequence of a normal gene is known, a mutant allele can be sequenced quite easily and quickly. However, sequencing is arduous, boring (very, very, very boring), and expensive, and there are many random polymorphisms in a cloned set of DNA molecules that don't disrupt the function of the gene and may be unrelated to the gene of interest. Such polymorphisms may make it very hard to find the really significant genetic alteration. The general guide here is that you should have a very good idea of where your gene is before you start to sequence.

Remember that that many mutational changes in DNA sequence have impacts on gene expression or gene function, but others do not. Many of the variants called "polymorphisms" were initially thought not to cause any alteration in gene function. As we learn more about the human genome, we are finding that some of the apparently harmless polymorphic alleles actually do cause functional changes. The inverse problem is also very real. Any particular difference in sequence between mutant and wild type *might* be the one causing the change in function, but it might not. That is why the kind of functional experiments described for evaluating the significance of the 1,000-bp viral deletion are SO important; the fact that the virus with the deletion does not cause the flu whereas other viruses with the complete 3.5-kb fragment do cause illness points us toward an intriguing hypothesis (i.e., that the gene defined by that deletion mutation is involved in causing the flu!). But we require many additional functional tests to prove that hypothesis and still more to allow us to determine what it is about that gene that is important for producing illness. No matter how logical our conclusions from examining DNA sequence appear to be, we really need some functional biological information to go with them. Such experiments are described below.

## REAL PROOF YOU HAVE CLONED THE TDF GENE

All of this information is fine, but the most compelling proof that you have your gene is an experiment in which you place the cloned normal gene back into a mutant cell or individual and restore a normal phenotype. In the mouse this can be done by transforming a specific type of cultured cells (called EC cells) with your cloned fragments. A variety of tools and eukaryotic viruses can serve as vectors to place your DNA molecule in the genome of a mouse EC cell growing in culture. You can then inject these cells into an early mouse embryo and place the embryo back into a female that has been tricked into thinking she is pregnant. (*Thank you Scott for not sharing the details of that part of the experiment!—JER*) A few weeks later you will be presented with a baby mouse, some of whose body comes from the host embryo and some of whose body is derived from the injected EC cells. If appropriate genetic markers are used in your host and EC cells, e.g., coat colors, you might be able to tell which part of a given tissue, such as the skin, came from host cells and which came from donor cells. These mice are thus mosaics and are often referred to as *chimeras*. If some of the injected cells helped populate the germline, you can now get your injected gene into succeeding generations and avoid having to study mosaics.

This same trick was used to put the cloned putative TDF gene onto one of the autosomes of the mouse. It was then straightforward to show that this gene alone caused XX fetuses to develop testis and subsequently male somatic sex as well. This proved that the cloned gene did indeed encode TDF and that the TDF gene was fully necessary and sufficient to confer primary male characteristics.

It should be noted that the TDF gene can do this wherever in the genome it is located, so why not use a simple sex determination system in which TDF is simply present on one autosome. Individuals with one copy of this TDF gene would be male; those with no copies would be female. Using this system, half the sperm of a male would carry TDF, and this would be male determining. The other half would lack TDF and be female determining. Quite obviously, homozygotes carrying TDF could never be produced and so the sex ratio would stay at 50:50. *(If it doesn't seem obvious, just think about it for a minute and you will get it.—JER)* Why not use such a simple system? Why not, indeed? Such a simple system would eliminate the need for X inactivation or for X–Y pairing. Why did evolution eschew such simplicity for what we have? We don't know. We can only tell you again what we have told you before, "Evolution is not Michelangelo and the Sistine chapel. It is a teenage kid with a broken car." Regardless of what might seem most elegant, it will do whatever works and thus the living world abounds in chicken-wire clutches.

What did we learn by cloning the TDF gene? The scientists who cloned the TDF gene actually learned quite a lot. They learned when and where this gene was expressed, and the analysis of its protein product is providing important insights into how this gene functions. It is hoped that you have learned the basic process by which human genes are cloned. It is a process we will return to in detail in each of the following chapters.

## WHAT IS THE HUMAN GENOME PROJECT?

We can't leave this chapter without at least noting that your tax dollars are committed to cloning the entire human genome, piece by piece, and then sequencing it. It is called the Human Genome Project. The purpose of the human genome project is to determine the complete DNA sequence of the human genome. Once that sequence is known, then scientists can spend more time studying the genes and other sequences found there and less time just trying to come up with the desired piece of DNA in the first place. By carrying out large-scale cloning experiments, by developing new faster methods for determining DNA sequences, by putting the resulting information into computer databases that are accessible through the Internet, and by developing new computer programs that can deal with the large volumes of information, the Human Genome Project is rapidly nearing its goal of completing the sequence very early in the 21st century.

# 13

# DNA Polymorphisms as Genetic Markers in Humans (and the Miracle of PCR)

The techniques introduced in the last chapter turn out to be critical not only for cloning human genes of a known position, such as TDF, but also for mapping genes within the human genome. It seems that differences in the DNA sequence of a given region of the human genome are quite common in our population. Thus one copy of chromosome 2 might possess the sequence ATTTCCGG at a given position along its length, whereas its homolog (the other copy of chromosome 2) bears the sequence ATTACCGG at the same site. This small change, a T to an A, may not have any functional consequences, and often it does not, but tools exist that allow geneticists to detect this change easily. Because such differences are relatively easy to detect, they soon became a common form of genetic marker during the first two decades after cloning was invented.

Obviously, one easy type of change to detect is one that alters a site at which a given restriction enzyme cuts. For example, if the restriction enzyme cuts at the sequence AAATTT, then a mutation changing the third A to a T will result in a sequence (AATTTT) that cannot be cut by this enzyme. Thus digestion of this region of DNA by this enzyme will yield a different pattern of DNA fragments depending on whether the sequence AAATTT or AATTTT is present at a given site. This is called restriction fragment length polymorphism (RFLP). Modern gene mapping began with RFLPs, and we will consider them first. *(Since then their popularity has been over-shadowed by something called microsatellite repeats that Scott will tell you about later. Even though most "real" human genetics these days centers around microsatellite repeats, it is still very important to understand the use of "RFLPs," as they are known, because even those of us who are die-hard*

---

*microsatellite repeat fans keep finding ourselves periodically having to admit that you just can't beat RFLPs for some purposes.—JER)*

## RFLPs AS A TOOL FOR GENETIC MAPPING

As noted earlier, the term RFLP refers to the fact that minor sequence differences, which can *eliminate* or *create* sites for specific restriction enzymes, are detectable in human populations. Consider the following example of two otherwise identical regions of DNA on two homologous chromosomes of a given male. One such copy of this region, the one he inherited from his father, possesses three sites for enzyme X, whereas the homologous region, the allele inherited from his mother, is missing the middle site. These two pieces of DNA correspond to the two copies of a given region of autosomal DNA present in this man.

```
        4 kb              2 kb
_____        The allele he inherited
   X                    X      X                  from his father
   1                    2      3

_____        The allele he inherited
   X                           X                  from his mother
   1                           3

        Probe A
```

The difference with respect to restriction site number 2 is really then just another case of allelism and will segregate just as will two alleles of any other genetic trait. If the polymorphism occurs on the DNA of an autosome, it will follow simple rules of Mendelian segregation; similarly, if the polymorphism is present on the X chromosome, it will display sex linkage.

To visualize this trait one need only possess a piece of cloned DNA homologous to the region in which the restriction site polymorphism occurs. For example, consider the cloned piece of DNA just diagramed and denoted as probe A. Probe A can then be used to visualize this difference in the following fashion:

Step 1. Obtain cells (usually blood) from this individual.
Step 2. Isolate DNA from those cells.
Step 3. Cut the DNA with restriction enzyme X.
Step 4. Run the DNA out on a gel.
Step 5. Transfer the DNA to a filter by Southern blotting.
Step 6. Hybridize the filter to labeled probe A and make an autoradiograph.

Suppose this man marries a woman who carries two different RFLPs that are recognized by probe A. These new RFLPs are diagramed below and result from the fact that her alleles carry restriction sites 4 and 5, which are not found in the allelic pair carried by her spouse.

```
                    2 kb
                                                    The allele she inherited
    X       X               X       X                 from her mother
    1       4               2       3
                    1 kb
                                                    The allele she inherited
    X       X       X               X                 from her father
    1       4       5               3
                  Probe A
```

As shown below (see lane 1), cutting the DNA from the husband with enzyme X, performing a Southern blot, and probing with probe A will result in an autoradiograph (a film image of the filter after hybridization) displaying 6- and 4-kb bands. (Remember that only those restriction fragments that overlap with probe A will show up on the Southern blot. Adjacent fragments won't hybridize and thus won't be visible.) The same analysis, as applied to the spouse's DNA in lane 2, will show bands at 1 and 2 kb. Lanes 3–8 display the restriction fragments identified by probe A for the parents and their six children.

|          | 1<br>Dad | 2<br>Mom | 3 | 4 | 5 | 6 | 7 | 8 |
|----------|------|------|---|---|---|---|---|---|
| 6 kb →   | —    |      |   | — |   |   |   | — |
| 4 kb →   | —    |      | — |   | — | — | — |   |
| 2 kb →   |      | —    | — | — |   | — | — |   |
| 1 kb →   |      | —    |   | — |   |   |   | — |

One can then easily see that each child recovered one and only one allele from each parent.

As markers for analyzing human inheritance, RFLPs have several enormous advantages over more classical phenotypic polymorphisms (such as eye color). First, they are easily assessed in any given individual. All one needs is a small amount of DNA. Indeed, as described later, even DNA from a single cell is sufficient (see discussion of using PCR to analyze RFLPs). Second, allelic differences in restriction sites (another way to describe RFLPs) are said to be *codominant*. Both alleles can be observed in the DNA of a given individual and thus the homozygote is easily distinguished from either heterozygote. Note that this is not true for recessive mutations that affect phenotypic traits. Third, such polymorphisms are very common in the genome. As more and more pieces of cloned human DNA become available, more and more restriction fragment polymorphisms are found. Fourth, unlike allelic differences that are revealed only by their phenotypic effects, the probes used to detect the RFLPs can quickly be physically mapped to specific chromosomal regions by *in situ* hybridization so that any genetic findings made with the use of this marker can be correlated to a physical location on one of the chromosomes. Thus it has been possible to construct a physical map of RFLPs along each of the 23 pairs of human chromosomes.

*(Actually, as noted in the last chapter, there are several different ways to map a given piece of DNA back to a location on a human chromosome. What is important here is not that you know how each method works but that you understand that the available methods keep advancing and changing to be faster, more sensitive, or more accurate and that each advance increases the rate at which we are finding things out. The rate at which new genes are being mapped to positions on chromosomes is just astonishing, and it's still increasing!—JER)*

### WHERE DO RFLPs COME FROM?

Over a period of several decades a large number of university, government, and industry labs have dedicated a vast amount of time to finding cloned human DNA sequences that recognize RFLPs that are common in the human population. Basically, these labs start out with libraries of random pieces of human DNA ligated into cloning vectors (viruses and plasmids) (libraries were defined in Chapter 12). These laboratories randomly tested huge numbers of such cloned fragments of human DNA for those sequences that recognized different sized DNA fragments in different people. In other words, they looked for pieces of cloned DNA that represented regions of DNA that have mutated or diverged rapidly since the beginning of the human species. Realize that the value of such a RFLP-sniffing probe depends on how variant that region of DNA is in the human population. A RFLP that is found very rarely, or only in very small or isolated populations, will probably have very little utility in mapping. *(Keep in mind that once you find a large family in which a disease or trait is segregating, you need RFLPs that work in THAT population. Whether they are present in some other population is irrelevant to you.—RSH)*

Years of work generated huge collections of such pieces of DNA that recognize very *polymorphic* (variable) sequences in the human genome. These collections are easily available from both private and commercial sources. Using the techniques described in this chapter, these RFLPs have been mapped with respect to each other to produce a detailed map of the human genome. Moreover, the technique of *in situ* hybridization (again see Chapter 12) allowed these people to map each RFLP onto a specific region of a given chromosome.

There are actually many other methods by which RFLPs can be placed on chromosomes. We aren't going to describe them all, but they all depend on the same basic concept: you ask whether your labeled, single-stranded probe hybridizes to single-stranded "target" DNAs which identify specific chromosomes. *(There is a whole world of assumptions wrapped up in the idea that you have sources of "tester" DNA on hand for which you know such a thing as its chromosome of origin. As with the rest of the Human Genome Project, the secret to this process of mapping a probe back to a chromosome is that someone else did a lot of work over a long period of time and at great cost to develop some reagent, such as a somatic cell hybrid cell line in which only chromosome 7 from a human is contained in a hamster cell line, allowing you to tell whether you have something from chromosome 7 based on whether it can hybridize to DNA from that cell line. The test isn't all that hard to do but starts from the BIG assumption that you have such reagents available.—JER)* The resulting efforts of the RFLP mappers were the construction of some of the first framework maps on which RFLPs had been localized at many positions along the length of each chromosome.

Back when maps were being built, primarily with RFLPs, the map was very sparse and there were a lot of "holes" in the map. It really took the development of the next kind of DNA marker, microsatellites, for a truly useful map to appear, densely saturated with markers assigned continuously along each of the chromosomes.

## MICROSATELLITES: A SECOND TYPE OF DNA POLYMORPHISM

As plentiful a source of genetic markers as RFLPs may seem to be, there is in fact a more useful set of genetic markers or differences called microsatellite repeat sequences. Because so many people use these markers, a lot of different jargony names, such as simple sequence length polymorphisms (SSLPs), have been developed. Microsatellite may not trip lightly off the lips, but it is better than saying "the marker formerly known as a short tandem repeat polymorphism." Microsatellites are short runs of repeated simple sequences, e.g., di- and trinucleotide repeats such as CACACACACA or TTGTTGTTGTTGTTGTTG, that are scattered at a huge number of places throughout the human genome. The curious thing about microsatellite repeats is that they appear to have been very mutable in the course of human history, at least in terms of the number of repeated units present at a given site. Thus, a given CACACACACA microsatellite sequence on chromosome 3 might have 29 copies of the CA repeat on the copy of chromosome 3 that I inherited from my father and

35 copies of the CA repeat on the homologous copy of chromosome 3 that I inherited from my mother. If we were to look at the same site in Catherine's genome, we might see 25 CA repeats at this site on her maternally derived chromosome 3 and 31 copies on her paternally derived chromosome 3. Thus, in just two people, we can find four different polymorphisms (alleles) for this one microsatellite array.

It should be easy to see why such polymorphic sites might make excellent genetic markers. These microsatellite arrays are present at very high copy numbers in the genome and are very easy to detect. Better yet, given the tens of thousands of CA repeats spread out around the human genome, you can usually find one right next door to any piece of DNA you have cloned, mapped, or otherwise want to study. The assay used to detect microsatellite repeat alleles is faster and easier than the assay used to detect traditional RFLPs.

### DETECTING MICROSATELLITE POLYMORPHISMS

Let's begin by imagining a microsatellite array that is flanked on both sides by unique sequences. Take a look at the following sequence consisting of a string of CAs embedded in an otherwise unique sequence *(unique means that this sequence is found in only one place in the human genome, i.e., NOT repeated—JER)*. At least in theory and assuming incredible levels of resolution, you might think that you could use hybridization to a Southern blot to tell the difference between fragment A and fragment B, which are identical except for the presence of three extra As and three extra Cs in the second DNA fragment.

A  __X_____CACACACACAC_____X__
B  __X_____CACACACACACACACAC_____X__
          Q _____

Realize that if we digest this DNA fragment with enzyme X, do a Southern blot, and probe with probe Q, the size of the band on the resulting autoradiogram will depend on the number of CA repeats in the microsatellite array. The more CA repeats there are, the larger the fragment will be. If you are thinking that this is just another type of restriction fragment polymorphism, you are entirely correct, so far. It is important to realize that this polymorphism is not due to a change in the presence of a restriction enzyme cutting site, it is due to a difference in the microsatellite repeat number. *(Frankly, if the alleles you want to distinguish differ by only six nucleotides in length, you are going to have problems telling any difference in fragment length between the two alleles if you use traditional RFLP technology with agarose gels and hybridization to Southern blots, but this technology can work well in some cases.—JER)*

(We should point out that you will get a very different result if you probe with the microsatellite array itself. This will hybridize to every CACACACA repeat in the genome. The result will be an autoradiogram that looks like a berserk supermarket bar code or even a solid black smear, depending on which simple sequence you use and how common it is in the genome. Eventually, we will show you how a similar experiment can play an important role in forensic analysis.)

So is that how we look at CA repeat lengths in the "real" world of the molecular biology lab? Sometimes, but not usually. Another technique came along in 1983 that has revolutionized the molecular genetics of the last decade. It lets you answer the question, "How many repeats on each of the two pieces of DNA?" It also lets you do so with the use of less target DNA from the patient, for less money, in less time, and with more gain of information. *(Rather like the Human Genome Project coming in ahead of schedule and under budget! Sound miraculous? Well, it was miraculous enough that this new technique earned Kerry Mullis a Nobel Prize! Read on for details.—JER)*

## POLYMERASE CHAIN REACTION

### THE MIRACLE OF PCR

PCR is based on the fact that DNA can be replicated in a test tube provided four or five things are added: (1) a DNA molecule to be replicated, known as the template; (2) primers, short stretches of single-stranded DNA corresponding to the ends of the region you wish to replicate; (3) DNA polymerase; and (4) nucleotides (bases) to be incorporated into the newly made copies. Suppose we wanted to copy the following DNA molecule:

```
5' _____ 3' (Watson)
   AAT ACGTT CGT AAT CGT CCGT ACT GGT T AT AT CGT T AT GCT AAAAC GGT A
   I I I I I I I I I I I I I I I I I I I I I I I I I I I I I I I I I I I I I I I I I I I I I I I I I I I I I I I
   TT AT GCAAGC AT T AGC AGGC AT GACC AAT AT AGC AAT ACGAT T T T GCC AT
3' _____ 5' (Crick)
```

We begin by denaturing this molecule into two separate single strands.

```
5' _____ 3' (Watson)
   AAT ACGTT CGT AAT CGT CCGT ACT GGT T AT AT CGT T AT GCT AAAAC GGT A

   TT AT GCAAGC AT T AGC AGGC AT GACC AAT AT AGC AAT ACGAT T T T GCC AT
3' _____ 5' (Crick)
```

Our primers are short stretches of DNA corresponding to the ends of the specific region that we wish to amplify. One primer is 5' TTCGTAA 3' and the other is 5' TTTTAGC. If these primers are hybridized (or annealed) to the single-stranded DNA, two types of molecules are formed:

5' ——————————————————————————————————————— 3' (Watson)
AAT ACGTT CGT AAT CGT CCGT ACT GGT T AT AT CGT T AT GCT AAAAC GGT A
                          I I I I I I I
                          3'  CGAT TTT  5'

and

        5' ——————— 3'
        TT CGT AA
        I I I I I I I
T T AT GC AAGC AT T AGC AGGC AT GAC C AAT AT AGC AAT AC GAT T T T GCC AT
3'                                                        5' (Crick)

The trick to PCR is that DNA polymerase will now extend the 3' end
of those primers by copying the template DNA. One of the secrets to
understanding PCR is to realize that polymerase will NOT make a copy
of one strand of DNA unless it has a primer like this with a 3' end from
which it can begin building or extending the copy strand. So the polymerase
cannot just copy any part of this piece of DNA that is present, but only
the parts downstream, or 3', of the primer. This is an especially valuable
feature because it means that you can put the entire human genome into
a test tube as the target for this reaction, and the ONLY piece that will
get copied is the piece in between the primers. So assuming that replication proceeds to the end of both molecules, the first cycle of replication
yields:

5' ——————————————————————————————————————— 3' (Watson)
AAT ACGTT CGT AAT CGT CCGT ACT GGT T AT AT CGT T AT GCT AAAAC GGT
I I I I I I I I I I I I I I I I I I I I I I I I I I I I I I I I I I I I I I I I I I I
T T AT GC AAGC AT T AGC AGGC AT GAC C AAT AT AGC AAT AC GAT T T T
3'                                                        5' (Crick)

and

5' ——————————————————————————————————————— 3' (Watson)
   T T CGT AAT CGT C CGT ACT GGT T AT AT C GT T AT GC T AAAAC GGT T
   I I I I I I I I I I I I I I I I I I I I I I I I I I I I I I I I I I I I I I I I I
   TTATGC AAGC AT T AGC AGGC AT GAC C AAT AT AGC AAT AC GAT T T T GCC AA
3'                                                        5' (Crick)

(Note: The newly replicated bases are in *italics*.)

We are now going to denature the two DNA molecules back into
four single strands, add the primers, and replicate again. Replication of
the two original strands will yield the two products just shown. But look
at what happens when we hybridize the same primers to the two new
strands:

        5'    TT CGT AA  3'
              I I I I I I I
*TTATGCAAGCATTAGCAGGCATGACCAATATAGCAATA*CGATTTT         (Crick)
3'                                                 5'

and

| | |
|---|---|
| 5'   TTCGTAA*TCGTCCGTACTGGTTATATCGTTATGCTAAAACGGTT*   3'   (Watson) | |

                      I I I I I I I
                   3' CGATTTT 5'

Now let us allow DNA polymerase to extend the 3' end of the primers by adding bases complementary to the longer strands again. We get

5'   _____  3'  (Watson)

    TTCGTAATCGTCCGTACTGGTTATATCGTTATGCTAAAA

    I I I I I I I I I I I I I I I I I I I I I I I I I I I I I I I I I I I I I I I I
    *TTATGCAAGCATTAGCAGGCATGACCAATATAGCAAT*ACGATTTT

3'                                       5'  (Crick)

and

5'   _____  3'  (Watson)

    TTCGTAA*TCGTCCGTACTGGTTATATCGTTATGCTAAAACGGTT*

    I I I I I I I I I I I I I I I I I I I I I I I I I I I I I I I I I I I I I I I I
    AAGCATTAGCAGGCATGACCAATATAGCAATACGATTTT

3'                                         5'  (Crick)

(Note: The newly replicated bases are underlined.)

We are going to run this same set of reactions again, but now let us worry about the two strands newly synthesized in the last reaction. (The two older strands will do just what they did before.) First, denature the DNA molecules and let the primers bind to the single strands:

5'   _____  3'  (Watson)

    TTCGTAATCGTCCGTACTGGTTATATCGTTATGCTAAAA

                          I I I I I I I
                      3' CGATTTT 5'

and

5'   TTCGTAA  3'

    I I I I I I I
    AAGCATTAGCAGGCATGACCAATATAGCAATACGATTTT

3'                                         5'  (Crick)

Okay, let the polymerase do its job one more time:

5'   _____  3'  (Watson)

    TTCGTAATCGTCCGTACTGGTTATATCGTTATGCTAAAA

    I I I I I I I I I I I I I I I I I I I I I I I I I I I I I I I I I I I I I I I I
    *AAGCATTAGCAGGCATGACCAATATAGCAAT*ACGATTTT

3'                                         5'  (Crick)

and

5'　　　────────────────────────────────────　3'　　(Watson)

TTCGT AAT CGT CCGT ACT GGT T AT AT CGT T AT GCT AAAA
I I I I I I I I I I I I I I I I I I I I I I I I I I I I I I I I I I I I I I I I I I I
AAGCAT T AGC AGGC ATGACC AAT AT AGC AAT ACGATTTT

3'　　　　　　　　　　　　　　　　　　　　　　5'　　(Crick)

(Note: The newly replicated bases are in *italics*.)

Realize that the replication of these two molecules will generate identical molecules whose ends are defined by the original primers. Those four daughter molecules will make 8 such copies in the next cycle, 16 in the cycle after that, then 32, 64, 128, 256, 512, 1,024, 2,048, 4,096, and so on. It will not take a great many cycles to generate an enormous number of identical copies of these molecules, all of which were derived from the single parent molecule with which we began. *(After 20 cycles of PCR amplification more than a million copies will have been made from one DNA molecule. Sometimes we do as many as 35 or 40 cycles! Each cycle takes just a matter of a few minutes to carry out, so the whole experiment is over in a few hours at most. I can remember before the first PCR machine was delivered and the students and post-docs had to hand carry the samples between water baths of different temperatures, over and over and over and . . . . —JER Haven't I heard that one before? It continues with something about walking to school in bare feet in 40 below weather at 4 a.m., up hill in both directions?—RSH)*

In theory, one can easily make enough of this molecule to see it as a single band on a gel! Enough copies can be made of it to clone the amplified DNA or to label it and use it as a probe. *(Lots of times, instead of growing up a clone and isolating the insert to make a probe, I just design primers that flank my region of interest and PCR amplify the piece I want to use as a probe.—JER)* In reality there are times when one may still have to do a Southern blot to visualize the resulting bands. For example, if you tried to use PCR to make a copy of a really large fragment, it would not be copied as efficiently as a short fragment, so you might end up with a low yield at the end of the PCR run. Then you might need the assistance of hybridization to be able to tell where on the gel the PCR-amplified fragment was, but the exposure times will be much shorter than with conventional RFLP analysis. The whole process is SO simple! Just mix the original DNA sample; the primers, one of them labeled with radioactivity or a fluorescent-colored tag (so you can detect where the DNA fragments are after you run them out on a gel to separate them by size); the polymerase; and the nucleotides in a small tube and stick them in a wonderful little machine called a thermal cycler or "the PCR machine." Scott's lab has two of these little gems. *(I don't want to guess how many your lab has, Julia.—RSH More than yours and it is still not enough! I cannot imagine having enough PCR machines because everyone in my lab uses PCR in most of their experiments.—JER)*

So where did those primers come from? Well, to do PCR you must know the sequence of the ends of the region that you wish to amplify. Such sequence information is easy to get by simply analyzing the first clone of this region that you obtained. You can then have primers defining the region you wish to amplify by either a lab at your company or university or by a private company. My lab can literally pick up the phone, call a company that makes primers, and have them in hand in 48–72 hours. They really don't even cost very much. *(Per primer perhaps, but the bill can still run up VERY fast when you start buying enough primers to "look" at positions throughout the human genome.—JER)*

At this point some of you may be wondering, "If you already know the sequence of this piece of DNA because you have already cloned it and determined the sequence, then WHY do you want to use PCR to amplify pieces of something you already have?" In many cases, we have a clone and we know its sequence, but we do not know what the sequence is in different individuals. Remember, the real question is not "what is the DNA sequence of the clone?" but rather "what is the DNA sequence, or what are sequence differences, in specific individuals in whom we are trying to follow some other genetic trait?" In most cases, the only value of this clone to us is to detect polymorphisms in a population, i.e., as a genetic marker to screen for mutations in a candidate gene. Another reason we might want to work with a fragment amplified by PCR is that we do not actually have the original clone ourselves. *(Perhaps some other researcher who has the clone has sequenced it. Instead of sending me the clone through the mail or by courier, my collaborator can just send me the primer sequences by e-mail and I am then free to amplify and "play with" this new DNA fragment.—JER)*

## USING PCR TO DETECT
## MICROSATELLITE POLYMORPHISMS

Suppose that sitting in the middle of the sequence that we amplified earlier was a $CACACA(CA)_n$ microsatellite repeat. Then the size of the resulting PCR product would depend on the number of copies of the CA dinucleotide repeat. Thus we could easily determine the CA copy number for both alleles of a given individual by doing PCR using primers flanking the microsatellite repeat. The result should be a mixed population of two fragments differing in size by the difference in the number of CA repeats. For example, imagine that allele one carries 50 copies of the CA repeat. You amplify a sequence that extends for 50 bp on each side of the micro-satellite. The result should be a 200-bp fragment. However, if the microsatellite repeat on the homologous chromosome carries 55 copies of the repeat, the resulting fragment will be 210 bp long. *(Some of the things we can do with PCR may be amazing, but the thing that always amazes me is that we can separate pieces of DNA that differ from each other by only 1 bp in length and we can distinguish those two bands from each other. Thus, if we wanted to compare the previously mentioned fragment between a case with 50 repetitions of the repeat and another fragment with 53 repetitions, even though the whole fragment differs by only 6 bp in length out of 150, we could distinguish them on a gel and know how long they are.—JER)*

## COMPARING MICROSATELLITE AND RFLPs AS GENETIC MARKERS

Both RFLP and PCR analyses of microsatellites can be used in genetic analysis, but microsatellites are far more commonly used. Why?

1. RFLP analysis requires large amounts of often precious human DNA. PCR-based microsatellite analysis can be done with minute amounts of human DNA.
2. RFLPs are often far less polymorphic than microsatellite repeats. One often has to be very lucky to find two different RFLPs for a given DNA fragment for a given family.
3. RFLP analysis is simply harder to do in some cases than microsatellite analysis. *(RFLP studies are often difficult or very finicky to run because human DNA preps are often reluctant to cut to completion with some restriction enzymes, and it is often hard to get a probe that is truly unique, i.e., that does not contain any repeated sequence material on it.—JER)* PCR-based microsatellite analyses avoid these problems. RFLP experiments take several days to do and the exposure of the autoradiogram can take 1–14 days. PCR can be done in an afternoon, with perhaps an extra day for the gel to run plus autoradiograph in some cases.
4. RFLP analysis requires that you test many restriction enzymes to even find out IF there is an RFLP where you want one. Microsatellite polymorphism can be found ALMOST anywhere in the DNA. Just the CA repeats alone occur in something like 50,000 copies and are often highly informative. Indeed, heterozygosity rates for individual repeats run into the 80–95% range for many CA repeats on the map.

## FORENSIC DNA ANALYSIS: A BRIEF DIGRESSION

Most of this book talks about DNA testing of people who want to be tested to resolve a question about their health care or people who have agreed to be tested to assist a research program aimed at understanding the underlying mechanisms of a genetic disease. However, these are not the only genetic mysteries solved by this kind of testing. Exactly the same kinds of assays—RFLPs and microsatellite repeat testing—get used by forensics experts to determine whether blood or sperm from a crime scene came from the accused individual or from the victim.

The basic principle of this test calls for examining multiple different polymorphic DNA markers that segregate independently of each other (are unlinked). The power of the test depends in part on the ability of the test agency to sample enough different markers to be sure that only one person could have given an identical set of alleles at every single marker tested. If you only test one marker for which there are only two alleles and you find that the crime scene sample and the sample from the accused are both homozygous for allele 1, what does it mean? Although it seems like the answer to this should be simple, it is not. What does it mean if 50% of the population is homozygous for allele 1? In that case, this result is not terribly meaningful. What does it mean if allele 1 occurs in less than 1% of the population, and homozygotes occur only once in every 10,000 individuals? Such a result might seem to mean more, but it still means that there are many individuals whose DNA would match that of the forensic sample. As you can see from these examples, it is actually very important to know much more than whether the two DNA samples match. It is also important to know things about the genetic marker being tested and about the population from which alternative matches might be drawn.

As you may have guessed as you read the last paragraph, just testing one genetic marker is not going to get you a definitive answer, as even a very rare allele of a genetic marker is likely to match many people in this very populous world of ours. What if you test another 10 independently segregating genetic markers and find that all alleles match for every marker, including another 5 very rare alleles? Well, if you test enough markers and find identity continuously, eventually you can arrive at an answer that suggests that instead of a 1/2 chance that the accused contributed the blood sample, or a 1/10 chance, or 1/100 chance, finally you reach a point where the denominator in the fraction is larger than the entire population of the earth! This starts to sound very convincing, but remember, it is all statistics. In the face of such odds, is there a way that someone else might have been the source of the blood sample?

What questions haven't we asked here? Does the accused have an identical twin. Yes? Right there you have put an end to the argument that NO ONE else on the face of the earth could have done it. In addition, other close relatives such as a sibling might have some very low likelihood of a match under circumstances that effectively exclude most of humanity. In one case in which a man's DNA sample matched the sample from the crime scene, the defense tried suggesting that the man's brother could have been the source of the DNA sample. It didn't work as it turned out; the brother had been far from the crime scene, out on a boat with witnesses at the time of the crime.

Here is the real clincher: So what if the DNA from the blood samples match? Does that prove that the accused did IT (whatever IT was)? Where did the two blood samples being compared come from? In one case an individual who was in fact guilty was initially cleared on the basis of a failure of his DNA to match with the sample from the crime scene. How could this happen? Does this mean that DNA testing doesn't mean anything? No. Let's examine what happened. All of the men who lived in the area where the crime was committed were requested to donate a blood sample to be submitted for testing. The fellow who was guilty got one of his buddies to donate a blood sample in his place. However, because the buddy lived in the target area, he also donated a sample in his own name. The forensics folks found themselves with an interesting tangle. None of the test samples matched the crime sample, but two of the test samples matched each other! Being smart forensics types, they quickly figured out that the guilty individual had not contributed a sample, but that someone had donated a sample in place of someone else in the target group. It didn't take them long to sort out who contributed twice and who had not contributed at all and to figure out that the fellow who had tried to skip out of donating was in fact guilty.

The converse of this situation is one that has certainly been discussed *ad nauseum* in the press. If someone did not do IT, but someone else plants a sample of their blood at the crime scene, a later test will of course show that their blood matches that from the crime scene! Does the match mean that the accused did it? No, but the match could be taken as evidence in favor of conviction if no one figured out the sample was planted. For this reason, forensic work that is carried out correctly uses safeguards and extensive documentation of every step in handling a piece of evidence. Forensics people have elaborate mechanisms for seeing to it that if you donate blood as a test sample that they have that sample in their control and can account for its whereabouts continuously after sampling. In a recent famous trial one of the issues raised was that the test blood sample from the accused had been handled in such a way that its whereabouts could not be accounted for continuously after it was drawn and before it arrived at the lab. One of the problems that faced anyone trying to evaluate the evidence was this: If the chance that anyone other than the accused produced the crime scene sample is less than 1 in some unbelievably large number, how does that compare to the unknown chance that anyone did something inappropriate with a misplaced tube of the accused's blood while it was unaccounted for? However, the fact that someone did not fill out some paperwork correctly is hardly evidence of sinister plots and planted evidence. Where is Perry Mason when we need him to sort out such things? Although many people worry about the validity of tests for matching DNA, and what the statistics mean, a lot of the testing is done with great care and precision, and we can feel confident in saying two samples match. After that, as for fingerprints, hair samples, handguns, footprints, or any other evidence, someone still has to figure out whether a match between the two samples means anything based on the other rules of evidence.

What other problems are there in forensic testing? First, the DNA isolated from a forensic sample may be very small in quantity, making the testing more challenging, both technically and in terms of the analysis. If you have to test more markers to increase the probability that an apparent match means something, what do you do if the crime scene sample is tiny and you cannot test very many markers? Second, crime scene samples may be in bad shape, depending on what has happened to them. Was the sample dried? Did it spill onto dirty ground? Was it present on a decomposing body? There can be a problem with trying to do tests on DNA that has broken down (been degraded) at the crime scene. Third, the blood or semen sample from the crime scene may be mixed with blood or other body fluids from the victim. The problem becomes much more complicated when trying to detect a match between the accused and the crime sample by "subtracting away" the allele sizes that were contributed by the victim (especially since some of the allele sizes found in the victim COULD be the same size as alleles found in the assailant, for which the statistics people have developed compensations).

Next, let us take a look at an interesting but extreme and unlikely situation that supposedly actually happened. What happens if someone has had a bone marrow transplant? The DNA produced by his blood cells now matches the DNA of the bone marrow donor and not his own DNA. If the legal system wanted to evaluate whether the man in question had committed a rape and asked the man to submit a blood sample so that DNA testing could occur (the usual procedure), the blood sample from the real assailant would not match the sperm sample from the real assailant! It appears that in that case someone smart finally figured out what had happened and compared sperm DNA from the accused to the sperm DNA from the crime scene sample.

Finally, let's keep in mind that comparisons of DNA samples can also lead to exoneration of someone accused. Once DNA testing became available, some men who had been in prison for rape were freed after it was determined that they didn't "match" the forensic sample (and others were confirmed as being rightfully imprisoned). Sometimes, in the course of trying to solve a crime, law enforcement agencies will take blood samples from many individuals who fit the description of the reported assailant.

Out of such situations arguments have arisen about the ethics of such situations. Men who had nothing to do with the crime who are required to provide a blood sample may be asked to consent to donate, but is it in fact consent when refusal to consent will be taken as suggesting guilt?

However, testing of such samples can clear many men who otherwise might remain under a cloud of suspicion. After a series of rapes (including one death) in a midwestern town, DNA samples from many men were tested. After the case was solved, the men wanted their DNA samples returned to them. The legal end of things wanted to keep the samples for reasons that you can probably imagine as well as I. After much confrontation and publicity, the samples were finally turned over. So the question arises, does society have any right to keep blood samples on men who have been found innocent? Once someone has been fingerprinted and later found innocent, the FBI does not tear up the fingerprint card. So there are two different questions for you to think about. Would it have been wrong for the legal system to keep those DNA samples? If it is wrong for DNA samples to be kept, is DNA somehow any different from other forms of evidence which, when retained by the law, have raised no outcry or protest that retaining fingerprints violates someone's rights?

We are not going to try to offer you any answers, but we want you to think very hard about the fact that although DNA can provide more precise answers than some other forms of evidence, the real bottom line is that there is nothing either more wonderful or more evil about DNA evidence as compared to any other forms of evidence. Should the handling of DNA as evidence be held to some different moral standard than other forms of evidence? (Did I hear a hesitant whisper of "maybe yes"?)

## OTHER IDENTITY CRISES

Criminal situations are not the only ones that call for comparison of allele sizes present in different DNA samples. You may have heard about paternity testing. It is possible, if you know the pattern of alleles present in the mother and the child, to tell whether the purported father of a child really is the father. Again, the answers arrived at are all based on probabilities, but the probabilities are frequently quite a compelling argument. *(Issues such as paternity can be incredibly important in civil litigation in people's minds and feelings. It's worth noting that many researchers and physicians suggest that in as many as perhaps 10% of births, the man listed as the father on the birth certificate simply cannot be the father of that child.—RSH)* Whether it is a criminal trial or a paternity suit, it is common for genetics experts to be called upon to explain to the court and the jury just what the findings are, how they can be interpreted, what alternative interpretations there might be, and what the likelihood is that someone has been ruled in or out as a match based on the tests that were done.

Examination of the alleles present in DNA samples has been used to answer other historical questions. When a grave was discovered that was purported to contain the remains of the Romanovs, the last Russian royal family, DNA samples from the bodies were compared to DNA samples from a variety of relatives of the royal family that had survived. It was concluded that the grave indeed held the Romanovs plus additional members of their household.

Other questions that have been asked include: Do patterns of alleles present in different indigenous people around the globe match the patterns of relationship predicted by anthropology and archeology? Somewhat. Can we all trace our ancestry back to one original "Eve" based on DNA differences in the mitochondria or back to one original "Adam" based on building a family tree out of DNA differences on the Y chromosome? Some researchers say "yes," whereas others say "no."

Every day it seems like someone comes up with some new clever question that can be answered by comparing DNA samples between individuals or between groups. As with forensic testing, the hard part is usually not in telling whether two samples match, it is in the interpretation of what the heck matching or not matching means.

# 14

# HUMAN GENE MAPPING:

## A GENERAL APPROACH

Any kind of inherited difference among different individuals can potentially be a genetic marker: a blood type, an eye color, even the shape of an earlobe. Most genetic markers these days are the types of DNA markers discussed in the previous chapters. However, the fundamental techniques of genetic mapping are the same whether your genetic marker is an RFLP, a microsatellite repeat, or your blood type. They all can be used to map genes such as disease-causing loci. To demonstrate this technique, let us begin by using RFLP analysis in the simplest possible case, one that meets two simple conditions: (1) We already possess a cloned piece of DNA corresponding to the gene of interest. *(Don't worry about where that clone comes from. Scott will explain in detail later; just trust us for a moment that you have such a piece of DNA that you can use as a probe for your Southern blots.—CAM)* (2) The disease-causing mutation (DCM), e.g., a small deletion that included the restriction site, created a new DNA polymorphism. With those two conditions met, we can move on.

We begin by considering a hypothetical mutation in the CF gene, the gene whose malfunction causes cystic fibrosis (see Chapter 15). In this case, the mutant allele can be observed as a 6-kb restriction fragment, in contrast to the 4-kb fragment characteristic of the normal or wild-type allele. *(Wild-type is common jargon for the genetic version that occurs "in the wild," which is to say the normal form of the gene, if such a thing can truly be said to exist.—JER)* Suppose then that two heterozygotes married and produced the children denoted as individuals 3 through 8 below. An autoradiograph resulting from digesting DNA from the members of this family and probing with a cloned piece of DNA from the cystic fibrosis gene is schematically diagrammed below.

Panel

| | 1 | 2 | 3 | 4 | 5 | 6 | 7 | 8 |
|---|---|---|---|---|---|---|---|---|
| | Dad | Mom | | | | | | |
| 6 kb → | — | — | — | — | — | | — | — |
| 4 kb → | — | — | | — | | — | — | — |

In this family, individuals 3 and 5 were born with cystic fibrosis; individuals 4, 7, and 8 are carriers of the disease; and individual 6 is homozygous for the normal allele. We could imagine much larger pedigrees, but no matter how large a pedigree analyzed, the 6-kb RFLP would always cosegregate with the cystic fibrosis mutation.

A similar analysis can be performed for sex-linked genes. Consider a female with the following two X chromosomal alleles:

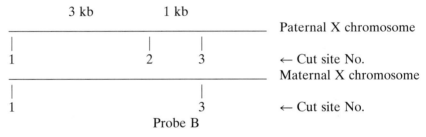

Probe B

In this case a piece of DNA corresponding to part of the muscular dystrophy gene is diagramed. Restriction enzyme site number 2 has been deleted by a 5-bp deletion that also knocked out the ability of the gene to produce a functional gene product. DNA from this woman's cells is cut with enzyme X, run on a gel, blotted onto filter paper, and probed with probe B. As shown in panel 2 of the following autoradiogram, this result is two bands, one at 1 kb and one at 4 kb. Suppose that this woman then marries a man carrying the normal allele of muscular dystrophy (i.e., an allele that possesses restriction site 2). *(So human geneticists really spend their time trying to figure out how to get people to select their mates based on genotype in case they are ever in a genetics study? Of course not, but the joke has come out of more than one geneticist's mouth when it turns out that someone with the 4-kb allele on both of her chromosomes is married to someone with the 4-kb allele on both of his chromosomes, thus making more than one generation of the family "uninformative."—JER)* The following autoradiogram reflects the restriction fragment sizes identified by probe B for the husband (panel 1), the wife (panel 2), and their five sons (panels 3–8).

Panel

| | 1 | 2 | 3 | 4 | 5 | 6 | 7 | 8 |
|---|---|---|---|---|---|---|---|---|
| | Dad | Mom | | | | | | |
| 4 kb → | | — | | — | | — | | — |
| 1 kb → | — | — | | — | | — | | — |

In this family, individuals 3, 5, and 7 will be born with muscular dystrophy, whereas individuals 4, 6, and 8 will be unaffected. Suppose a similar mating had produced six daughters (offspring 9–14 in the following diagram). Although none of these daughters would be affected, it would be easy to determine which ones are carriers for the disease.

Panel

| | 1 | 2 | 9 | 10 | 11 | 12 | 13 | 14 |
|---|---|---|---|---|---|---|---|---|
| | Dad | Mom | | | | | | |
| 4 kb → | | — | | — | — | | | — |
| 1 kb → | — | — | — | — | — | — | — | — |

Obviously, individuals 10, 11, and 14 will be carriers for the disorder, whereas individuals 9, 12, and 13 are homozygous for the normal allele and thus their sons will not be at risk. *(The method of tracing DMD mutations just presented is described for teaching purposes only. In reality, researchers studying Duchenne muscular dystrophy have developed another kind of assay that uses PCR to search for small deletions in the DMD gene. By using multiple pairs of primers, each from a different region of the DMD gene, they are able to amplify many different parts of the gene simultaneously in one PCR reaction. This test is supposed to be able to detect 98% of deletions that are located within the DMD gene.—JER)*

## WHAT IF THE DISEASE-CAUSING MUTATION DOES NOT CREATE A LARGE AND EASILY IDENTIFIABLE DNA POLYMORPHISM?

In most cases, we are not fortunate enough to already possess a clone corresponding to the gene of interest. In fact, obtaining such clones is the holy grail of human genetics. In this case we must content ourselves by finding genetic markers that lie close to the gene of interest.

Okay, so plan A for mapping human genes is really simple:

1. Find a family or families where the disease, and thus the gene, is segregating. Big families are better than small families, but we can cope either way. Get blood, and thus DNA, from *every* single living man, woman, and child in that family. (Family reunions are great for this purpose.)
2. To find the gene for a recessive autosomal disorder, you need to follow the segregation in that family of a very large number of the genetic markers, in this case microsatellite repeats, located throughout the genome (with the markers located 10–20 map units apart across the whole genome for which the complete length is about 3,300 map units) to look for a microsatellite repeat that cosegregates with the

disease-causing mutation (DCM). In the case of a recessive mutation, the affected individual is either a homozygote (has a defect in both copies of the gene) or is a *compound heterozygote* (i.e., someone that has mutations in both copies of the gene, but the mutations are not identical).

As shown in Fig. 14.1, if both copies of the disease gene came from the same ancestor (in cases of consanguinity or inbreeding), then the affected individual would actually be homozygous for the specific gene defect or mutation. They would also probably be homozygous for some of the microsatellite repeat markers that are closest to the disease gene. Thus the genetic marker would be homozygous in all/many of the affected individuals and present in all the folks who are carriers.

*(Sometimes homozygotes for a DCM DON'T express the disorder or get the disease. This reflects a poorly understood phenomenon called no penetrance or reduced penetrance.—JER)*

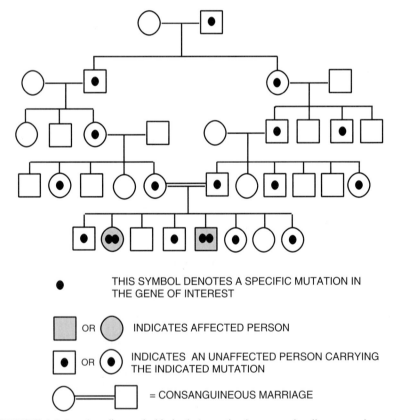

THIS SYMBOL DENOTES A SPECIFIC MUTATION IN THE GENE OF INTEREST

OR ◯ INDICATES AFFECTED PERSON

⬛ OR ◯ INDICATES AN UNAFFECTED PERSON CARRYING THE INDICATED MUTATION

◯═⬛ = CONSANGUINEOUS MARRIAGE

FIGURE 14.1     A pedigree of a kindred segregating for a recessive disease-causing mutation in which the homozygotes are the result of identity by descent.

Please remember that the marker is not actually the gene, just a landmark near the gene helping you to recognize that you are in the right neighborhood. The farther away from the gene itself, the less likely that the marker will also be homozygous. As shown in Fig. 14.2, in a family in which the parents of the affected person aren't related, i.e., they don't carry defective copies of the gene that are identical by descent (IBD), the affected person could get two defective copies of the gene. However, in this case the defective copy from mom might have a completely different DCM from the DCM present in the defective copy carried by dad. This is the case of the compound heterozygote alluded to above. So, if the two copies of the disease gene came from unrelated ancestors, then the background "fingerprint" of allele sizes at surrounding markers could be quite different.

*(For a variety of reasons, mapping of recessive disease genes is so much easier in families where inbreeding—marriage of close relatives—increases the chances that the affected chromosomal region will be recognizable in part by the stretch of homozygous markers surrounding the disease gene itself.—JER)*

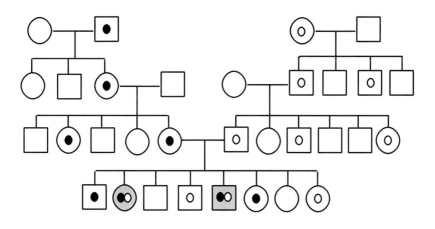

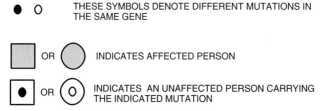

● ○    THESE SYMBOLS DENOTE DIFFERENT MUTATIONS IN THE SAME GENE

▢ OR ◯    INDICATES AFFECTED PERSON

[●] OR (○)    INDICATES AN UNAFFECTED PERSON CARRYING THE INDICATED MUTATION

FIGURE 14.2   1. A pedigree of a kindred segregating for a recessive disease-causing mutation in which the "homozygotes" carry two different mutations in the "disease gene" that arose in different ancestors.

Obviously, if you are looking for a dominant trait, you look for a marker allele that is only present in affected individuals and absent in everyone else! *(I know this sounds like a lot of work, but the labs that do this stuff are basically drowning in money, equipment, and very talented personnel.—RSH Can you tell that Scott is jealous?—CAM)* You should work out for yourself the way in which a marker associated with a sex-linked trait would segregate.

These approaches work really well as long as the available collection of genetic markers contains at least one sequence that is sitting right on top of the mutation of interest, so close that the site defined by the microsatellite repeat and by the disease-causing mutation virtually never recombine. An example of such a mapping is presented in Chapter 19. The technique also works if the microsatellite repeat is separate from the gene of interest, but so close that recombination virtually never occurs between the DCM and the genetic marker (see the section on linkage disequilibrium).

What happens if the closest microsatellite marker in the collection you test is just *close?* Well then the closely linked microsatellite repeat allele will usually follow the DCM in the pedigree, but not always. Sometimes it will recombine away from the DCM. That's okay, all we need to do is remember what we learned about recombination. *(I know that "close" is only supposed to count in horseshoes and thermonuclear warfare, but recombinational mapping is another example where simply "close, very close" will do.—RSH "Oh, what I wouldn't give for close, just close," says the lady who is part way through a genome scan in which she must continue to determine genotypes until she scores a "hit."—JER)*

## RECOMBINATIONAL MAPPING

Consider the example of a female who is doubly heterozygous for both an X-linked disease-causing mutation (denoted dcm) and an X chromosomal DNA sequence that is defined by an RFLP. The following diagram shows that the MSR-A (microsatellite repeat-A) allele lies on the same X chromosome as the disease causing mutation, whereas both MSR-B and the normal allele of disease-causing mutation (denoted DCM) are on the other X chromosome. If there were no recombination, any ovum that carried MSR-A would also carry dcm and any ovum that carried MSR-B would also carry DCM.

| MSR-A | dcm |
|-------|-----|
| MSR-A | dcm  O |

| MSR-B | DCM |
|-------|-----|
| MSR-B | DCM  O |

However, there is a substantial amount of meiotic recombination in human beings. As can be seen in the preceding diagram, an exchange event occurring between these two genes can produce both an egg carrying MSR-A and DCM and an egg carrying MSR-B and dcm. *Please remember that the probability of exchange increases as physical distance between the two markers (in this case the microsatellite repeat sequence and the DCM gene) increases.*

- Physically close genes will recombine rarely.
- Physically distant genes will recombine frequently.

To illustrate, suppose the female whose X chromosomes were just diagramed marries a man whose X chromosome carries MSR-B and DCM. The autoradiogram for the MSR and the disease status of both parents and their six sons (individuals 6 through 8) is shown:

|  | Panel | | | | | | | |
|---|---|---|---|---|---|---|---|---|
|  | 1 | 2 | 3 | 4 | 5 | 6 | 7 | 8 |
|  | Dad | Mom | | | | | | |
| MSR-A | — | — | | — | | — | | |
| MSR-B | — | — | | — | | | | — |
| Affected | No | No | Yes | No | No | No | Yes | No |

As noted, without recombination, any ovum that carried MSR-A would also carry dcm and thus produce an affected son. Similarly, any ovum that carried MSR-B would also carry DCM and produce a normal son. Individuals 3, 4, 6, 7, and 8 fit with this prediction. However, individual 5 carries both MSR-A and carries DCM (i.e., he is unaffected). This individual must have resulted from the following recombination event. The frequency with which such recombinant gametes will be produced will decrease as one finds other MSR markers that are closer and closer to the disease-causing mutation.

| MSR-A | | dcm |
|---|---|---|
| MSR-A | | dcm O |

| MSR-B | X | DCM |
|---|---|---|
| MSR-B | | DCM O |

The frequency of recombination can be measured simply by counting the number of recombinants and dividing by the total number of offspring. In the case just described, there were six progeny, one of whom was recombinant. The recombination frequency is thus 1/6, or 16.6%. By convention, each percentage point of recombination is said to equal one *map unit.* Thus the MSR tested here and the dcm gene are said to be 16.6 map units (also called centimorgans) apart.

In those cases where the MSR and the dcm are far apart, the map distance will equal 50 map units, or 50% recombination. [For some of you this statement will be an intuitively obvious restatement of independent assortment. For the remainder of you who are skeptical, I present the following proof. Suppose we repeated this analysis using MSRs from a gene on another chromosome (denoted MSR-2a and MSR-2b). We will consider those instances in which the MSR-2a allele came in from the same parent as the dcm allele and the MSR-2b allele came in from the parent with the DCM allele. These double heterozygotes will produce MSR-2a dcm, MSR-2b dcm, MSR-2a DCM, and MSR-2b DCM gametes at equal frequencies. Because the number of recombinant gametes, MSR-2a DCM and MSR-2b dcm, will equal the number of so-called nonrecombinants or parental types, then the recombination frequency will equal 50%. It may help to draw this out if you are still confused.]

The purpose of MSR mapping is to find those MSRs that are as tightly linked to the disease-causing mutation as possible. The problem lies in measuring small frequencies of recombination in an organism (humans) that has very few offspring per couple. The human geneticist is faced with analyzing small families wherever she/he can find them. Fortunately, there is a statistical tool called the *LOD score,* which allows us to sum together the results of many small families. Defining LOD scores rigorously requires using a good deal more math and statistics then is possible here. A brief attempt at such a definition is presented in Box 14.1 for those who like such things. For the rest of us, suffice it to say that such a method exists and it allows us to say things like "this MSR and the disease-causing mutation are 10 map units apart" with high confidence.

Fine, suppose you find a microsatellite repeat that is approximately 10 map units from the disease causing mutation. Not good enough! What you need to do now is to find other microsatellite repeats that are closer to the one on which you made your first observation that show even tighter linkage (i.e., less recombination) to the disease-causing mutation. You and your army of postdoctoral fellows need to keep this game up until you have several microsatellite repeats that sit right on top of your disease gene. Fortunately, there are many microsatellite repeats and, just like Doritos, they'll make more.

### THE PROBLEM OF COUPLING

In the process of attempting to map a gene you must look at each set of parents and each set of children. For each microsatellite repeat you test you must determine whether each child carries a recombinant chromatid or a nonrecombinant chromatid. Your estimate of linkage will depend on just how many recombination events are observed. It is important for you to realize that such an analysis depends on knowing which alleles are on which ho-

## BOX 14.1 A MORE RIGOROUS DEFINITION
## OF LOD SCORES

LOD scores are meaningful only for a given frequency of recombination (called $\Theta$ or theta). A LOD score is defined as:

LOD $= \log_{10}$ [(the probability of obtaining the observed progeny if the two genes are linked with a recombination frequency of $\Theta$)/(the probability of obtaining the observed progeny if the two genes are unlinked)].

One then goes through each pedigree and computes the probability that you would have found so many recombinant and so many nonrecombinant offspring if the real value of $\Theta$ were the value being tested. One then carries out the same calculation to ask what is the chance that you would have observed the same data if there really were no linkage ($\Theta = 0.5$). One then divides the first probability (that of linkage for the value of $\Theta$ that is being tested) by the second probability. Then you take the $\log_{10}$ of that ratio. In other words, if someone told you that for two genes and for $\Theta$ equal to 0.1 map units, the LOD score equals 3, you would know that the odds are 1,000 to 1 in favor of these data resulting from two genes 0.1 map units apart rather than a statistical fluke in the segregation of two genes that are in fact unlinked. (Why? Remember, $\log_{10}$ of 1,000 is 3.)

The LOD score then simply gives a probability of obtaining a given set of progeny for any given map distance. In practice one lets a computer calculate the LOD scores for all values of $\Theta$ ranging from 1/10,000 to 0.5. The value of $\Theta$ that gives the highest LOD score is considered to be the best estimate of the true recombination frequency.

The wonderful thing about LOD scores is that as more and more small families are studied, data can be compiled simply by adding their LOD scores together. Remember that the log of the product of two numbers can be calculated simply by adding the logs of the two numbers. For a given value of $\Theta$ and three families, the LOD score can be calculated by adding the LOD scores for that value of $\Theta$ for each of the three families. This is essentially the same mathematical operation as taking (the probability of obtaining the observed data for family 1 if $\Theta$ is the same number) multiplied by (the probability of obtaining the observed pedigree for family 2 if $\Theta$ is the same number) multiplied by (the probability of obtaining the observed pedigree for family 3 if $\Theta$ is some number) and so on.

In the final analysis, a LOD score of 3 or greater is usually considered to be the minimal value that provides clear evidence for linkage.

mologs in the parent. This is to say that suppose you have a father who is heterozygous for the disease-causing mutation and for two different alleles (a and b) at a given microsatellite repeat (MSR2). The progeny he produces is interpreted very differently depending on whether his genotype is

$$
\begin{array}{ll}
\text{MSR-2a} & \text{dcm} \\
\hline
\text{MSR-2b} & \text{DCM}
\end{array}
$$

or

$$
\begin{array}{ll}
\text{MSR-2a} & \text{DCM} \\
\hline
\text{MSR-2b} & \text{dcm}
\end{array}
$$

The relationship of the two alleles on the homologous chromosomes is called the *coupling relationship*. How we determine the coupling relationship can be easy or complex. For example, for a sex-linked pair of markers, if we know the father's genotype, we know the coupling relationship in his daughters. *(If I were writing your final, and for those of you at U.C. Davis I probably will, I would ask you to explain to me why that last statement is true.—RSH)* For autosomal markers, however, it can be harder. Suffice it to say that we have ways of finding these things out. Moreover, when we can't figure it out, we have statistical tricks for coping with our ignorance. *(They aren't real pretty but they work.—RSH)*

As you go through your family testing lots of markers, you finally find several markers that lie close, within a few map units, to your gene. How do you put them in order?

### PUTTING TWO CLOSELY LINKED MICROSATELLITE REPEATS AND THE GENE INTO A LINEAR ORDER

Is it obvious to you by now that meioses in which three markers are present tell you not only map distances but gene order as well? For example, consider the following triple heterozygote for three tightly linked genes (AaBbCc). We begin by arbitrarily postulating that B is to right of A. (The technique would work equally well if we postulated the reverse order, A to the right of B, but we need to choose one order to begin the analysis.) So starting with A to the left and B to the right, if the C gene lies to the right of both the A and the B genes, then all Ab recombinants will carry the c allele, whereas all aB recombinants will carry C.

$$
\begin{array}{ccc}
\text{A} & \text{B} & \text{C} \\
\end{array}
$$
— — — — — — — — — — — — — — — — — — — — — — — — — —
$$
\begin{array}{ccc}
\text{a} & \text{b} & \text{c} \\
\end{array}
$$

Conversely, if the gene order was C–A–B, then all Ab recombinants will carry the C allele, whereas all aB recombinants will carry c.

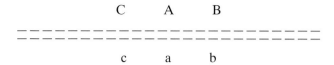

If the order was A–C–B, then some Ab recombinants will carry the c allele and some the C allele; similarly, some aB recombinants will carry the c allele and some will carry the C allele.

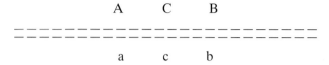

You don't need a lot of recombinants to pull this off; a handful or so may suffice, which may be hard to find and rare. This is the best part of doing human genetics. Interesting families often call you up and make appointments to see you or someone else!

## THE BASIC PROCESS OF HUMAN GENE HUNTING

If we put everything together, the basic steps in human gene mapping are as follows.

1. Find a family, or better yet families, where your disease or trait is segregating. This may not be easy and researchers are pretty possessive of the families they do find. If your competitor just found a family in which the trait of interest is segregating in four generations of living members, odds are she/he is NOT going to call you up and volunteer to give you their names and addresses. *(Actually, one of the underlying principles here is that the family belongs to itself and decisions about participation really are the province of the family and that we, the researchers, do not control who the family members interact with. Some families throw themselves into a research study with great enthusiasm, some participate while hardly seeming to care, and some contact different researcher groups, establish Web pages giving information about their disease, and take a very activist role in the whole research process.—JER)*

2. Use a large collection of microsatellite probes to search for markers that cosegregate with the disease or trait. *(The primary collection that we use in screenings is about 350 markers!—JER)*

3. Once the first evidence of linkage has been found, narrow in on the gene by testing more markers in this interval.

4. Position the disease gene BETWEEN two closely linked markers.

5. Obtain a chromosome walk spanning the interval between the two nearest flanking markers.

The next chapter will describe exactly how this was done for the cystic fibrosis gene.

## SPECIAL TECHNIQUES IN HUMAN GENE MAPPING

### THE CONCEPT OF LINKAGE DISEQUILIBRIUM

The method described earlier is a perfectly acceptable method for mapping human genes. However, it does make a big difference whether one knows the exact genotype of each parent and it helps if one knows which allele of each gene is on each homolog. There are, however, some shortcuts that allow one to rapidly examine large pedigrees for each microsatellite repeat, the disease-causing mutation, and to deal with some of the less than ideal situations where you do not have all of the information you want or just the right combination of study participants.

Suppose, for example, that one had a small set of families who share a common ancestor thought to have been affected by the disease. This kind of effect may be found in populations that originate from a small founding population. If one of the founding members of this kindred or population carried the genetic defect, then the disease might occur at a higher frequency in this population than in the general population (but not necessarily). Moreover, affected individuals in such a population have an increased chance of being descended from the original founder.

Such populations exhibit a *founder effect*. In such a case, it is possible that the disease-causing mutation may be marked by a specific microsatellite repeat, a microsatellite repeat sufficiently close to the disease-causing mutation that recombination events are few, if they occur at all. Thus affected individuals will preferentially carry one particular microsatellite repeat allele that is present on the particular chromosome on which the original mutation event occurred or on a chromosome of the individual who immigrated into this population and brought the mutation along.

This association between specific alleles and other genes on the same chromosome is referred to as *linkage disequilibrium*, a difference in the frequency with which a given pair of alleles turns up in an affected population when compared to the normal population. What one does in practice to exploit or search for linkage disequilibrium is to examine families in which a given disease gene is segregating, using microsatellite repeats at as many sites in the genome as possible. Those microsatellite repeats that are unlinked to the disease-causing mutation, e.g., those that map onto another chromosome, will segregate independently of it. However, in those cases where the microsatellite repeat lies very close to the disease-causing mutation, the affected individuals are more likely to carry this allele.

Keep in mind here that the probability of a recombination event taking place over the course of a "historical" length of time is determined both by the distance between the two items on the chromosome and by the number of generations passed. However, because it is a matter of probabilities, there is always a chance that if a recombination event did take place, it could have taken place early in the history of this population as well as late. If the recombination event was a very unexpected event close to the gene and early in the time course we are examining, then the level of disequilibrium will not be as high. If no recombination events took place between this microsatellite repeat and the gene until later in the time period, the linkage disequilibrium may be quite pronounced.

Taking advantage of linkage disequilibrium in a population has worked on situations in which apparently unrelated people actually share an ancestor. As discussed in the next chapter, this worked in the case of cystic fibrosis because something more than 70% of CF patients in the United States all carry a specific allele, the delta F508 mutation, and appear to be descended from one ancestor. Markers closest to the mutation itself are the least likely to have been separated, or "traded away," through a recombination event back in history, as the chance of a recombination event is proportional to the distance between two markers. When the cystic fibrosis gene hunters tested markers in the vicinity of CF, they found alleles that were more frequent in American CF patients than among the general population.

What if you can't find a large population or kindred descended from a small number of people, one of whom carried the mutation you want to map? What do you do then? Huh, what do you do then? *(Give up and work on flies.—RSH And see what happens to your funding, Scott.—JER)* Actually, because you only get the benefits of linkage disequilibrium in some situations, it is beneficial if you have some other tricks up your sleeve. Fortunately, there are other ways to look at alleles near a candidate genetic location in a population where many of the mutations are independently derived and the apparently unrelated families do not secretly share a common ancestor. In the next paragraph we examine one such way.

### USING SIB-PAIR ANALYSIS TO MAP A GENE

Basically, sib-pair analysis is a method in which you look for the rate at which affected siblings share the same allele(s). (Recall that Dean Hamer used this method to map a possible gene for sexual orientation, as described in Chapter 8.) The value of this method is that it does not require that all, or even most, affected individuals in a population carry the same allele of the disease-causing mutation or the same flanking microsatellite repeat alleles. Rather this approach is based on the idea that affected siblings in the same family will share the same disease allele(s) and the same tightly linked microsatellite repeat alleles. The particular disease-causing allele in question may be different in sib-pair A vs. sib-pair B, but it turns out to be an issue of the rate of sharing. Thus if a marker is heterozygous in 95% of individuals in the population and we find that the disease-causing chromosome in each sib carries the following alleles for a given MSR:

| Sib | Allele | Sib | Allele |
| --- | --- | --- | --- |
| A1 | 2 | A2 | 2 |
| B1 | 7 | B2 | 7 |
| C1 | 13 | C1 | 13 |
| D1 | 9 | D2 | 9 |
| E1 | 11 | E2 | 11 |
| F1 | 7 | F2 | 7 |
| G1 | 5 | G2 | 5 |
| H1 | 2 | H2 | 2 |
| I1 | 6 | I2 | 4 |
| J1 | 12 | J2 | 12 |
| K1 | 10 | K2 | 10 |

and so on, we begin to wonder if this MSR might lie close to the disease-causing mutation.

If instead we had seen

| Sib | Allele | Sib | Allele |
| --- | --- | --- | --- |
| A1 | 2 | A2 | 7 |
| B1 | 7 | B2 | 3 |
| C1 | 13 | C1 | 1 |
| D1 | 9 | D2 | 5 |
| E1 | 11 | E2 | 11 |
| F1 | 3 | F2 | 7 |
| G1 | 10 | G2 | 5 |
| H1 | 2 | H2 | 2 |
| I1 | 6 | I2 | 4 |
| J1 | 12 | J2 | 6 |
| K1 | 10 | K2 | 8 |

we suspect that the MSR under consideration is nowhere near the disease-causing gene.

Sib-pair analysis has some lovely properties. The best of these is that because you only look at affected sibling pairs, you don't have to worry about other environmental or genetic influences required for a given trait to be expressed. *(Oh dear, well you DO have to worry about something called phenocopies, individuals with a disease that looks just like the one you are studying but with a different underlying cause. I work on a disease that has both genetic and environmental forms, and we always have to worry that even our careful efforts to exclude environmental cases during the medical exams may not always succeed. Fortunately, if you have an idea of the probability that any one person in your study might be a phenocopy, there are things you can do with your calculations to take that into account.—JER)*

## CONSANGUINEOUS MARRIAGES AND
## GENE MAPPING

In recessive diseases there is this really cute effect that allows you to select for families with consanguinity by recruiting families from parts of the world in which cultural patterns encourage the intermarriage of close relatives. If information is collected from distantly related individuals, remote cousins perhaps, for whom all four parents share a common ancestor, you make the assumption that whatever alleles they carry in the immediate vicinity of the gene will be homozygous, with the probability of being homozygous related to the distance from the gene and the probability for a recombination event having occurred during that often small number of intervening generations. In real research terms, the effect is this: For a normal genome scan, each member of a large family must be genotyped, and each individual must have their test information run in a separate lane so that each person's genotype can be determined individually.

Thus, the analysis of recessive diseases in inbred families is straightforward; as shown in Fig. 14.3A it takes three lanes per marker (including a water control) to screen the entire family! In one lane you put the PCR product produced when a pool of DNA from all of the affected individuals is used as a target. In lane 2 you put a similar pool of PCR products from all of the unaffected at-risk individuals. Lane 3 gets the water control (a sample with no DNA in it). Thus, affected individuals, who should all carry exactly the same allele on both chromosomes, should produce one robust band at the site of the affected allele. Unaffected relatives will display a range of band sizes representing many different alleles. The water lane should be blank. In the affected lane the allele band will generally be dark and heavy, and in lane 2 the same band size will appear as there are carriers among the unaffected who carry that band size along with another. If the marker you are looking at is just a bit farther away, most of the affected individuals will be homozygous for the affected allele, but some of them will be recombinants who are heterozygous. As a result you will see a very dark line in lane 1 for the affected allele with an occasional faint band at the positions that correspond to the other allele sizes. So instead of having the screening of a marker take up 40 lanes on a gel, it takes up 3 lanes and you can get a dozen or more markers on one gel. Imagine being able to do a genome scan and FIND a gene with one-tenth the work of anyone

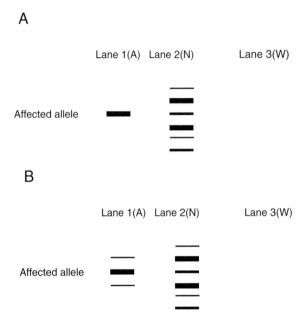

FIGURE 14.3   A method of using a MSR marker to map a gene for a recessive trait in an inbred family. (A) The result when the MSR markers actually defines or lies within the gene of interest. (B) The result when the MSR lies a short distance away from the gene.

going after an autosomal dominant gene through standard practices of slogging through it on a one individual–one lane basis.

## WHAT DOES THE NEWSPAPER MEAN WHEN IT ANNOUNCES THAT A GENE HAS BEEN FOUND?

In some cases, "found" means that the gene has been mapped, that its location on one of the human chromosomes has been identified. Why would this be a big deal? (1) Once we know where the gene is located we can move ahead with a positional cloning approach to obtaining a copy of the gene. (2) Once we know where a gene is located that causes the disease in question, we can start asking questions about whether this is the only gene that causes the disease or whether there are multiple, different genes that can all give the same symptoms. (3) We can carry out presymptomatic diagnosis for some members of large families in which the mapped gene causes a disease, although just a map location does not allow for development of any general diagnostic test that could be used on everyone who wants to be tested for a mutation in the disease gene. It also does not tell you anything about the gene product or the underlying processes of the disease. This kind of find makes headlines because it often is the first breakthrough on efforts to find the gene in question and offers promise that more findings will follow quickly. *(The more dreadful and/or more common the disease the more likely it is to make the headlines. Are you surprised?—JER)*

In some cases "found" means that the gene has been cloned. This is a much more important event because once the gene is cloned, its DNA sequence can be determined. (1) Once the sequence is known we can design presymptomatic diagnostics tests that can be used on anyone at risk for the disease. (2) We can study the function of the gene product and learn more about the underlying causes of the disease. (3) We might even be able to put copies of the gene back into cells in which the defect is the inability to make functional copies of the gene product. This process is called gene therapy and is discussed futher in Chapter 22. The science of gene therapy is still in its infancy as scientists work to figure out what works best for getting copies of genes into cells, getting them into the right cells, getting those gene copies to make the gene product, etc. It is not yet known to what extent gene therapy will really turn out to be useful in routine clinical situations. We are not there yet, but it is one of the directions that cloning is taking us.

# 15

## CYSTIC FIBROSIS

*One of the early breakthroughs in the application of positional cloning (the cloning of something based on its location in the genome) was the cloning of the gene that leads to cystic fibrosis if both copies of the gene are defective. At the time of this victory, many of the hard-won resources of the Human Genome Project were not yet available and clone by phone was not an option. Large numbers of people worked long hard hours to clone this gene, and among them was a college student with cystic fibrosis who worked in one of the labs that made the breakthrough. Medical advances relative to CF had made it possible for him to grow up and attend college, but as the other scientists watched him alternate between attending classes, working in the lab, and dealing with his illness, it was clear that much more medical advancement was needed.*

*The tale of the cloning of the CF gene was really a tale of determination and a tale of hope, as shown in the flood of mail that began to reach the researchers after the announcement of their find: letters of thanks, letters of hope, and letters from small children writing to say thank you for being given the opportunity to grow up. As more and more genes are cloned, we all become a bit more blasé about it. Yep, one more gene—fifth one in the newspaper this week. But for every gene that is cloned there are people like those letter writers who have been waiting for someone to come up with their particular breakthrough, to bring them, if not a cure, then at least the hope of one.—JER*

Cystic fibrosis (CF) is one of the most common genetic diseases in the United States and Europe. The disease, which is the result of loss of function of both copies of the CF gene, is a potent killer of children, primarily as a result of chronic respiratory infections. Many studies demonstrate that the primary defects are pulmonary obstruction, bacterial infections of the trachea, pancreatic insufficiency due to blockage of the secretory ducts, and reduced male fertility or sterility, although not all CF patients have all of these problems. It is inherited as an autosomal recessive mutation, and a sample pedigree is shown in Fig. 15.1.

Among Caucasians, the incidence of the disease is approximately 1/2,000 births, and carriers (heterozygotes) are found at a frequency of 1/22. Some investigators have proposed that the high frequency of these mutations in the Caucasian population results from the possibility that infants and young children heterozygous for mutations in the CF gene were more resistant to the epidemics of diarrhea that once ravaged Europe. In this sense, mutations at the CF gene may be considered analogous to mutations in the sickle cell gene, a severe blood disorder in homozygotes for which we see resistance to malaria in heterozygotes. *(What Scott is driving at here is that mutations need not necessarily be "bad." In the context of both CF and sickle cell disease, heterozygosity for the mutation can confer a real survival*

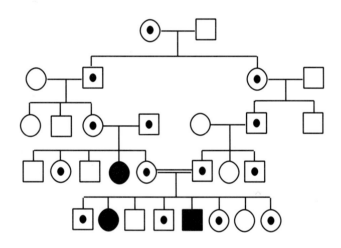

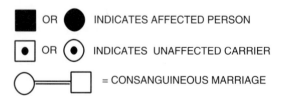

FIGURE 15.1    A pedigree displaying the mode of inheritance for cystic fibrosis.

*benefit relative to an infectious disease, whereas homozygosity can be deleterious or even fatal.—CAM)*

## MAPPING OF THE CYSTIC FIBROSIS GENE

Because of the high frequency of this disorder, attempts to map the CF gene have been ongoing for quite some time. The first success occurred in 1985 when the CF gene was found to be linked to a mutation, called PON, that caused a detectable alteration in a protein in the blood. Unfortunately, nobody knew where the PON polymorphism mapped in the genome, so that didn't help much. More significant success came later that year when the CF gene was found to be linked to a RFLP marker on chromosome 7q. Using more markers, and the tricks described in the previous chapters, researchers were able to draw the following map and to show by *in situ* hybridization that CF was in 7q31-32 (which means bands 31 and 32 of the q or long arm of chromosome 7).

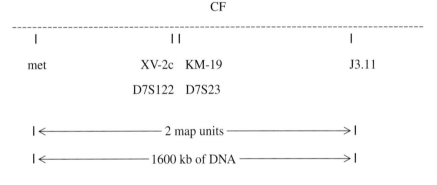

CF

| met | XV-2c  KM-19 | J3.11 |

D7S122  D7S23

|← ——————— 2 map units ———————→|

|← ——————— 1600 kb of DNA ———————→|

### CLONING THE CF GENE

Markers XV-2c and KM-19 are very close to the most common mutation in the CF gene, i.e., the vast majority (90%) of CF chromosomes carry a specific set of alleles of XV-2c and KM-19 carried by only 25% of the general population.

This genetic mapping suggested a straightforward strategy for cloning the gene.

1. Clone the entire region of DNA that spanned those markers. To clone the region from KM-19 to J3.11, chromosome walking was carried out. This region turned out to be greater than 1,500,000 bp in length! Clearly the gene was in there somewhere, but where? Realize that as you walk along the chromosome there are no neon signs telling you that you have found the gene of interest. Unfortunately, there were no obvious structural differences, such as large deletions or insertions between the DNA of

normal and mutant individuals. *(Please recall our discussion of chromosome rearrangements and their effect on DNA structure in the last chapter.—RSH)*

2. Look for sequences that were conserved during evolution. It's a good bet, although not always a guarantee, that gene sequences are important enough to be conserved during evolution. So real genes, unlike the spacer DNA sequences that separate them, should be conserved in evolution. This can be tested by performing a Southern blot using DNA from males and females of various organisms (called a zoo blot). The researchers were lucky in this case. They found five conserved regions. Two of these regions could be ruled out because researchers had found recombinants between polymorphisms within these regions and the CF mutation. A third was dismissed because it did not define a functional gene.

Techniques exist to sequence any given DNA molecule. Note that if a region of DNA does contain the gene of interest (as opposed to being a piece of junk or spacer DNA that separates genes), then the sequence of that region should contain a region of base pairs capable of encoding a protein. This is called an *open reading frame* (ORF). ORFs are long runs of bases that can specify amino acids and are not interrupted by stop codon sequences in frame with the open reading frame of the amino acids. The last two regions of conserved DNA met the three criteria: They were conserved, they had an open reading frame, and they had not been ruled out by a recombination event.

3. Look for sequences that are expressed in the right tissues (in this case lung and pancreas) and at the right times during development. If a given region of DNA is your gene, then it should be expressed (transcribed) in the appropriate sex and tissue and at the right time during development. Looking at gene expression in various tissues can be done easily using a technique called *Northern blotting.* In this technique, samples of isolated RNA are run on a gel and then the size-separated mRNA molecules are blotted onto filter paper, just like a Southern blot, only using RNA instead of DNA. The filter paper is then probed with your piece of cloned genomic DNA. If a band appears on the autoradiograph, the messenger RNA derived from that piece of DNA is present in the cells from which the original RNA was extracted. Many samples can be run on adjacent lanes in a single gel. This allows the comparison of RNA samples from many different tissues or from the same tissues from different individuals. It also makes it possible to compare RNA samples from normal and mutant individuals. In the case of the cystic fibrosis gene, this technique worked well. As diagramed in Fig. 15.2, one of the two possible candidate genes was expressed at the right time and in the right tissues.

4. Sequence the gene or a cloned DNA copy of the mRNA transcript (called a cDNA) from homozygous mutant individuals and compare that sequence with the normal gene. Finding the exact molecular position of several mutations that destroy gene function provides good evidence re-

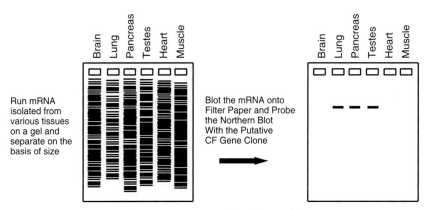

**FIGURE 15.2**   Northern blot analysis.

garding the whereabouts of the gene. There are two ways to do this: sequence the genomic DNA or sequence cDNA. So the work begins: Just sequence the genomic DNA and cDNAs from mutants and normal individuals and hope, hope really hard, that you find a consistent and sensible difference.

Recall that a good laboratory can sequence perhaps as many as several hundred kilobases of DNA in less than a month or so. If any of the sequence is already known (and if you have enough equipment and money), DNA sequencing can go much faster. For example, if you know the sequence of the normal gene a mutant allele can be sequenced quite easily and quite quickly. However, sequencing is arduous and expensive, and there are many random polymorphisms in a cloned set of DNA molecules that don't disrupt the function of the gene and may be unrelated to the gene of interest. Such polymorphisms sometimes make it very hard to find the really significant genetic alteration.

In the case of CF this technique worked beautifully, and in 1989 a collaboration between scientists in the United States and Canada used the sequences of CF mutations to prove that they had cloned the CF gene. The gene turns out to encode a large membrane protein called CFTR (cystic fibrosis transporter) that is involved in the transport of Cl⁻ (chloride) ions into and out of cells. It turns out that most (~66%) of the mutations of CF mutations in the U.S. Caucasian population are due to a 3-bp deletion that removes the amino acid phenylalanine at position 508. The remainder of CF mutations — more than 550 other much rarer mutations have been identified — are scattered throughout this rather large gene that spans over 250,000 bp of genomic data.

As might be expected, many of these mutations prevent the production of a protein product or produce a product not functional enough to associate with the membrane. (Realize that being able to at least associate with the membrane is a *sine quo non* for a protein like CFTR that must function as an ion transporter!) However, some missense mutations produce altered forms of the CFTR protein that is still inserted into the membrane, but displays weaker or reduced activity. Homozygotes for these mutations often lack the pancreatic problems associated with the more severe alleles. In at least one other case, a disorder known as *congenital bilateral absence of the vas deferens* (CBAVD), which appears in approximately 1–2% of infertile males, often appears to result either from heterozygosity for strong alleles of mutations in the CFTR gene or from double heterozygosity for two different and weaker alleles of this gene. Given that male sterility is commonly observed in individuals homozygous for stronger alleles of the CFTR gene, this result is perhaps not too terribly surprising.

5. Look to transformation as the final proof that the CF gene has cloned. Put plainly and simply, transformation rescue is the ultimate proof that the identified gene is really the disease gene. Cultured cells from patients with CF cannot make the CFTR protein and so they cannot transport certain kinds of salts across these membranes. Scientists have been known to spend their lives in labs figuring out how to measure things like how much salt a cell can eat or excrete in an hour. So we can see a measurable defect in these cells. If these cells are transformed with a normal CFTR gene, they regain the ability to transport salt.

As we will discuss in detail in Chapter 19, scientists are also trying to put the CFTR gene into lung cells from children with CF. If this works, it will be a dramatic success for the pubescent field of genetic medicine. There is a lot of work in this area and the first efforts are underway. This is a best case scenario for gene therapy in many ways, and most of us are optimistic that it will work in the near future. For right now, much of our effort focuses on developing simple tests for carrier detection that will allow us to identify heterozygotes within the population more efficiently.

# 16

# MAMMOTH GENES:
# MUSCULAR DYSTROPHY
# AND NEUROFIBROMATOSIS

*When we originally designed this book, all of these introductory para-
graphs were supposed to be fully fictional. Indeed, Catherine has been working hard on a story about a child with Duchenne muscular dystrophy (DMD).
For me, Duchenne muscular dystrophy is an illness for which I can abide
little fiction. My brief encounter with two boys dying of this disease has given
this illness far too stark a reality. Back when I was a high school student in
the late 1960s, I did a brief stint of volunteer work for the Muscular Dystrophy
Association. I did what a high school student could do. I answered phones
for the telethon and visited two brothers who were both suffering from
this disease at a local convalescent home. The reality of their illness defies
description; this is truly a horrid disease in which alert young people slowly
waste away to death in their late teens. However, my memories of this experience are of hope and dignity. Those two boys wanted to talk of just one
thing or, more correctly, of one person. They both idolized Elvis Presley.
For his part, Mr. Presley had gone to some great lengths to return their
affection. He had flown them to one of his concerts in Las Vegas and met
with them before and after. There had been cards of best wishes and, as I
recall, a phone call or two from Mr. Presley. Despite what this disease was
doing to their bodies, Mr. Presley's kindnesses had made these two boys feel
quite special and, as one of them told me, even quite lucky.*

*Because DMD is one of the most common and best known genetic disor-
ders, the point of this chapter is to introduce you to the genetics and molecular
biology of this disease. However, as you may by now have come to suspect,
there is also another agenda, namely to introduce a new genetic concept or
two. In this case the basic science revolves around the fact that the gene*

*which, when mutated, causes Duchenne muscular dystrophy is the largest
known human gene. Perhaps not surprisingly, very large genes have very
peculiar problems, especially when it comes to being large mutational
targets.—RSH*

The diploid human genome consists of approximately 6,000,000,000 bp
of DNA. Our current best guess is that human beings have only some
50,000 to 100,000 genes. *(Only 50,000 to 100,000? Try searching through
this collection of genes to find something and you will quickly realize that
it is a number too large to really visualize. Because the number of genes is
so large and because each gene represents a large complex of information
rather than a single data point, the growing amount of information on these
human genes has required the development of new technologies and computer
algorithms to allow us to work with information on such a large scale.—JER)*

According to Victor McKusick, these genes can be divided into roughly
five groups based on their sizes. *Small* genes, such as those for the blood
proteins α- and β-globin and insulin, range in size from 800 to approximately
4,000 bp. McKusick's *medium* gene category includes genes ranging from
11,000 to 45,000 bp in length. [As you can imagine, writing these large
numbers out will get tedious as the gene size increases. So we will revert
to using a shorthand that was introduced earlier (i.e., 1,000 bp equals 1
kilobase pair or kb). Using that nomenclature, this class of genes may be
said to range in size from 11 to 45 kb.] This medium class includes such
genes as those encoding the collagen and albumin proteins. McKusick's
next class, *large* genes, is represented by the 45-kb phenylalanine hydroxy-
lase gene. The fourth class, which he refers to as *giant* genes, ranges from 160
to 250 kb and includes the cystic fibrosis gene discussed in the last chapter.

Two genes in McKusick's final class of genes, which he refers to as
*mammoth* genes, are the subject of this chapter. This class includes a gene
called DMD (for Duchenne muscular dystrophy) that encodes a protein
known as dystrophin. Mutations in this sex-linked gene, which is approximately 2.5 million bp long (or 1.5% of the length of the X chromosome),
cause the diseases known as Duchenne muscular dystrophy and Becker
dystrophy. This gene is bigger than the entire genome of the bacterium
*Haemophilus influenzae,* which is now completely sequenced! It is SO
big that before it was cloned it was believed that Duchenne and Becker
dystrophies had to be caused by separate genes because they mapped to
places on the X chromosome that were millions of base pairs apart. It
appears that the map position comparison was done between a Duchenne
family that had a mutation near one end of the gene and a Becker family
with their mutation near the other end of the gene. Because it had not yet
been conceived that genes would be as big as the DMD gene, the automatic
presumption was that they MUST be different but neighboring genes. The
second member of this mammoth class is a gene called NF1, which when
mutated gives rise to a disorder known as neurofibromatosis.

## WHY ARE SOME HUMAN GENES SO BIG?
## (OR, REMEMBER "INTRONS?")

Well, first off, it isn't because they are encoding giant or mammoth messenger RNAs or proteins. Table 16.1 compares the gene size, the number of introns, and the size of the final mRNA for several human genes. Throughout this discussion remember that the GENE is the entire DNA structure that includes the promoter region with its on and off "switches." In contrast, the final mRNA molecule is a completely spliced and processed transcript that is ready to be used in the translation process. However, even then, only part of the mRNA corresponds to the coding sequence that contains the actual information needed to make the protein gene product. By any standard of comparison, the term mammoth seems truly appropriate for the DMD gene. True, the DMD gene does encode a 17-kb messenger RNA, which is about 10 times the size of average mRNAs. But the DMD gene is 2500 kb in length; less than 1% of the gene can be accounted for by the size of the transcript. Similar inequalities can be seen for the smaller gene classes as well. So why are these genes so big?

As presaged in Chapter 5, much of the length of these larger genes is taken up by noncoding sequences (*introns*) that are interspersed with parts of the coding sequence (*exons*). As shown in Fig. 16.1, the full length of the gene, composed of exons and introns, is transcribed to produce a full-length transcript. The introns are then spliced out of this initial transcript to produce the final mRNA molecule that is transported to the cytoplasm to be translated. Splicing takes place within the nucleus and is a very carefully executed and regulated process designed to make sure a proper message is delivered to the cytoplasm.

Although the genes in McKusick's small gene class have only 2–3 quite small introns, those in the larger classes can have anywhere from 10 to 70 introns of varying sizes. *(I was floored when I heard about one that has 100 introns!—JER)* We really don't have a terribly clear idea of just what

TABLE 16.1    Characteristics of Several Human Genes[a]

| Class | Example | Gene size (kb) | mRNA size (kb) | No. of introns |
|---|---|---|---|---|
| Small | β-Globin | 1.5 | 0.6 | 2 |
| Medium | Albumin | 25 | 2.1 | 14 |
| Large | Phenylalanine hydroxylase | 90 | 2.4 | 12 |
| Giant | CFTR | 250 | 6.5 | 26 |
| Mammoth | DMD | 2500 | 17 | ~70 |

[a] Modified from McKusick *Mendelian Inheritance in Man* (1992).

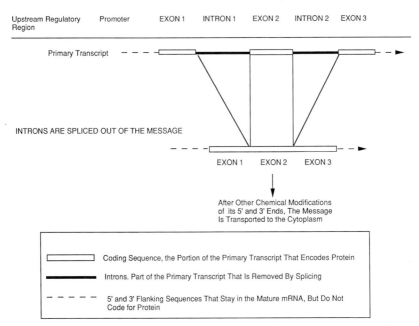

FIGURE 16.1    A schematic drawing of a gene and its transcript showing introns and exons.

introns do or why they are there. Clearly, some introns, which is especially true for first introns within some genes, contain important sites for regulating the activity of that gene. In other cases, the organism can use a trick called *differential splicing,* as shown in Fig. 16.2, to pick and choose between alternative sets of exons, depending on what part of the coding sequence it needs in different tissues and at different times during development. This allows a single gene to produce different forms of a given protein at different times in development, but the function of most introns remains unclear.

Oddly enough, those of us who work on the genomes of higher organisms sometimes find entire genes within the introns of other genes. For example, Scott works on a gene in the fruit fly *Drosophila melanogaster.* He has known for some time that this gene and several other genes are contained within the intron of a much larger gene of unknown function, and now a new gene has been discovered within the third intron of the original gene of interest. This gene is within the intron of a gene (*nod*) that is itself within the intron of a larger gene. *(Many human geneticists joke about the introns and intergenic regions being garbage DNA or spacers, but we need to keep reminding ourselves that just because we haven't figured out what something does doesn't mean it does nothing. Work on organisms like fruit flies by Scott and others keeps plugging holes in our knowledge about the*

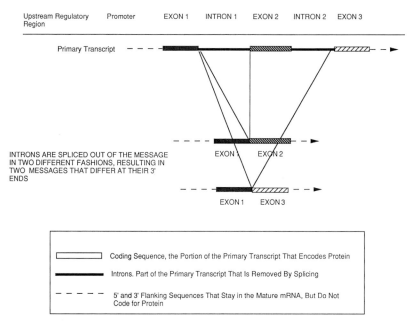

FIGURE 16.2　Differential splicing.

*genomes of complex creatures, and it is a frequent surprise and delight to find that the mysteries they have solved offer the answers to questions that have human geneticists puzzled.—JER)*

I know this view of some groups of tightly associated genes arranged inside each other's introns like stacks of wooden Russian dolls is a bit more complicated than the simple picture that emerges from studies in lower organisms. It is, unfortunately, a reality of the genomes of higher organisms. Why did things evolve this way? We can't tell you because we don't know. We can only reiterate what we have said before: "Evolution is not Michelangelo and the Sistine chapel. It is a teenage kid with a broken car and no money. Who cares what's pretty, just do whatever works!"

Introns may have arisen because they carried some selective advantage, such as the possibility for additional levels or transcriptional regulation or differential splicing, or other advantages that we do not yet fully understand. Alternatively, they may have been created by rogue genetic elements called transposons that can move themselves throughout genomes in a parasitic fashion. As long as such insertions could be spliced out of the transcript before the message was sent to the cytoplasm for translation, they might have carried with them no obvious disadvantage. Either way, introns created space in the genome that has become available for the evolution of new genes.

It is possible that having spacers within genes lets the cell use information from existing genes as building blocks for the assembly of new genes that have some functions from each of several previously existing genes. If a new gene function is needed in a cell, imagine the difficulties of trying to "evolve" a whole new gene from some random stretch of DNA by mutating individual bases with any efficiency or soon enough to do any good or without just producing something nonfunctional. However, with building blocks that can be rearranged, the cell has the option of making a "new" gene out of copies from pieces of several different genes. The cell doesn't have to carry out any very precise assembly process to stick the gene pieces (what we call exons) together because joining the building blocks can be a fast-and-dirty process that happens out in the "garbage" DNA of the introns with no danger of interrupting the open reading frame of the coding sequence if an imprecise joining is made. No, this isn't something our bodies do overnight to meet some new environmental demand. Evidence shows that over the very long time frame of evolution, new genes have come into being that consist of an assembly of pieces, with one piece resembling an exon in a cancer gene, another piece resembling an exon from an enzyme, and another piece resembling an exon from a gene that influences early development. *(My favorite example is a gene I am studying in which one motif—a sequence of amino acids associated with a particular function—found in a gene that affects the level of pressure within the eye is also found as a small part of a different protein in nasal mucus and again as a small part of a VERY different protein that is a receptor for a toxin from the black widow spider. Each of these genes shares this one motif but has very different sequences in the rest of the gene. So is this scenario in which the cell builds new genes out of copies of pieces of existing genes a fantasy? A fairy tale? It remains to be seen, but I am betting that the truth will turn out to be a little bit of each of the theories discussed here.—JER)*

## DOES LARGE SIZE AFFECT THE GENETIC BEHAVIOR OF THESE GENES?

Imagine a gene like DMD that is so large that it occupies more than 1% of the length of the chromosome on which it resides, in this case the X chromosome. This represents a huge target for mutation. Indeed, such genes are extremely mutable and the types of mutations observed commonly include chromosome aberrations such as translocations and deletions. In fact, males bearing loss-of-function mutations at the DMD gene represent 1/3,500 live births, and at least one-third of such males are the result of newly arisen mutations. Indeed, the rate of mutation at the DMD gene is estimated to be 1 new mutation in every 10,000 gametes, a rate 10 to 100 times greater than that observed for most human genes.

Similarly, the human disease neurofibromatosis is the result of mutations in a gene called NF1, which is also huge (greater than 300 kb). Like DMD, the NF1 gene has a mutation rate of 1 in 10,000. The rest of this chapter will focus on the genetics and molecular biology of these two genes.

## DUCHENNE MUSCULAR DYSTROPHY

Duchenne muscular dystrophy is a well-known fatal disorder that results in death in the late teens due to muscle wastage and deterioration. Because the DMD gene is X linked, most affected individuals are males. These males are normal at birth, but develop muscle weakness at age 4–5. These males are normally confined to a wheelchair by their early teens. By their late teens, these individuals usually succumb to either respiratory or cardiac failure. We now know that the DMD gene encodes a protein called dystrophin that is required for muscle maintenance and to prevent muscular atrophy. Affected males lack this protein and thus succumb to progressive muscular atrophy.

Most carrier females show no obvious symptoms, despite the fact that half the nuclei in their muscles express the mutant allele. This is because muscles are composed of very large cells that arose by the fusion of many individual cells. These cells each contain hundreds to thousands of nuclei per cell. On average, half of these nuclei produce dystrophin (i.e., they have inactivated the mutant-bearing X), and this level of functional dystrophin production appears to be sufficient. Given that X inactivation is random, one might imagine, at least in some cases, that a female embryo might inactivate the normal X in a large fraction of those embryonic cells that will go on to produce muscles. Such a thing shouldn't happen very often, but then this is a fairly common genetic disease. Indeed, some 8% of carrier females show some significant muscle weakness. *(It also goes without saying that an XO female, a woman with Turner syndrome, would also express the full illness if she carried the DMD mutant, just as a 46 XY male does.)*

The DMD gene had been mapped originally to the short arm of the X chromosome by standard linkage analysis. A more precise mapping of the DMD gene to the short arm of the X (band 21) was made possible by finding several very rare females who showed no signs of Turner syndrome but did exhibit full-blown Duchenne muscular dystrophy. In several of these cases, the affected females were heterozygous for an X autosome translocation. Although the autosomal breakpoints of these translocations varied from case to case, the X breakpoint always corresponded to band Xp21 on the short arm of the X chromosome (see Fig. 16.3).

Recall that a translocation results from the breakage of two nonhomologous chromosomes (in this case, the X and an autosome) and subsequent rehealing such that the proximal element of the X chromosome is capped

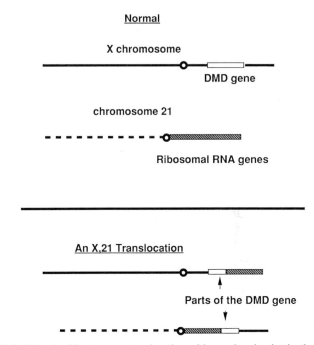

FIGURE 16.3   An X-autosome translocation with one breakpoint in the DMD gene and one in the rRNA gene on chromosome 21. Includes "blow-up" of the breakpoint regions on both chromosomes showing DMD gene sequences contiguous with rRNA gene sequences.

by the distal element of the autosome and vice versa. Thus, these females carry a normal X chromosome, a normal autosome, and the two chromosomes that resulted from the translocation. When the breakage events that created the translocation occur within a gene, or genes, they disrupt those genes and result in a loss-of-function mutation. *(RNA polymerase, the enzyme that carries out transcription, can do lots of neat tricks, but it can't jump. By splitting a gene into two parts and moving one part to a new chromosome, you have killed that gene. The RNA polymerase molecule simply has no way to leap to another site in the genome in order to complete transcribing this gene.—RSH)* The chromosome and position of the autosomal breakpoints of these translocations varied from female to female, but the X breakpoints were always in the same place, band Xp21.

So why are such translocation-bearing females affected? After all, they have a normal X chromosome in addition to the X involved in the translocation. *The problem lies in the fact that all of the cells in the bodies of these females arose from embryonic cells in which the normal X chromosome was inactivated!* Recall that if one of a female's X chromosomes is involved in a translocation involving an autosome (i.e., an autosome and an X chromosome have exchanged some material), the normal X will be inactivated in all of her cells. Why? Think back to what happens during early embryonic development, when X inactivation occurs. Half the cells will shut off the normal X and the other half of the cells will shut off the X chromosome involved in the translocation. In doing so, they will also shut off some part of the autosome involved in the translocation as well. That's a cell lethal event. All cells that try to shut down the translocation-bearing X will die. Thus, by cell selection in the early embryo, all cells that shut down the translocation chromosome died. All of the living cells will contain an inactivated normal X and an active translocated X. When females heterozygous for this translocation are born, all of them will have inactivated the normal X. They have muscular dystrophy because the X chromosome breakpoint of this translocation disrupts the DMD gene. The lack of the product of the DMD gene, dystrophin, causes Duchenne muscular dystrophy.

*(Don't feel dismayed if you have to reread this paragraph to "get it." Scott thinks in terms of chromosomes and I often have to do a diagram on paper to understand things that he finds obvious!—JER)*

### CLONING THE MUSCULAR DYSTROPHY GENE: METHOD 1

One of these translocations was fortuitously broken in the DMD gene on the X chromosome and in a ribosomal RNA (rRNA) gene on chromosome 21. Realize that this translocation results in DNA fragments that possess ribosomal gene sequences fused to parts of the DMD gene. Thus if one could make a clone library from these cells and probe that library with sequences corresponding to the ribosomal RNA gene, then it may be possible to fish out clones that carried the DMD sequence as well. Ribosomal RNA genes are very abundant in most genomes; in fact, such genes were among the very first genes to be cloned.

Investigators used such a cloned ribosomal gene to screen a library of DNA made from the translocation-bearing patient. In that library, clones corresponded to the breakpoints of the translocation that contained DMD gene sequences contiguous with rRNA gene sequences. These sequences could be identified easily by *in situ* hybridization as sequences that hybridized both to sites on chromosomes 13, 14, 15, 21, and 22, where the ribosomal genes reside, and to band Xp21 on the short arm of the X chromosomes. Using these fragments of the DMD gene, these workers could thus find the DMD sequences in clone libraries from normal individuals and begin walking in both directions to isolate the entire gene.

There is a lesson here about the importance of very, very rare individuals with unusual mutations and/or phenotypes in doing genetics, human or otherwise. Females with full-blown DMD are very, very rare. Yet the careful study of just these females made both mapping and cloning the DMD gene much easier, just as XY females bearing deletions for the TDF gene made cloning the TDF gene possible. Frequently, one of the keys in the use of a rare individual to solve a genetic problem is an observant physician who realizes the unusual, important nature of this individual and arranges the patient's participation in a genetics study. One of the greatest frustrations for the human geneticist is that we KNOW that there are patients out there who have some unusual set of characteristics that could provide a breakthrough, which could help many other patients in the long run, but finding them in the office of some clinician in a small town in Texas or the middle of New York City can be almost impossible. One of the ways that some genetic studies are starting to deal with locating unusual patients, or even just expanding the number of people participating in the studies, is through Web pages that provide information about the disease and invite participation in studies.

### CLONING THE MUSCULAR DYSTROPHY GENE: METHOD 2

An alternative approach was also successful in getting the DMD gene. This technique also depends on the study of a rare individual. In this case the individual was a boy with no family history of any disorder who was found to be affected simultaneously with Duchenne muscular dystrophy and with two other disorders associated with the loss or inactivation of genes known to be tightly linked to the X chromosomal DMD gene: retinitis pigmentosa and X-linked chronic granulomatous disease. Such cases, where a single deficiency removes several essential genes and thus produces a complex set of phenotypes, are often referred to as *contiguous gene syndromes* because they are presumed to reflect the simultaneous loss of two or more closely linked genes. In this case, the boy in question carried a small but cytologically visible deficiency including band Xp21. Kunkel and collaborators pulled off a really cute and very difficult trick called *subtractive hybridization* to isolate the region encompassed by this deletion and thus to get an entry point into the DMD gene.

They mixed a small amount of restriction enzyme-cut, denatured DNA (sample 1) from an unaffected person with buckets of denatured DNA from cells of this boy (sample 2). (Please recall that the term "denatured" means that the DNA molecules have been separated into single strands.) They then allowed the DNA from both sources to "renature" back to being double-stranded DNA and selected only those DNA molecules in which both the Watson and the Crick strands came from "normal" DNA.

Realize that most of the normal DNA sequences will reanneal with DNA from the deletion-bearing cells as the deletion DNA is in 250-fold excess. Only those DNA strands not present in the deleted boy's DNA will have to seek out their own counterparts. Not only did one of the clones they recovered hybridize to Xp21 by *in situ* hybridization, but it was also absent in the DNA of several other children carrying new deletion mutations at the DMD gene. By using these clones to perform a chromosome walk *(a rather long and complex walk because there was so much territory to cover!—JER)*, Kunkel and collaborators were able to isolate clones corresponding to the entire DMD gene.

## STRUCTURE AND FUNCTION OF THE MUSCULAR DYSTROPHY GENE

Using cloned DNA from within the DMD gene, the investigators walked out in both directions, using translocations and deficiency breakpoints as signposts to tell them how far they had gone and which direction they were going. It turns out that the DMD gene is just plain gigantic, mammoth even (over 2.5 million bp in length). The DMD gene contains at least 70 introns and produces a 17-kb mRNA transcript that encodes a huge protein known as dystrophin. It takes the cell more than 12 hours to produce the DMD RNA transcript that then gets spliced to produce the 17,000-bp mRNA. The DMD gene is transcribed most abundantly in skeletal and cardiac muscle and, to a lesser extent, in the brain. Although the dystrophin protein is under intense study, we still do not know a great deal about its exact function or about the other proteins with which it interacts.

Perhaps not surprisingly, large deletions, ranging from those removing just one exon to cytologically visible deletions that remove the entire gene, account for more than 60% of the loss-of-function mutations that cause DMD. Most, if not all, of these deletions are sufficient either in their extent or structure to completely prevent the cell from making dystrophin, i.e., the small deficiencies observed in DMD patients apparently either alter the reading frame of whatever message is produced or block proper splicing. Even in those cases that do not appear to delete most or all of the gene, the predominant theme in DMD is the failure to make the dystrophin protein.

*(Before Scott moves on I want to talk a little bit about the study of the unusual patient with DMD, chronic granulomatous disease, and retinitis pigmentosa mentioned earlier. In the course of cloning the DMD gene, the same set of studies led to the cloning of the neighboring gene for chronic granulomatous disease. By that time everyone was presuming that all three diseases in this boy had resulted from the deletion that caused his muscular dystrophy. Only recently has it been shown that the gene for retinitis pigmentosa that maps to the same area of the X chromosome is close to DMD but is NOT inside the deletion in this boy's DNA. Here is an important point: Just because several diseases are present in the same person does not mean that they are the result of the same underlying cause. The co-occurrence of*

*several diseases in one person can sometimes provide us with a powerful tool for research, but it can also really trip us up, as happened in this case. There are actually more than 20 genes in the human genome than can cause retinitis pigmentosa. Because there was no family history information for the boy in question, we do not know whether his retinitis pigmentosa was the result of a mutation in one of the other 20-plus genes (maybe even the one next to DMD) that was coincidentally present or whether something about the alteration to the chromosome structure in the region around the DMD deletion might have affected expression of the neighboring retinitis pigmentosa gene without actually removing it or changing its sequence. I know people who are working hard to solve this particular mystery.—JER)*

With a gene as big as DMD, might one also expect to find mutations that don't completely knock out the gene, mutations that might permit some level of dystrophin production? Indeed, one does find such mutations; such mutations appear to produce a muscular weakness disorder with a very different phenotype known as Becker muscular dystrophy. The symptoms of Becker muscular dystrophy are very similar to DMD, and the mutations that produce Becker muscular dystrophy map within the DMD gene. Becker muscular dystrophy is first picked up clinically at 6–18 years of age, and patients often aren't confined to a wheelchair until 25–30 years. They live to 40–50 years of age and often produce children. Indeed, the ability of men with Becker muscular dystrophy to produce children is reflected in the fact that most cases of this disorder are inherited. Unlike DMD, where approximately one-third of the cases are due to new mutations, less than 10% of Becker muscular dystrophy cases result from new mutations.

Like DMD, Becker muscular dystrophy results from a lack of functional dystrophin. However, most mutations that lead to Becker's dystrophy arise from either single base missense mutations or from small deletions that do not disrupt the reading frame of the protein. Thus, unlike DMD patients, they produce a defective dystrophin, but they produce dystrophin with at least some level of activity! Even this very small amount of dystrophin activity appears to significantly deter muscle wastage and thus greatly ameliorates the phenotype, at least in comparison with Duchenne muscular dystrophy. *(I always think of the difference here as being DMD, NO gene product, and BMD, a gene product that doesn't work very well.—JER)*

Please note the crucial lesson here: Mutations at the same locus can produce different phenotypes, depending on the type of mutation and its position within the gene. *This is to say that the phenotype is not the consequence of the damaged gene, but rather the result of the failure of that gene to make a functional product.* The effect of producing an altered product may be quite different from the effect of producing no product at all. This point has been made before in the chapter on cystic fibrosis and when we talked about the fact that mutations in the AR gene *can* produce androgen insensitivity syndrome (testicular feminization) or spinal bulbar atrophy, depending on the type of mutation. *(They now say that some mutations in the AR gene can cause prostate cancer!—JER)* It is a basic concept in human genetics: Not only does it matter which gene got damaged, but also how that gene was damaged.

## PRENATAL DIAGNOSIS AND GENE THERAPY
## FOR MUSCULAR DYSTROPHY

In those cases where the DMD mutation results from a deletion (more than 60% of the cases), prenatal diagnosis is fairly straightforward. Even small amounts of fetal tissue are sufficient to determine whether the fetus is male and whether a functional DMD gene is present. In the other 40% of cases, microsatellite repeat-based linkage analysis may be required. Either way, accurate prenatal diagnosis is possible for more than 90% of known carrier females. There are also reasonably good genetic and chemical tests with which to screen for carrier females. Obviously, no amount of such prenatal diagnosis will ever simply eradicate this disease; new mutations are just too common. However, in females who are potential carriers for either of the two dystrophies such tests may provide prospective parents with the information they desire.

It is hoped that we will soon be able to cure this disease by just injecting functional DMD genes into the muscles of affected children. Unfortunately, the very large size of the exons alone is making this task Herculean indeed. Still, people are trying hard to build copies of the DMD gene in which introns have been spliced out to produce minigenes to be put back into muscles. The fact that patients with even low amounts of dystrophin of intermediate function, as is the case for patients with Becker's muscular dystrophy, show a less severe course of the disease than do patients with DMD suggests that even getting very small amounts of functional dystrophin produced may be quite therapeutic. Sooner or later something along these lines is going to work, but it may take a while. We will discuss the basic approaches for this technique in a later chapter.

For now, we want to conclude this chapter by briefly presenting just one more mammoth gene disease, neurofibromatosis 1, with a similar set of genetic problems. Our focus is not to just load you up with descriptions of diseases, but rather to convince you that the principles we have presented earlier are applicable to a large number of human diseases.

## NEUROFIBROMATOSIS 1: ANOTHER REALLY
## BIG GENE

Neurofibromatosis is a relatively common disorder of the nervous system whose phenotypic expression can range from "café-au-lait" spots on the skin to large fibromatous tumors of the nervous system. Although these tumors are usually benign, malignancies resulting in death are not uncommon in these patients. In some cases, the tumors or masses can be quite large and disfiguring. The frequency of this disorder is approximately 1 case per 3–5,000 people. Although the severity of phenotypic expression can vary within families, all cases of this disorder appear to be due to mutations in a gene known as NF1, which has been mapped to the long arm of chromosome 17. As is the case for DMD, the NF1 gene is quite large, greater than 300 kb. Unlike DMD, which is recessive and sex linked, mutations at the NF1 gene are dominant and show a pattern of segregation that is characteristic of autosomal dominant mutations.

As is the case for muscular dystrophy, the mutation rate at the NF1 gene is quite high, approximately one new mutation per 10,000 gametes. Again, approximately one-half of the cases of NF1 appear to reflect new mutations. Presumably, this very high mutation rate simply reflects the very large target size of the gene. Not surprisingly, some of these mutations include chromosome aberrations, such as translocations, but point mutations, including a nonsense mutation (a mutation that interjects a stop codon into the coding sequence), have also been found.

We understand a lot about the function of the protein product of this gene in the cell. The NF1 gene is a member of a class of genes known as tumor suppressor genes, which will be described in detail in the next chapter. The function of the NF1 protein is to control cell growth, and thus the phenotype of mutations at this gene, large masses of tissue with a heightened risk of malignancy, is at least partially explicable. We will use this transition to move into the next chapter and begin our discussion of genes and cancer.

# 17

## GENES AND CANCER

*Gwen stared down on the freshly dug grave where her sister Michelle would be buried. Michelle was Gwen's eldest sister: She was only 36 when she died. Gwen turned to look at her mother's grave which lay beside Michelle's. Gwen lost her mother 10 years ago to the same affliction: breast cancer. If Gwen were to walk 30 yards down the cemetery lawn, she would come across her grandmother's grave; she too died of breast cancer. Gwen felt numb. She turned around to see her other older sister Sandy who was just 29 and had recently been diagnosed with breast cancer.*

*A month or so later Gwen visited Sandy in the hospital. Sandy was about to undergo surgery to have both breasts removed: a bilateral mastectomy. The doctors had warned Sandy that it was likely that the cancer had already spread. That day their family physician, Dr. Morris, took Gwen aside to have a talk about her future.*

*"Gwen, breast cancer may run in your family. You are 19 years old, and presently show no signs of the disease, which is not to say that you will not be affected someday in the future. There is at least a chance that your mother and sisters developed breast cancer because they carried one of a number of genes that predisposed them to develop cancer. If that is true, then there is a good chance that you could develop cancer as well. Gwen, you might well carry that gene." Dr. Morris folded his hands in his lap, "Gwen, I think you should see a medical geneticist, I think that you should be tested."*

*That night Gwen couldn't sleep. The next day she made an appointment to see the medical geneticist recommended by Dr. Morris, but the thought of being tested for some cancer gene was just too scary to think about. She had read articles in magazines and newspapers about "cancer genes" being found by scientists, but she wasn't sure what such things meant. What would it mean to her? Did she really want to know? What if the test was positive? What would she do? She knew that it was possible to have both of her breasts surgically removed before the cancer could develop and spread. Was she desperate enough to take that step? Would it protect her if she did?*

*Gwen finally went to the appointment. It turned out that a cancer-promoting mutation in the BRCA-1 gene was segregating in her family. DNA tests using blood from Gwen and Sandy and tissues preserved from her mother and sister showed that all of them carried the same mutation. Gwen tested positive: Her risk of developing breast cancer was very high. In the months that followed, there were many more appointments with the medical geneticist, with the genetic counselor, and with oncologists and surgeons. She asked many questions and heard many answers. Some of the answers had to do with genetic testing for the BRCA-1 breast cancer gene that is discussed in this chapter. Some had to do with the odds of an individual inheriting this mutation developing cancer. Some of the answers had to do with discussions of the uncertainties about how much protection a double mastectomy might offer her. In the end, the choice had to be hers.*

*A few months later, Gwen dropped by her sister Sandy's house for coffee. She walked into Sandy's living room and slumped into the chair by the window. Sandy looked at Gwen and asked, "What are the doctors recommending?"*

*"The surgeon wants me to have both of my breasts removed . . ." Gwen's eyes filled with tears.*

*"To prevent what could happen in the future . . ." Sandy said.*

*Sandy sat in silence before reaching out to Gwen. Gwen came to the side of the couch and took Sandy's hands in hers. Sandy looked carefully at her little sister, hating the thought of Gwen having to follow in her footsteps but knowing it was not her decision to make. "It has to be your choice, but whatever you choose, I will be with you all the way."—CAM*

The preceding story is about cancer. Cancer is a disease that results from just one of the cells in your body overriding the normal controls of cell division. Most of the cells in your body are not supposed to divide anymore. There are some notable exceptions, such as the cells that make up the villi of your intestines, the cells that comprise your bone marrow, the cells in the lowest layer of your skin, and your germ cells. The basic pathway by which each cell must divide is diagramed in Fig. 17.1. As can be seen, a cell begins division in a resting or preparatory state called G1. Once it commits to cell division it then enters the S phase, where DNA replication occurs, another preparatory phase called G2, and then enters the mitotic division (M). For the most part, your cells are permanently parked in a sidetrack of the cell cycle called G0 (pronounced gee zero). Sitting quietly in this stage, your cells indefinitely forego the possibility of division in favor of a stable commitment to execute their particular functions.

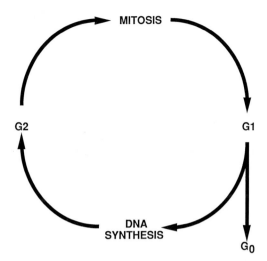

FIGURE 17.1    The cell cycle.

Sometimes, however, *a single cell* overrides that inhibition. It loses that commitment to just "being" and begins to divide. Perhaps not very rapidly at first, but nonetheless it begins to divide. This is the start of a *tumor,* the beginning of the most awful word in our language, *cancer.* Most of the time, the daughter cells descended from one miscreant cell will divide slowly, staying together in a dense, well-defined, and tightly bordered mass that can usually be removed by a competent surgeon. Such tumors are referred to as *benign.* However, sometimes the daughters of those cells also lose their inhibitions regarding the invasion of normal neighboring tissue and begin to spread throughout the organ or tissue in which they arose. Such tumors are described as *invasive.* In some cases these tumor cells, now committed to rapid unbridled cell division, find their way through the lymph or bloodstreams into other sites or tissues, establishing new sites or nodes of tumor formation. This movement to new sites is called *metastasis.* Invasive or metastatic tumors are referred to as *malignant.* Malignant tumors can and often do kill human beings. It is obviously in our interest to understand how they arise.

The most important sentence in the preceding paragraph was "sometimes, however, a single cell loses that inhibition." Tumors begin from single cells. They arise as a consequence of mutations within that cell in one of two types of genes. Many types of tumors arise as a result of mutations in genes called *tumor suppressor genes.* For purposes of simplicity, tumor suppressor genes will be divided into two classes: those genes whose protein products are *cell-cycle regulators,* which act as guards to prevent cells from dividing, and those whose products are essential for *DNA repair.*

Mutations that inactivate cell-cycle regulators, or "guard protein" genes, prevent those guard proteins from being produced, or from functioning normally. As a consequence, the ability of the cell to stay in G0 is compromised. In the absence of such proteins the cell is at risk. Mutations that impair certain DNA repair systems impair the ability of cells to properly replicate their DNA without making errors and/or impairing the ability of the cell to repair DNA damage done by the environment. The loss of these repair proteins not only makes every cycle of DNA replication a mutagenic event, it also allows damage to the DNA of the cell to escape repair or to be repaired improperly. Thus the probability of a cell acquiring a cancer-causing mutation increases drastically. ·

Cancers can also be promoted by mutations in genes called *proto-oncogenes.* Proto-oncogenes are genes whose products stimulate cellular proliferation. In this case, cancer-causing mutations are mutations that activate these genes in tissues where they should stay repressed and non-functional. In many, and probably most, tumors, multiple mutations are required for these tumors to develop into a full-blown malignancies.

As you read the last paragraph it is hoped that at least some of the following questions occurred to you.

1. Where do these mutations come from, and how many mutations are required to create a full-blown malignancy?
2. Do the same mutations cause tumors in all cell types?
3. Are cancer-causing mutations ever inherited and, if so, does that mean that it might be possible to predict who will eventually get certain types of cancer?
4. Does that mean that things that cause mutations, such as radiation and certain chemicals, can also raise the risk of cancer?

These questions are addressed by considering a hereditary form of cancer called *retinoblastoma.* Retinoblastoma is a cancer of the retinal cells of the eye that is most commonly diagnosed in young children. In certain families, retinoblastoma is inherited as a simple autosomal-dominant mutation (see Fig. 17.2). These cases of inherited retinoblastoma make up about 40% of

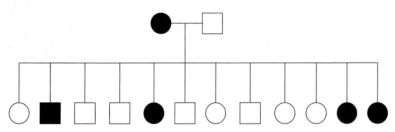

FIGURE 17.2    Pedigree for retinoblastoma.

the total cases of retinoblastoma uncovered each year. Children exhibiting this inherited form of the disease usually exhibit multiple tumors in both eyes. The remainder of the cases are sporadic and usually manifest themselves as only a single tumor in one eye. Because retinoblastoma results from mutations in a tumor suppressor gene, an understanding of the genetics of this disease will provide several crucial insights into the genetic basis of cancer. Most notably, the discussion of retinoblastoma will open a discussion of the general category of genes called tumor suppressor genes. We begin by discussing the inherited form of retinoblastoma.

## RETINOBLASTOMA AND TUMOR SUPPRESSOR GENES

### THE GENETICS OF INHERITED RETINOBLASTOMA

Inherited retinoblastoma is usually manifested earlier than the sporadic form. This inherited form is the result of a mutation (Rb) that maps to the long arm of chromosome 13. When a child inherits the mutant Rb gene (Rb) from one of the parents, they are likely to develop tumors in both eyes at an early age. In this case we can point our finger at one mutation in just one gene, say that mutation causes cancer, and we would be right.

Why do only some retina cells form tumors? If every cell of this child carries the one copy of the Rb mutation, how come every cell doesn't initiate tumor formation? The answer lies in the fact that the normal copy of this allele, obtained from the other parent, is sufficient to control cell division and prevent tumor formation.

*(Wait a minute, Scott! So you are saying that Rb is dominant when one thinks about pedigrees and recessive in terms of individual cells?—CAM Exactly right, Catherine.—RSH So how come* any *tumors are formed? Don't all the cells have the normal copy of Rb that they inherited from the other parent?—CAM)*

The answer to that question came from the mind of one of the smartest people I know, Dr. Alfred Knudson, a scientist at the Institute for Cancer Research in Philadelphia. Dr. Knudson figured out that the retinal cells of Rb/+ heterozygotes only become competent to form tumors when they lose that normal (+) allele of the Rb gene, i.e., they must carry two loss-of-function alleles of the Rb gene in order to form a tumor! In these children the first hit was inherited and the second hit happened in a retinal cell after the child was born (see Fig. 17.3). This is called the *two-hit hypothesis.* To form an eye tumor, cells have to knock out both good copies of the Rb gene. Cells throughout the eyes and bodies of these children already have one bad copy, but tumors will only occur in those cells that lose the other copy as well. The inherited Rb mutation is thus dominant in terms of

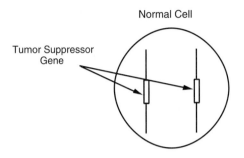

Normal Cell

Tumor Suppressor Gene

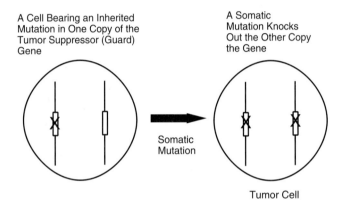

A Cell Bearing an Inherited Mutation in One Copy of the Tumor Suppressor (Guard) Gene

A Somatic Mutation Knocks Out the Other Copy the Gene

Somatic Mutation

Tumor Cell

Knudson's Two-Hit Model for Tumor Formation

FIGURE 17.3    The "two-hit" hypothesis.

pedigree analysis: Virtually all offspring that receive this mutation will develop tumors. However, at the cellular level it is recessive, i.e., tumorigenesis is observed only in an occasional cell when the normal allele on the normal homolog is either deleted or mutated.

## HOW CAN CELLS LOSE THE NORMAL RB GENE?

Realize that mutations in any gene, including the normal allele of Rb, do occur at a very low frequency (approximately 1 per 100,000 or 1,000,000 cells/cell generation) during the process of DNA synthesis in each cell cycle. Many cycles of cell division are required to produce millions of cells in the retina, called retinoblasts. At each of those divisions was a very low risk of mutating the normal (+) allele of the Rb gene. So even though the somatic mutation frequency is low, it will happen given enough cycles of cell division. As a consequence, several cells in each retina will endure such mutations and thus be left without a functional Rb gene. The resulting cells have lost both copies of a potent tumor suppressor gene. This process of losing the remaining normal allele is referred to as *loss of heterozygosity* (LOH). It is a common event at many sites in the genome in human tumors. *(There are other means to lose the normal gene in addition to simple mutations: mitotic recombination, chromosome loss, and reduplication. If you know what these terms mean, great. If you don't, well it really doesn't matter; however, if you want you could go find your professor right now and make her/him explain them to you in much detail. I'll bet she/he would like that.—RSH)*

It is important to realize that the normal product of the Rb gene plays a crucial role in the nondividing cells in the retina. This protein, cleverly called the RB protein, is required to block cells from starting the mitotic cell cycle. In its absence, cell proliferation begins and the cell starts on the path to tumorigenesis. It turns out that inherited and sporadic retinoblastoma differ only in that in the inherited form only the second event needs to occur, whereas in the sporadic cases two independent mutations must knock out both copies of RB in the same cell. This might seem to be a rare event, but it does occur. It should occur at a frequency of $1/10,000,000,000$ to $1/1,000,000,000,000$ cells. However, recall that there are $10,000,000$ retinal cells per person. Thus it should be observed in something like 1 in every 1,000 to 1 in every 100,000 humans. (Indeed the observed frequency of spontaneous tumors is $\sim 1/40,000$.)

Another example of an inherited tumor, which, like retinoblastoma, behaves as a dominant in pedigrees but also requires inactivation of the normal allele at the cellular level, is *Wilm's tumor*. This tumor develops in one or both kidneys. The Wilm's tumor gene has been cloned and studied at the molecular level. Not surprisingly, like the Rb gene, the product of this gene normally functions to repress the functions that allow cell division. As pointed out later, similar findings have been made for inherited forms of breast and colon cancers.

*(Both retinoblastoma and Wilm's tumor are due to mutations in genes that encode what I call guard proteins. This term comes not from distinguished scientists, but rather in response to one of my daughter Tara's board games. The purpose of the game is for all players to* cooperate *to rescue a sleeping princess. One can only wake and rescue the princess by killing, bribing, or drugging the seven guards that surround her. (This is a child's game?) You can think of a cell's commitment and/or ability to divide as the sleeping princess. She can only wake and begin to move the cell into division when the guard proteins have been inactivated in one way or the other.—RSH)*

So do other genes/proteins have to be inactivated to get to a full-fledged tumor? Probably, but we are not sure about what other events are required in the retina cell for tumor formation. However, we can be somewhat more specific with respect to a rare hereditary form of colon cancer known as *familial adenomatous polyposis* (FAP). Like the tumors just described, FAP is inherited as a simple autosomal dominant, and the first step in the formation of the tumor(s) is the loss of the normal allele of the FAP gene on the long arm of chromosome five. Again the protein product of the normal FAP gene appears to be required for the control of cell division. However, in this cancer the further progression of the tumor from a small polyp to a malignant tumor can be divided into clear stages that can be distinguished by a pathologist. A study of these various stages reveals that the development of an invasive and metastatic tumor requires multiple new mutations.

Thus, although the formation of the early polyp appears to require only one mutation, additional mutations in other genes (most notably the gene encoding a protein called p53; see later) are required for that early tumor to become a dangerous malignancy.

How come children with the inherited form of retinoblastoma don't get tumors in cells outside of the eye? They do, as some of the children suffering from the inherited form of this disorder are also at risk for some other kinds of cancers later in life, especially those treated with radiation therapy. However, these people are not at risk for all types of cancers. The same is true for the inherited form of Wilm's tumor. As you have probably guessed from our discussions in the previous two sections, there appear to be different types of guard genes/proteins that protect different types of cells in various tissues.

Those kids with the spontaneous form of retinoblastoma shouldn't be at risk for other tumors, as most of their cells have two normal RB genes, right? Correct, both for the sporadic form of retinoblastoma and for the sporadic form of Wilm's tumor.

Are there other inherited forms of cancer associated with mutations in these "guard genes?" Two more will be discussed, but the number seems to increase weekly.

### LI–FRAUMENI SYNDROME AND A GENE THAT ENCODES A PROTEIN CALLED p53

One of the most important guard proteins in our cells is known as p53. This protein plays a crucial role both in preventing unwanted cell division and in regulating the response of the cell to DNA damage. More than 50 types of cancer have been shown to carry new mutations in the p53 gene; indeed, more that 70% of all colon–rectal cancers carry mutations in this gene. The same is true for many other kinds of tumors. So what happens if one inherits one defective copy of the p53 gene? The answer is a hereditary disorder known as Li–Fraumeni syndrome, in which heterozygotes are at risk to develop a wide variety of different tumors in different tissues of the body, including sarcomas and tumors of the ovary.

### BREAST CANCER AND A GUARD GENE
### CALLED BRCA-1

In the last decade it has become increasingly clear that some 4–10% of cases of breast cancer could be explained by mutations of either of the two genes BRCA-1 and BRCA-2 (see Fig. 17.4). Mutations at these genes also appear to predispose their carriers to other types of cancer as well. Like the tumor suppressor genes described earlier, BRCA-1 and BRCA-2 both appear to play crucial roles in blocking uncontrolled cell division. Heterozygotes already are missing one copy of this gene; in tumors the remaining normal allele has also been inactivated. As was the case for FAP, loss of both copies of either BRCA-1 or BRCA-2 alone is probably not sufficient for tumor progression. Analysis of DNA from tumors often shows a loss of heterozygosity for multiple other sites in the genome.

As you might imagine, finding a gene that predisposes an individual to breast cancer has powerful implications for testing individuals with multiple cases of breast cancer in their families. As an example of testing within an affected family, we need to look no further than the story with which Catherine began this chapter. The test to which Gwen's doctor alluded was in fact a genetic test for mutations at the BRCA-1 and BRCA-2 genes. If such a test was available to Gwen, it *might* be able to provide her with considerably more information on which to base her decision about surgery. *(We hope that you will take just a few minutes to reconsider Gwen's dilemma in light of this test. Suppose she found that her mother and sisters did carry a mutation in the BRCA-1 gene, would she want to be tested? Suppose she was positive and chose the route of surgical prevention, would she want to test her future children, and, if so, when? What might she tell them?—RSH If she didn't test her future children prenatally, would she be allowed to test*

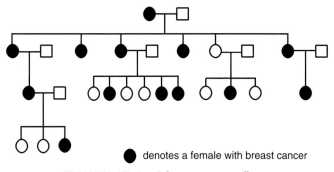

denotes a female with breast cancer

FIGURE 17.4    A breast cancer pedigree.

*them after they were born?—JER In Scott's classes, he asks students to write creative essays about this sort of dilemma. Perhaps you could find some way to carry this issue away from this book as well.—CAM)*

We do need to note, however, that although such a test may have substantial value in many families, it should also be kept in mind that the frequency of spontaneous or sporadic breast cancer is sufficiently high that in many cases the finding of several individuals with breast cancer simply reflects a chance clustering of events rather than a genetic predisposition. The finding of BRCA-1 and BRCA-2 genes has also revealed that mutations in these genes may be substantially more common in some populations or ethnic groups than in others. This finding raises the issue of whether it might be worthwhile to consider a general screening of those populations where a specific mutation in the BRCA-1 and BRCA-2 may be more common.

Do all the guard genes defined by cancer-causing mutations encode cell cycle control proteins? No, some like BRCA-1 appear to encode proteins that regulate the transcription of other genes, most notably genes that control the cell cycle. Another such example of this phenomenon is von Hippell syndrome, a hereditary predisposition to brain tumors. In this case the mutated gene is a known component of the enzyme complex that transcribes genes into mRNA.

So, we've begun to answer some of the questions posed at the beginning of this section. Yes, cancer can be inherited; yes, we can diagnose the inherited gene underlying these tumor syndromes by DNA analysis very early in life; and yes, various cancer-causing mutations may be cell-type specific, whereas others, like mutations in the p53 gene, may be a general characteristic of many types of tumors.

Are there other kinds of mutations, in addition to those in genes that encode guard proteins or their transcriptional regulators, that can cause tumorigenesis? While we are on the subject, where do all those additional mutations come from? We discuss just that in the next section.

## COLON CANCER, SKIN CANCER, AND MUTATIONS IN DNA REPAIR GENES

As you read the previous pages, you may have begun to wonder just how all of these mutations, which are necessary to produce a full-blown malignancy, can arise within the progeny of that single misbehaved cell. Well, to some extent, it is simply a matter of large numbers of cells and strong selection in the tumor for those cells that divide most rapidly. However, it turns out that tumors may often begin with mutations in genes that impair the ability of cells to repair their DNA and thus to fix either errors that occur during the replication process or DNA damage that occurs as a result of exposure to environmental mutagens.

The relationship between defective DNA repair and carcinogenesis is best illustrated by a hereditary form of colon cancer called *hereditary nonpolyposis colon cancer* (HNPCC). Researchers have found two sets of families segregating for HNPCC. One set of families allowed them to map a gene that predisposed individuals to HNPCC to chromosome 2, and the second set of families allowed them to map a second cancer-causing mutation to chromosome 5. Analysis of tumors from both families revealed an unusual genetic instability in the tumor cells. This instability was most easily manifest as the expansion or contraction of sequences in the human genome called *microsatellites* (see Chapter 12). Microsatellites are short runs of repeated simple sequences such as di- and trinucleotides (e.g., CACACA-CACACACA) that are scattered throughout the genome. Although the number of repeats at each site was constant in normal cells of these individuals, cells in the tumor might carry huge expansions of the repeat (e.g., CACACACACACACACACACACACACACACACACACA) or contractions of the repeat (such as CACACACACA).

Clearly, something was wrong in tumor cells, something that interfered with the cell's ability to replicate these microsatellite sequences. However, did this repair defect have anything to do with the development of a malignancy? In order to answer these questions and to determine the biochemical defect in these tumors, several investigators turned their attention to a more tractable genetic system: yeast cells. Yeast cells are an incredible "model system" for asking questions about basic cellular functions; it is easy to clone genes and to isolate mutants in this organism. Much of what is known about a variety of basic cellular processes has been learned by clever experiments on this very simple organism. Indeed, investigators, including Tom Petes, Mike Liskay, and Richard Kolodner, quickly found *repair-deficient mutations* in yeast (mutations that made the yeast cell hypersensitive to even very low doses of agents that could damage DNA) that showed the same instability for microsatellite repeats as did tumor cells. Whereas human microsatellite sequences cloned into yeast were normally replicated stably, in yeast cells bearing these mutations, these investigators saw rapid amplification and deletion of the microsatellites. The genes defined by these repair-defective mutations turned out to encode enzymes involved in a DNA repair process called *mismatch repair.*

Mismatch repair is one of a number of cellular processes that repair the damage done to DNA both by exogenous agents (such as sunlight; environmental radiation; carcinogens in the air, water, or food) and by everyday life in the cell. Mismatch repair deals with a number of such types of damage, especially errors caused by or during replication. In the absence of a functional mismatch repair system, those errors that do occur during replication are not repaired but rather incorporated into one of the two daughter strands following the next replication (see Fig. 17.5). The resulting daughter strand will thus carry a new mutation at the site of the original replication error. Because such replication errors are not infrequent, cells that cannot achieve error correction by mismatch repair will have greatly increased mutation rates. (Please note that errors occur throughout the genome during replication, not just at the microsatellites. However, errors do seem to be more frequent when DNA polymerase attempts to replicate such sequences and thus are easier for us to detect. In the absence of mismatch repair, new mutations will occur at a high frequency throughout the entire genome!)

## DNA REPAIR FUNCTIONS AS
## TUMOR-SUPPRESSOR GENES

As these yeast mismatch repair genes were being discovered, investigators were rapidly closing in on mapping the human colon cancer genes on chromosomes 2 and 5 (Fig. 17.6). Much to everyone's delight, but perhaps to no one's surprise, the human versions (homologs) of two of the yeast genes involved in mismatch repair (MSH2 and MutL) turned out to map to exactly those two intervals. Further experiments revealed that the colon cancer-causing mutations on chromosomes 2 and 5 were, in fact, mutations in these genes and, in the tumors themselves, the normal allele of these genes often appears to be deleted. *(Think about what we just said for just a minute. Basically, we are talking about the Knudson "two-hit" model again. Individuals inheriting a cancer-causing allele of one of these genes are at*

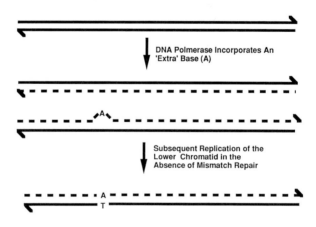

In the Absence of Mismatch Repair, Replication Errors
Can Result in New Mutations

FIGURE 17.5    Faulty repair of a microsatellite sequence.

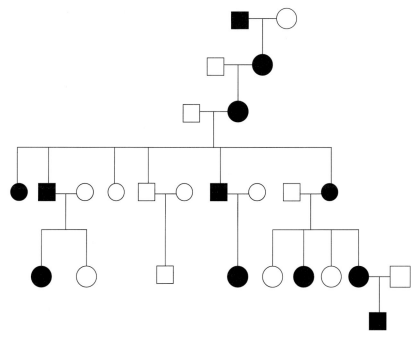

FIGURE 17.6    A pedigree for HNPCC.

*high risk for HNPCC. The "second hit" is the mutational "knockout" of the normal copy of the MSH2 or MutL gene in one or more of these cells, which creates a cell with no functional copies of these genes, a cell that can no longer accomplish mismatch repair. That cell is now going to experience a high frequency of new mutations every time it replicates its DNA.—RSH)*

Why? Why should a defect in DNA repair cause cancer? Think about it for a minute. If you have an error-prone system for replicating DNA, one that cannot repair the occasional errors made during replication or spontaneous damage to DNA, then every round of replication is a mutagenic event. Every round of replication gives you a chance to lose another tumor suppressor gene by mutation. Every round of replication makes the loss of control of cell division more inevitable. If DNA repair defects really make a cell susceptible to cancer, then shouldn't defects in other repair systems also make cells more cancer prone? Yes! Indeed, there are a number of human diseases in which individuals are demonstrably repair defective and highly cancer prone, including xeroderma pigmentosa, Bloom syndrome, ataxia telangiectasia, and Fanconi's anemia. Each of these diseases results in homozygosity for inherited recessive mutations in genes required for various aspects of DNA repair. For example, patients with xeroderma pigmentosa are defective in various aspects of a process called excision repair, whereas children with ataxia telangiectasia are deficient in a process that forces cells with unrepaired DNA damage to stop dividing until they can repair their DNA. In each of the disorders, mutations in genes whose protein products are required for DNA repair predispose their bearers to develop tumors. How? Well, the best guess is that these repair-deficient disorders effectively raise the mutation rate and, in doing so, increase the probability of mutating the tumor-suppressor or guard genes.

Wait a minute, isn't the function of all of those tumor-suppressor proteins simply to keep the division functions of the cell from being activated? Couldn't the cell bent on becoming a tumor get around this by a mutation that activates a proliferation-inducing gene (one that starts some component of the cell cycle) right under the nose of those guard proteins? Yes, unfortunately, and that story is found in the following section.

## THE END RUN

Dominant tumor-promoting mutations that *activate* genes are called proto-oncogenes and push the cell past the guards and into the division cycle.

Chromosome rearrangement and instability are hallmarks of tumor cells. Indeed such rearrangement may play a crucial role in the initiation of some tumors. One of the best examples is chronic myelogenous leukemia. Greater than 90% of such tumors carry a specific translocation involving chromosomes 9 and 22 that is not present in normal cells of these patients. In this case a normally inactive gene (called *abl*) on chromosome 9, which acts to *promote* cell division, is moved by translocation to lie downstream from a very powerful set of promoter elements located on chromosome 22 (see Fig. 17.7). After translocation, these promoter elements, which are always on in lymphoblast cells, inappropriately turn the *abl* gene on in white blood cells. As a result, these white blood cells begin to constitutively enter the division cycle! Burkitt's lymphoma, another form of cancer, results from a similar mechanism.

Please note that there is a crucial difference between such rearrangements and the inherited tumor-promoting mutations discussed earlier. These rearrangements are dominant and are found only in tumorous cells. They are NOT present as a "first hit" throughout the body. Thus they are dominant on the cellular level but are not transmitted as a dominant disorder within a family.

## SUMMARY

It should be obvious by now that cancer is in reality a genetic disease that results from one or a series of mutational events. Not surprisingly

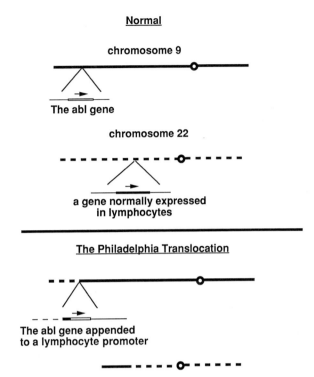

FIGURE 17.7   The Philadelphia translocation (small arrows indicate the direction of transcription for the abl and lymphocyte genes).

then, agents that cause mutations such as the ultraviolet rays in sunlight or the tars in cigarette smoke behave as potent carcinogens. Indeed the ability of a chemical to induce mutations in bacteria is now used as a good first assay of the ability to that chemical to induce tumors! This test, called the Ames test, has provided a fast and inexpensive method for screening the many compounds that we eat, inhale, wear, or rub on our bodies each day.

So do yourself a big favor and use sunscreen this summer. Ultraviolet radiation from the sun is a potent mutagen. Your cells have excellent repair systems to cope with the damage this radiation causes, but those systems are not perfect and new mutations will occur if your cells are exposed to too much sunlight. The right combination of such mutations, and one cell is on the path to skin cancer (*melanoma*). We remind you that skin cancer can be deadly!

It should also be obvious that many cancers reflect a multistep mutational process. Early in their development many tumors may not yet have acquired the mutations that render them invasive and possibly metastatic as well, which makes them far easier to treat and excise. This is why early detection is so incredibly important. Pay attention to any change in a mole on your skin. If you, are a woman, learn how to do breast self-examinations, and see a doctor *immediately* if you find something. If you are a man, learn how to do testicular self-examinations. Pay attention to any usual discharge or bleeding. If you have questions about these or any other early warning signs of cancer, ask your doctor or call the American Cancer Society. We know that we are preaching here, but frankly *your* life depends on it!

# V

# INTERACTIONS OF GENES

# AND THE ENVIRONMENT

We have tried at several points in this text to draw crisp distictions between genotype and phenotype. This section illuminates another aspect of that connection: the interaction between genotype and various environmental influences in creating a phenotype. We begin with the traditional treatment of this subject, a discussion of multifactorial inheritance. We then consider possible interactions and contributions of the environment and of genetics with respect to criminality. We end by discussing the genetics of a human virus (the AIDS virus) and the human genetics of resistance or susceptibility to AIDS.

# 18

# MULTIFACTORIAL INHERITANCE:

## TOWARD FINDING THE GENES FOR

## MANIC DEPRESSION AND

## SCHIZOPHRENIA

Up until now we have focused on those traits that are specified by single genes. We have also limited our discussion to those cases in which a given genotype specifies a given phenotype, regardless of extraneous or environmental factors, and those cases for which there are discrete traits that can be simply defined. Unfortunately for most geneticists, the traits in which we are most interested do not fit these criteria. Most human characteristics are not simple yes/no traits, but are continuous traits that show some pattern of distribution.

As shown in Fig. 18.1, height and weight are excellent examples of this sort of trait. People cannot simply be classed into "plus/minus" trait categories in terms of height or weight, but rather each of us falls somewhere on the continuum. Some of those at extreme ends of the weight and height continuums may well be suffering from pathological conditions, but it is difficult to draw a simple line between "normal" and "affected" individuals. Moreover, both of these traits, while strongly genetically determined, are also clearly dependent on the environment. These traits require a more sophisticated view of the mechanism by which a given genotype produces a given phenotype.

### MANY TRAITS ARE SPECIFIED BY MORE THAN ONE GENE

It is ludicrous to even suggest that a trait as complex as height or intelligence or athletic ability could be determined by a single gene. Indeed, the height of an individual is determined by the additive interaction of a large

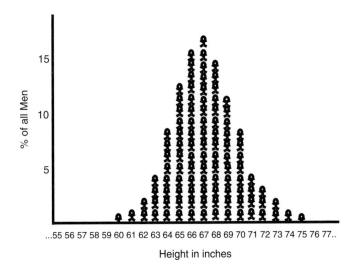

% of all Men

...55 56 57 58 59 60 61 62 63 64 65 66 67 68 69 70 71 72 73 74 75 76 77..

Height in inches

FIGURE 18.1    A histogram for variation of height. Adapted from Harrison et al., *Human Biology*, second edition (Oxford University Press, 1977).

number of genes. This phenomenon is referred to as *polygenic inheritance* and reflects the fact that a very large number of gene products go into determining one's height. Although the contribution of some of these genes, such as those coding for growth hormones and their receptors, may be more important than others, the final height of an individual will, in part, reflect the sum of the activity of many genes. This point can be made explicitly with the following bit of whimsy.

Let us imagine a highly fictitious (at least I think it's fictitious) example. The football-quarterback trait, which confers the ability to accurately throw an oblong ball to a receiver who is 50 yards away and moving fast, while dodging a whole bunch of very strong guys who want the ball for their very own, reflects the additive effects of genes encoding skeletal structure, musculature, hand–eye coordination, agility, speed, vision, reflexes, and intelligence. Further suppose that for each of these 500 or so genes there are athlete alleles, normal person alleles, and, worst of all, couch-potato alleles. One could then imagine that an individual exhibiting the quarterback trait would be homozygous for athlete alleles at most or even all of these genes. Conversely, couch-potatoes — like myself — well, you get the point. The rest of us, with some combination of various alleles, might well fall on a continuum between those two ends of the spectrum.

Okay, so given that a continuous trait might require the action of many genes, would you expect each gene to contribute to the phenotype equally or might some genes contribute more than others? Frankly, we don't know the answers to those questions yet. However, later in the chapter, some approaches to these questions will be discussed. Before we can begin to discuss such issues, however, we need to realize that for many such quantitative traits, the final phenotype may well reflect an interaction of genes with the environment.

## MANY TRAITS REFLECT THE INTERACTION
## OF THE GENOTYPE WITH THE ENVIRONMENT

It should be obvious that if we took identical twins and raised one under conditions where ample nutrition and health care were available, while the other suffered the ravages and diseases of famine, the two twins might well grow to different heights. This is not to say there isn't some strong genetic predisposition to grow to a given height, but rather that the developing human being is constrained both by its genetic instructions and by available resources.

To continue the analogy drawn earlier, even an individual possessing the right combination of alleles required for the quarterback trait will never make it to the Super Bowl or a Hanes commercial without proper nutrition, health care, training, and encouragement during high school and adolescence. However, it is highly unlikely that any amount of training, coaching, and nurturing could create the quarterback phenotype in an individual who lacked the requisite genetic makeup.

So how do we ever sort out the relative contributions of genetics and environment for any trait, much less for quantitative or polygenic traits? The answer is a trick referred to as calculating the *heritability* for a given trait. Measurements of heritability, which usually rely on twin studies, are used to determine how much of the variation for some trait in some population is genetic and how much is environmental. A discussion of the manner in which heritability is computed in twin studies, and the hazards intrinsic to using heritability estimates, is provided in Box 18.1.

## SOME TRAITS MAY REQUIRE A THRESHOLD
## NUMBER OF DELETERIOUS ALLELES

Imagine that there was a trait that was not truly continuous. *Cleft palate* is a good example. Although the severity of this trait varies between affected individuals, it is not a continuous trait within the population. Babies are either born with a cleft lip or palate or are normal. Yet cleft palate does seem to "run in families" and follows the rules for multifactorial inheritance that are listed in the next section.

We think of cleft palate and other similar disorders such as *anencephaly* and *spina bifida* as being *threshold traits.* These are traits in which affected

## BOX 18.1. HERITABILITY MADE SIMPLE

Heritability measures the proportion of phenotypic variance that results from additive genetic variance. Heritability (H) is defined in terms of two values, $g$ and $e$, such that $g$ represents the contribution of additive genetic factors with variance $G$, and $e$ represents the contribution of additive environmental factors with variance $E$. We can then approximate H by

$$H = G/(G + E).$$

The values $G$ and $E$ are frequently difficult to estimate in human beings. The best approach is to compare identical (monozygotic or MZ) and fraternal (dizygotic or DZ) twins (with both members of any given DZ twins being of the same sex, if possible). The assumption is that the variance observed in MZ twins reflects only environmental variance ($E$), whereas the variance in DZ twins reflects both genetic and environmental variance ($G + E$). Thus the genetic variance can be estimated by subtracting the variance observed in MZ twins from that observed in DZ twins: $(G + E)_{DZ} - (E)_{MZ} = G$. This allows us to calculate H as

$$H = \frac{\text{(variance in DZ twins)} - \text{(variance in MZ twins)}}{\text{(variance in DZ twins)}}.$$

### CRUCIAL POINTS CONCERNING ESTIMATES OF HERITABILITY

1. In those cases where MZ twins are always truly identical or concordant, the variance in MZ twins will approach or equal zero. In such cases H will approach or equal 1. This is the case for Down syndrome.

2. If the environment changes, then H can change as well without any change in the genetic structure of a population.

3. A measurement of H is useful only for a given population at a given time.

4. H values cannot be compared between populations. One can easily imagine, for example, two populations with the same genetic composition being exposed to very different environments. One can see how as the environmental influence on a trait may change, so will the estimate of heritability, without any change in the gene pool.

5. It is not straightforward to calculate H for discontinuous traits. In this case, it may be possible only to demonstrate a significant increase (via the $\chi^2$ test) in the concordance between MZ twins as compared to DZ twins or, less meaningfully, siblings.

6. The method of calculating heritability presented here in fact really only works well for calculating heritability in populations of twins. Estimates of heritability for large populations require similar, but more statistically complex, methodologies.

individuals are thought to carry more than a certain threshold number of "deleterious or advantageous" alleles that are required to create a phenotype, as shown in Fig. 18.2. Forming the top of the palate during embryogenesis requires that two masses of tissue in the head of the developing fetus arch up over the tongue and the mouth cavity and fuse properly to form the upper palate and the two sides of the upper lip. Those tissue movements surely require the products of many genes acting in concert. These movements also must occur in a very short window of fetal development. Clearly, some alleles might produce defective proteins that retard this process. As long as they don't retard it too much and the arch of the mouth is built before the window of time is closed, things will be fine. However, if more deleterious alleles are added to the genome of this embryo, progressively more impairment of movement is observed. Finally, as shown in Fig. 18.2, in true "the one straw or allele that broke the camel's back" form, one too many proteins is impaired, and the arch fails to be completed before the temporal window closes.

One of the more interesting consequences of proposing such thresholds is that it becomes easy to suggest that thresholds might be altered by environment, such that two identical genotypes might display different phenotypes in different environments.

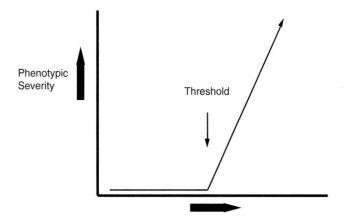

Number of Deleterious Alleles

FIGURE 18.2    A schematic view of the threshold model.

If all of this seems less solid or more confusing than simple Mendelian inheritance, it is. The understanding of polygenic traits and of genotype–environment interactions is becoming an increasingly interesting and profitable avenue of inquiry in human genetics. Multifactorial inheritance presumably underlies some of the more clinically important human traits, such as susceptibility to several major diseases or illnesses (heart disease, stroke, diabetes, schizophrenia; see later). Some data also support the notion that alcoholism may be an example of multifactorial inheritance. It is also possible, as noted earlier, that this type of inheritance might play a role in establishing some crucial aspects of personality as well. Indeed, several studies have argued that there may be strong genetic components to several aspects of personality. For these reasons, it is worth considering just how one would recognize multifactorial inheritance in a pedigree.

## THE RULES FOR
## MULTIFACTORIAL INHERITANCE

As is the case for autosomal recessive and autosomal dominant traits, there are some rules that let you determine that a given trait is best explained by multifactorial inheritance.

1. Although the trait obviously runs in families, there is no distinctive pattern of inheritance (autosomal dominant, autosomal recessive, or sex-linked) within a single family. In other words, when nothing else makes any sense, start thinking about multifactorial inheritance. Nonetheless, a few rules are helpful in identifying a trait whose expression reflects multifactorial inheritance. These are:
   - The risk to immediate family members of an affected individual is higher than for the general population.
   - The risk is much lower for second-degree relatives (aunts, uncles, grandchildren, etc.), but it declines less rapidly for more remote relatives. This latter point is a hallmark of multifactorial inheritance and distinguishes it from autosomal recessive inheritance.
2. The risk is higher when more than one family member is affected. Traits reflecting multifactorial inheritance are controlled by the number of "deleterious" or "advantageous" alleles segregating in the family in question. If there are several affected individuals within a family, the odds go up that a large number of "deleterious" or "advantageous" alleles are segregating within the family. Obviously, in such a family the risk to subsequent children is increased when the parents are consanguineous.
3. The more severe the expression of the trait, the greater the recurrence risk. Presumably, the severity of the phenotype is proportional to the

number of deleterious alleles distributed over a large number of genes carried by that individual. An individual with many such alleles will on average pass on half of those alleles to their children and share half of them with their siblings. Imagine that some trait, such as cleft lip or cleft palate, was governed by alleles at 400 genes and that "normal" and "bad" alleles exist at all of those genes. Assume that any 50 such "bad" alleles distributed among 400 genes are sufficient to produce a phenotype and that the severity of the phenotype gets worse as the number of "bad" alleles increases. Thus, an individual carrying 60 such "bad" alleles will be mildly to moderately affected. On average, he will pass only 30 "bad" alleles onto his children, which is usually too few to cause a phenotype. However, a more severely affected individual with, say, 150 "bad" alleles will usually produce children with 75 such alleles. Such offspring are very likely to be affected.

4. Identical (or monozygotic) twins need not always be concordant with respect to the trait. Indeed, the concordance can fall anywhere between 100% and the concordance observed between siblings. This reflects the influence of environment on the expression of the trait, especially in cases where the number of alleles is just over the threshold at which characteristics of the trait can occur.

Obviously, most of these rules work best with traits that are to some degree discontinuous. However, the rules can be applied, with some degree of difficulty, to fully continuous traits such as height. In the following pages, we will turn our attention to some of the more controversial examples of traits that are presumed to reflect multifactorial inheritance.

*(A word of warning: Just because a trait exhibits some of the characteristics listed above does not prove that it is due to multifactorial inheritance. One could easily imagine that being a successful actor or getting into medical school might also "run in families." In matters such as these, caution in proposing any kind of genetic basis for a trait is not just a good idea, it's essential.—RSH and JER)*

## SOME EXAMPLES OF COMPLEX HUMAN TRAITS THAT MAY (OR MAY NOT) BE EXAMPLES OF MULTIFACTORIAL INHERITANCE

We have previously considered the possible genetic contributions to sexual orientation back in Chapter 8. Recall that by carefully selecting a study population designed to reduce the contribution of both other environmental variables and other genes, Dean Hamer and colleagues were able to provide evidence for a gene on the long arm of the X chromosome that may predispose some males to become homosexual. However, also recall that even if Hamer's data are taken at face value, this gene can only explain a fraction of homosexual men. Hamer and colleagues stress that other genes, and certainly strong environmental factors, contribute to the process by which male sexual orientation is established.

We will see a similar set of arguments in the next chapter when we consider the genetic contribution, if any, to violent aggression. However, several other notable examples have presented themselves in the scientific literature and in the news media. At least two of them, mental illness and the genetics of IQ, deserve mention here.

### SCHIZOPHRENIA, MULTIFACTORIAL INHERITANCE, AND QUANTITATIVE TRAIT LOCI

Schizophrenia is a true peril of our species! Affecting up to 1% of the population, it constitutes a major human health problem. The concordance rate in monozygotic twins has been estimated to be from 45 to 60%, compared to 10–20% for dizygotic twins or siblings. If one parent is affected, the risk to the child is again 10–20%; however, if both parents are affected, the risk to offspring increases to 40–60%.

It is clear to clinicians that there is more than one disorder called schizophrenia. Unfortunately, differentiation of subtypes vexes even the best clinicians. The subtypes apparently reflect at least two, and quite likely more, genetic anomalies. It is clear that schizophrenia in one kinship reflects the action of a major locus on chromosome 5. Other kinships, in which schizophrenia behaves as an autosomal dominant trait, do not suggest linkage to chromosome 5.

A similar conundrum exists for *bipolar affective disorder* (*manic depressive psychosis*). Again, this disorder affects about 1% of the population and the best explanation is multifactorial inheritance. The concordance rate for monozygotic twins is approximately 70–80%, whereas the concordance rate among dizygotic twins is only 15%. In some of those large kinships that have been studied, linkage has been found to markers at a site on chromosome 18 in some families and to markers on other chromosomes in others. However, in many other families there is no evidence for a major gene in chromosome 18. Things are clearly quite complicated. One problem here is diagnosis. For example, people with a genetic predisposition to bipolar affective disorder might exhibit this trait, but they might also be alcoholic, depressed, or normal. Since alcoholism and depression are common, and since these disorders may arise for other reasons, it is sometimes hard to say just who is "affected" and who is not in a given family. Regrettably, there are no simple laboratory tests for these disorders. Imagine trying to do linkage analysis when folks can't even agree on which family members are or are not affected!

## THE GENETICS OF IQ AND THE POLITICS OF RACE

Every once in a while, at a social gathering, such as a cocktail party, someone that Scott does not know or, more importantly, someone who does not know him will discover what he does for a living. When he tells them that he is a geneticist, they will lean over to him and, in a whisper, ask, "Do you think that intelligence is hereditary?" We worry a good deal about the people who ask this question, and there have been far too many of them. First of all, why do they whisper? They wouldn't whisper if they were asking about heart disease, height, obesity, or color blindness. Scott always tells them, "I believe that intelligence is inherited to some degree." He reminds them that human beings can't fly and don't swim very well by the standards of other more aquatic mammals such as seals or dolphins. Even the best of us cannot outrun a gazelle or outclimb a bear. Our hides are thin, we have lousy claws, and the acuity of our senses of sight and smell don't match those of many other animals, especially at night. We are the dominant critter on *terra firma* because, and only because, of our intelligence. We developed language and a method to pass what we learn and know onto both our young and our whole species. We create and build tools. We work, hunt, and kill well in groups. All of this requires the hard-to-define quality, or large set of qualities, called intelligence.

The current best guess is that more than half of our 50,000 to 100,000 genes are primarily expressed in our brains. It would be foolish to imagine that allelic variations in those genes don't exist or that such variation is without phenotypic consequences (i.e., that this variation doesn't explain some of the differences we see among us for various aspects of that elusive thing we so crudely call intelligence). Moreover, studies in lower organisms, such as fruit flies and even mice, are quickly identifying genes that regulate such processes as learning, memory, aggressive behavior, and various other aspects of cognitive functioning. Mutants in these genes are expected to have predictable affects on these processes.

Even in human beings, we have already described several genetic conditions, such as Down syndrome and the fragile X syndrome, that are associated with mental retardation. We have also noted the behavioral problems or anomalies sometimes associated with Klinefelter syndrome. Clearly, there are genes whose proteins are required for the proper functioning of our nervous systems. Those genes, like all genes, are expected to be mutable, and there is every reason to believe that in people, as in animals, these allelic differences will have phenotypic effects.

However, when Scott talks about such issues, he is also quick to note that whatever genetic differences do exist (and they may be strong or subtle), they are very likely well spread throughout the human population such that no group is likely to have a monopoly on such differences. As noted later for IQ, the variation that can be observed for various aspects of intelligence and functioning within the human population appears to be observed in all subgroups of that population. He also tells these people that intelligence is a case where the final phenotype reflects an interaction of many genes with the environment in a way so complex that we have yet to develop any greater understanding of the genetics of intelligence than we have of the meaning of the term "intelligence" itself.

Still, the people with the whispered questions about intelligence don't like that answer; they don't like it one bit. They are not satisfied with this answer because it doesn't respond to the question they really wanted to ask. They wanted to talk about reports in the news arguing that the heritability of IQ scores is high (and for some populations it is) and reports noting that there are differences between average IQ scores for the various racial groups in the American population. They are making an argument that if IQ is heritable, and if IQ values differ between two populations, then the difference must be genetic. They then conclude that people of one race or another are biologically smarter than other people. One can imagine the horrid social consequences of accepting such a view.

Speaking solely on scientific grounds, there are at least two major things wrong with this argument. First, recall that heritability cannot be compared between two populations. This is because the proportion of the variance in IQ scores that is due to environmental factors in the two populations may be very different, even if the genetic structure of the populations is similar or identical. This central tenet of genetics tells you that you cannot use heritability to compare the genetic structure of two populations. Nothing in available data argues for a genetic difference with respect to IQ between human populations. That differences in average IQ scores between races really exist or are biologically meaningful even if they do is arguable. Even if such differences do exist, there is NO reason to believe that they are reflective of differences in genotype between races. This caveat is especially important given many studies that suggest a strong environmental component to performance on IQ tests. Moreover, at least some of the early studies of heritability and IQ may have been examples of scientific fraud.

Second, although IQ scores may well measure some aspects of intelligence, exactly which aspects of intelligence they measure is an area still open to dispute and investigation. More importantly, there are also types of intelligence, such as creativity, intuition, or some types of abstract reasoning, that are not addressed by IQ tests. Clearly, IQ tests are around and may be useful in identifying and helping people whose scores fall toward the low end of the curve. But exactly what the tests measure in individuals toward the middle and upper ends of the curve is unclear. Moreover, the variation for IQ within any population group assayed is quite wide and various populations overlap significantly, despite differences in the mean.

Before we can meaningfully discuss the relationship of IQ to genotype and of genotype to intelligence, we need a better understanding of just what an IQ test measures and a better set of definitions of intelligence. We will also need to free this issue from the trappings of racism that have shrouded it for the last two decades. As stated earlier, we have no doubt that there may be genetic differences relating to various cognitive processes; we just don't think that such a hypothesis is addressed by the studies described earlier.

The next chapter addresses an equally vexing issue, that of the genetics of violent aggression in humans. Many of the issues considered here should stay in your minds as we continue.

# 19

## THE MONOAMINE OXIDASE A GENE AND A GENETIC BASIS FOR CRIMINALITY?

*In 1997 in California a man was tried and convicted for the grisly murder of a young girl. During the trial, this individual persisted in inflicting further pain on the victim's family. He also displayed both a total lack of remorse and an appalling lack of respect for our society and its laws. During the coverage of the trial, a local television station aired a report in which someone who had known this man for quite some time claimed that "he was just wrong from the beginning, just a bad seed." Odd words those, "bad seed" and "wrong from the beginning," words suggestive of an inborn lack of whatever keeps most of us from committing such atrocious acts. I listened to those words and nodded quietly in assent. Later, I started to think about what I meant by that judgment. Did I really believe that this man was born that way? Was he, to use the parlance of the tabloids, "a natural born killer?" Did I believe that he was missing gene products essential for building the parts of the human psyche that proscribe most of us from such behavior, genes that make empathy, caring, and guilt possible? Did I believe that such genes even existed? Did I believe that there truly was such a thing as a genetically predestined "bad boy" or "bad girl" from birth, and thus that mutations in certain genes might play a role in criminality? And,* much more importantly, *IF I did believe those things, did I believe them because there was solid genetic evidence to support such ideas or did I believe them because that belief or prejudice is ingrained in our culture? What follows in this chapter is an attempt, however unsatisfying, to sort out some of the answers to these questions.—RSH*

The idea that criminal behavior might have some of its origin in genetics received its first real support from studies in the late 1960s regarding XYY males. As should be realized by now, such males arise from nondisjunction at meiosis II in their father and should be hyperploid for only the very small number of genes carried by a Y chromosome. One really doesn't expect such males to be greatly different from XY males.

However, the surprising finding, reported by Dr. Patricia Jacobs in 1968, was that XYY males made up more than 3–5% of some prison populations as compared with 0.1% of the general population. More strikingly, if attention is restricted to male prisoners over 6 feet tall, the frequency of XYY males has been estimated to be as high as 10–20%. You can imagine how this finding was interpreted. People imagined that males were "naturally more aggressive" anyway and that adding that extra Y chromosome simply pushed things over the edge, resulting in hyperaggressive violent "super males."

As you can imagine, this observation and such misbegotten ideas stirred a significant amount of social interest and controversy. *(I recall a professor in college telling me that this observation was so "politically unacceptable that it simply couldn't be true."—RSH)* This observation was used as a defense in several murder trials, although, to the best of our knowledge, never successfully. A major newspaper once also incorrectly identified a notorious murderer in the midwest as an XYY male, creating a misconception that is accepted even now among some individuals. Several cities created prospective "treatment" programs to identify such males at birth and to "intervene" psychologically early during their development.

The problem with connecting the XYY karyotype with criminal behavior is that people simply didn't really look at the data. It is true that XYY males are taller than most, usually over 6 feet tall. It is also true that they are overrepresented in some penal populations, but it isn't true that all XYY males end up as criminals. Indeed, greater than 90–95% of XYY males lead normal lives without serious encounters with law enforcement. True, 5–10% of these guys do end up in prison, but it surely is not a general part of the phenotype. More interestingly, these men are not always, or even usually, incarcerated for violent offenses. Rather they are more often jailed for repetitive violations of probation agreements, possession of stolen property, writing bad checks, etc. The simple conclusion is that having an extra Y chromosome *does not* make a man more violent. It may predispose some men to get into trouble with the law, but it does not make them more violent.

Once these facts were widely known, the idea of an "XYY syndrome" linked to criminality became a bit of a scientific pariah. Indeed, in a rather unscientific survey of five basic college genetics textbooks lying around my office, I could find no mention of overrepresentation of XYY males among prison populations. This lack of coverage is quite common despite the fact that the basic observation is highly repeatable. There are two possible reasons for this rather conspicuous silence.

The first reason for not discussing the overrepresentation of XYY males in penal settings may be that we don't understand *why* any XYY males are abnormal. I've heard colleagues wave their hands when a savvy student will ask a question along these lines and reply, "Well, these men tend to be taller and maybe. . . ." Maybe what? Other workers have suggested that the large body size of XYY males may lead authority figures, such as teachers, employers, or parents, to expect more of these people and that some XYY males may deal with this stressor by becoming aggressive. (*Speaking as someone who is 6'3", I'm a bit sensitive to this "tall is bad" idea, but to the extent that being very tall early in life may be intimidating to others, these people may have a point.—RSH*) I am more intrigued by the suggestions made by some workers that a small fraction of XYY males may exhibit slightly lower IQs or learning difficulties. One could imagine that it is the frustration caused by such deficiencies that induces the difficult behavior later in life. Perhaps these problems underlie the behavioral problems observed in a small fraction of XYY males.

Why should there be any problem at all? One possible explanation arises from the observation that XYY males appear to produce sperm bearing only a single X or Y chromosome. They do not appear to produce XY- or YY-bearing sperm at detectable frequencies. They also show normal fertility. Moreover, some cytological analyses of these males suggest that the majority of their meiotic cells carry a single Y chromosome, perhaps indicating a selection in the male germline for cells that have lost one of the two Y chromosomes. If this is true, then the gametes produced by these males may arise primarily from the occasional XY cells produced in the male germline. Perhaps such males suffer from some testicular or hormonal abnormalities arising due to the loss or death of XYY spermatogonia and their replacement by XY germ cells or perhaps the presence of some Y chromosome gene in two doses exerts a phenotypic effect too subtle for us to understand.

The second reason for the collective silence of the textbooks on this matter may be that the whole issue became "politically incorrect." One simply didn't want to imagine that something as complex as criminality might be genetic. Indeed, the debunking of the violent XYY supermale myth seemed to add steam to the arguments of those who do not want the words "genetics" and "behavior" used in the same sentence. Nonetheless, as described in the next section of this chapter, a far stronger case has now been made. In this case, violently aggressive behavior has been shown to result from a specific mutation in a known gene.

## GENETICS OF VIOLENT AGGRESSION IN A FAMILY IN DENMARK

Figure 19.1 displays a pedigree for a family in the Netherlands studied by Brunner and colleagues. Indicated males in this family were often subject to seemingly unprovoked and uncontrolled violent outbursts. These aggres-

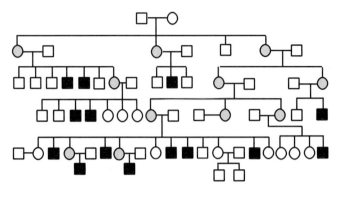

■ denotes a male exhibiting borderline mental
retardation and violent aggression

○ denotes a carrier female

FIGURE 19.1    Brunner's pedigree.

sive episodes ran the gamut from ranting and shouting, to exhibitionism, plus serious crimes such as rape, arson, and assault with deadly weapons. (One of these men forced his sisters to undress at knife point. Another man raped his sister and then later, while incarcerated, attacked the warden with a pitchfork. A third member of this family attempted to run over his employer with a forklift.) All of the affected males were mildly retarded with an average IQ of only 85 (100 is normal). Although there were no affected women, sisters of affected males frequently gave birth to affected sons.

All of this evidence pointed strongly not only to a genetic basis for this trait, but indeed to a sex-linked mutation underlying this behavior pattern. To test this possibility, Brunner and collaborators set out to determine if they could map the responsible mutation. As you might imagine, they identified a number of sequences on the X chromosome for which RFLP polymorphisms were common and followed the segregation of those polymorphisms through the pedigree. To their delight, one such polymorphism, which mapped within a known gene, showed very tight linkage to the violent aggression trait. This gene, called MAOA, encodes a protein known as monoamine oxidase A that is required to break down molecules known as neurotransmitters in the brain.

Neurotransmitters are small molecules that facilitate communication between cells, known as neurons, that comprise your neurological system. As shown in Fig. 19.2, although signals within a neuron are carried electrically from one end of the cell to the other, a given neuron communicates with the next neuron in the sensory or motor pathway by releasing neuro-

neurotransmitter molecule

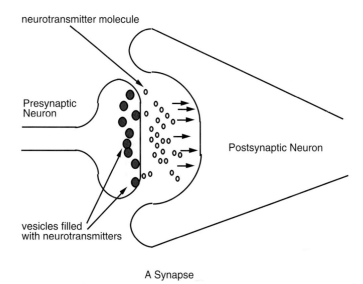

Presynaptic
Neuron

Postsynaptic Neuron

vesicles filled
with neurotransmitters

A Synapse

FIGURE 19.2    Neurons and neurotransmitters in synapses.

transmitters into the small space between these cells, known as a synapse. The second neuron absorbs the neurotransmitters, triggering it to fire an electrical signal along its length, and so on.

Obviously, the presence of neurotransmitters at the synapse needs to be very tightly controlled if the nerve cells are to function properly. They must be released rapidly by the stimulating cells and absorbed and/or degraded by the responding neuron. One of the enzymes required to break down neurotransmitters is MAOA. A variety of biochemical studies on urine samples taken from these males indicated markedly abnormal metabolism of neurotransmitters that are normally broken down by MAOA, including dopamine, epinephrine, and serotonin. Very high levels of these compounds were found in the urine of these males, consistent with an inability of these males to break down these compounds.

As diagramed in Fig. 19.3, in a subsequent paper, Brunner and colleagues demonstrated that affected males in this kindred carried a point mutant in the eighth exon of the MAOA gene that changes a glutamine codon to a stop or termination codon. This nonsense mutation was not found in the MAOA genes carried by unaffected brothers of affected males; obligate female carriers were also found to carry one normal and one mutant allele, although they were phenotypically normal. Thus, there was a precise correlation between whether males carried this nonsense mutation and whether they expressed violent aggressive behavior.

How is the defect in MAOA correlated with the violent outbursts exhibited by these men? Realize that among their many roles, neurotransmitters

MAOA GENE

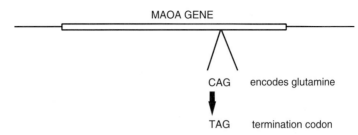

CAG    encodes glutamine

TAG    termination codon

FIGURE 19.3    A point mutant in the MAOA1 gene.

function as part of the body's "fight or flight" response to threats or danger. In most of us, as the levels of these transmitters increase in our brain in response to various stresses, they in turn are broken down by MAOA. Thus most minor stimuli produce only transient increases in neurotransmitter levels. One can then imagine that the degradation is greatly impaired in males lacking the MAOA enzyme, thus allowing levels of transmitters to rise far in excess of an appropriate level. Indeed, in several cases, crimes committed by these males closely followed traumatic family events, such as the loss of a loved one. It is then possible that even small stress might overstress these males to the point where violent outbursts become almost beyond their control.

Most workers now seem to accept Brunner's conclusion that this mutation largely explains the violent phenotype seen in this family. Nonetheless, a number of scientists have raised the issue of just what other environmental influences might be at play within this family or community and whether other genetic effects might also be at work (i.e., are there other mutations segregating in this family that influence the effect of the MAOA mutation?). The underlying basis of this question is our curiosity regarding just how general this effect may be? How many other people have the same, or a similar, mutation in the MAOA gene? Are those people prone to the same behavior? Are they in prison?

These are serious questions and the answers are difficult to obtain. You might be wondering why we don't just start sequencing the MAOA gene from every serious criminal who will stand still long enough to let us draw blood. There are several reasons why such studies are both difficult ethically and incomplete scientifically. First, there is a serious problem of "informed consent" when doing this sort of research with inmates. Can people in prison truly give free informed consent (imagine what type of pressures they might feel to agree even if they don't want to)? Second, suppose we did find inmates bearing such a mutation. Would that constitute a legal defense? Might the governor be more likely to pardon a condemned man if he thought the crimes were driven by his genes? Third, as with XYY males, would finding a higher fraction of inmates with such a mutation than expected based on the frequency in the general population really prove "cause and effect."

Not surprisingly, the response to Brunner's work is one of cautious interest. Researchers are trying to obtain animal (mouse) strains that lack this mutation, are looking for other families like the one studied by Brunner, are looking carefully at the multiple side effects of many drugs that impair neurotransmitter breakdown, and are very carefully beginning the necessary population studies.

Still, the question in most of our minds is: Just what do we do if all, or even some, of this pans out? Would we, as a society, screen male babies or fetuses for this mutation? If we did such screening, what action would we take? Would bearing this mutation be a cause for termination of a pregnancy? One text reports that at least one news commentator argued for sterilization of affected males and carrier females. If your first response to these questions is like mine, you are wincing and saying, "Oh no, we wouldn't do anything like that." However, remember that screening programs were created in several cities for XYY males. Worse yet, people tried to devise monitoring or treatment programs to diffuse the aggressive tendencies of these males. *(Talk about a great way to create a self-fulfilling prophesy!—CAM)* Moreover, I once asked 20 students in my senior seminar class if they would choose to terminate a pregnancy if they knew that the male fetus carried the Brunner MAOA mutation. The answer was an overwhelming "yes." Stop and think what your answer might be, then ask yourself why.

It would be irresponsible of us to close this chapter by leaving you with the wrong impression that the defect observed by Brunner is likely to account for any more than a small fraction of criminal behavior. So the question arises: How many other genetic influences might there be? Unfortunately, we cannot answer this question. Even more unfortunately, the issue has led to some rather careless speculation. I once listened in horror while a professor told a class in medical genetics that he thought that virtually all criminal behavior was genetic, that the likelihood of dying in a hail of bullets over a bad drug deal was as genetically influenced as other human traits such as blood clotting and color vision. There is currently no hard data to support such an assertion. Moreover, as was said in the discussion of the genetics of IQ, or lack of genetics thereof, in the last chapter, at least some of these discussions seem to be more about social politics than science.

*(Still, this is an issue that is not going to go away. To lay our prejudices on the line, we suspect that while most criminal or violent behavior has its roots in environmental causes such as child abuse, hunger, drug addiction, and seemingly hopeless poverty, there will be more cases like the MAOA mutation. We will find people, like the man alluded to in the beginning of this*

*chapter, who truly are missing something, something necessary, something genetic. Our society is going to have to find ways to cope with the crime and punishment of such individuals. Their mere existence and their crimes challenge the basic and cherished concepts of free will and individual responsibility. How could we punish someone if they cannot prevent themselves from committing a crime? Perhaps in those cases we will refocus our interests as a society from punishing and incarcerating to treatment and medical control. Perhaps. . . .—RSH)*

# 20

# GENETICS OF THE HUMAN
# AIDS VIRUS

*"I want you to get a blood test," Joan said.*

*"A what?" Vince replied.*

*"A blood test. We should be able to trust one another fully, and knowing that both of us are HIV negative is the first step in the right direction."*

*"HIV? You have got to be kidding! You don't really think I have AIDS do you? Look at me . . . I look and feel great! Besides, drug users and homosexuals get AIDS, not people like us.*

*"That is not true. HIV is becoming very common in people just like us! We've been using condoms for 8 months now and I want to go on the pill. Besides, we have talked about marriage too. Doesn't that mean monogamy from now on? So let's just be sure it is absolutely safe for both our sakes, okay?" Joan asked.*

*"Usually when marriage is the topic, the first step with most women is a big rock that shines on the finger and you only want a blood test? I guess I should be happy. Does this mean the ring is out?" Vince joked.*

*Joan slugged him playfully and Vince smiled. He would do it, for her. That morning, while driving to his apartment, Vince composed a mental chronology of his past sexual history: "Let's see . . . I am 29. I started having sex when I was 16 years old. All of the women were clean . . . no IV drug users. I have never done IV drugs or had any homosexual contact . . . so what's the worry? I am athletic, strong, and never get sick. Other people get AIDS; I will be fine."*

*By the time Vince had arrived at home, he had fully convinced himself that he was not an HIV carrier. Four days later, Vince and Joan were at the local community clinic getting their blood tests. They had to wait two weeks for the results. During the two weeks, the thought of being HIV positive went through Joan's mind several times, what she would do, how she would handle it, etc. Vince thought of it only when Joan mentioned it.*

*"What will we do if one of us has it?" Joan asked.*

*"That's not an issue, Joan. Neither of us have it. So don't fret. We'll just go in and get the negative results and get on with our life together."*

*"If I have it, I want you to know that I don't expect you to stay as my boyfriend or future husband. I don't want to expose you to it. I want you to stay as my friend, but . . . ." Vince hushed Joan with a kiss.*

*"It's not an issue, we are going to be fine." Vince smiled.*

*After the waiting period, Joan and Vince returned to the clinic. "So, what do we do?" Vince asked.*

*"Well, we go in and the counselor will hand you a piece of paper with your results. Then, if you're negative, you leave."*

*"What, if you're positive, they make you stay," Vince joked.*

*"No, but people who test positive usually want to stay and talk to the counselors about their options and what they should do next," Joan answered.*

*Joan and Vince were called in. It was four and a half hours before they left the clinic.*

This chapter is going to take a bit of a detour from talking about human genetics. Instead we are going to turn our attention to a virus called HIV that infects human beings and, in doing so, gradually destroys their immune system. The result is a disease called acquired immunodeficiency syndrome (AIDS). We decided to devoted an entire chapter to this disease for two reasons. First, discussing AIDS and HIV lets us tell you about viruses and their genomes in general. In that sense, it is hoped that it will broaden your understanding of genes and proteins. Second, AIDS is the plague of the current era, the one most likely to have an impact on your life in one way or another.

## ACQUIRED IMMUNODEFICIENCY SYNDROME

In 1982, the Center for Disease Control (CDC) first used the acronym for the acquired immune deficiency syndrome, best known as AIDS, to describe what seemed to be a new disease that needed to be monitored. What was known? They knew that it was lethal and that there was no cure for this virus. No one was quite sure how quickly it would spread if an individual carries HIV in his/her blood; he/she may not have AIDS and, as previously stated, may show no signs of sickness. However, an HIV carrier will eventually develop AIDS and become increasingly ill.

Scott once described AIDS as a "corrupt police force." If the city you live in has a police force composed of criminals, who will protect you? Your immune system is your built-in police force. When your body encounters infections (criminals), your body creates protective antibodies (police officers) to attack and destroy the infection. The HIV virus is like a criminal that possesses the strategy to eliminate your immune system (police force) and leave you almost defenseless.

In a general sense, AIDS is the result of the ability of the HIV virus to attack and destroy a specific type of white blood cell, known as a T lymphocyte for short. Because T cells play a critical role in mediating our bodies' immune response, their gradual ablation in infected people results in a rapid deterioration of their immune response. The ever-weakening immune system makes them easy targets for infection by many of the bacteria and viruses that live around and within us all. Indeed, it is these opportunistic infections, often causing pneumonia, infections of the brain, or similar disorders, that usually result in death.

In the intervening years, we have developed better drugs to fight the opportunistic infections and, better yet, have also developed several sets of drugs that actually act to slow or prevent viral replication. These findings have allowed the development of effective AIDS treatment protocols that seem likely to greatly extend the life span of many, if not most, infected individuals. We have also found a very small number of people who survive HIV infection and who appear to be resistant to HIV infection. *(At least some of the latter group of people are resistant because of a mutation in a human gene that encodes a component of the attachment or "docking" protein complex that the virus uses to attach to the cell. People lacking or carrying structural variants of these proteins appear to be immune to infection. Again, not all mutations are deleterious.—RSH)* The discovery of such people may provide important clues for both preventing the disease upon exposure (imagine a drug that blocked the attachment site) as well as assisting in the development of effective vaccines.

## THE HISTORY OF AIDS

HIV was a previously unknown virus until 1984. It was isolated in human blood and found to selectively attack a specific group of white blood cells that were crucial to the body's immune response. HIV has now been recognized as the direct cause of acquired immunodeficiency syndrome or what we call AIDS. It is assumed that HTLV-III, HIV, was first introduced into the United States in the 1970s but not recognized until 1981. No one is quite sure just how new the virus is or exactly where it came from. There are speculative theories on how it originated and some interesting rumors as well.

It has been speculated that the disease originated in Africa at least 10 years before it arrived in the United States. It is not clear whether this virus originated in humans or animals. If the virus existed for some time, there may have been some type of genetic alteration that made it more pathogenic, but no data supports this theory.

It is also theorized that the AIDS virus is a human variant of a virus that is also found in monkeys. The virus found in the macaque monkey, which closely resembles AIDS in both structure and growth characteristics, is called simian immunodeficiency virus (SIV). When SIV is present in the blood of healthy wild African green monkeys, the virus is not disease producing. However, when the disease is passed onto another species, it is pathogenic. So what is the problem with the idea that HIV recently originated from SIV, or another similar monkey virus? Well, there is at least some evidence to suggest that HIV has been around for a very long time. Certainly, the fact that its discovery in this country was recent does not mean that it is a new virus. So at this point we just don't know if there is any truth to the idea that HIV arose from a related monkey virus in recent history.

If HIV is not a new virus and has not jumped recently into the human species via an animal reservoir and if it is not a mutation of an already pathogenic virus, how did it suddenly appear in the human species so rapidly? Perhaps it didn't appear so suddenly. Perhaps it has existed for thousands of years within a specific and extremely isolated population in Africa. The isolated population may have adapted to the virus and been able to cope with the severity of the infection. As time passed, the demography in Africa changed, and people have left tribal life and headed for the larger populated cities, thus perhaps carrying the virus with them.

## VIRUSES

Think back to the days of when you were in grammar school. I remember being carefree, playing with my friends, learning to read, watching cartoons, and catching every flu, cold, and virus that was brought into my school. I also remember going to the doctor's office and pleading with my mother not to take me there because they were always giving me shots to, as my mother said, ". . . make [me] all better." I recall breathing a sigh of relief when the doctor told me I had a virus because that meant "lots of rest and no shot in the bottom." Now I understand why I never got a shot when I had a virus, as the shots of our childhood were usually antibiotics, effective against bacteria but not against viruses. There are a few medicines available that can be used to fight specific viral infections, such as herpes and AIDS (see later), but these are few and far between.

A *virus* is a noncellular organism that can only reproduce within a host cell. A virus contains genetic material, either DNA or RNA, encapsulated in a protein coat. The genetic material may be double or single stranded, depending on the virus. One of the key things about viruses is that they are very efficient, stripped down organisms that carry along as little genetic "baggage" as possible. Cellular organisms can be infected with a virus, with the cell serving as a kind of life-support system or incubator that provides resources the virus needs but does not have on its own. Viruses come in various shapes, sizes, and compositions. Viruses have specific mechanisms that allow them to infect host cells and direct the machinery of the cell to produce progeny viruses.

AIDS is classified in a unique subgroup of viruses called *retroviruses,* which are found not only in humans but animals as well. Retroviruses do not always cause disease in their hosts; in some cases the host may lack the machinery and energy-generating capabilities to manufacture progeny viruses. Like all viruses, retroviruses exploit the cells of living organisms to perform growth and division for them. The retrovirus infects a cell by attaching itself to and penetrating a cell. Once the virus has entered the cell, the host's genetic machinery reads the encoded material of the virus's genetic blueprint and reproduces it. A cell may die during virus replication, but before it does, the viral chromosome(s) has already been released into the host's system.

Before the AIDS virus was discovered, only two retroviruses had been found in human beings: the T-cell leukemia viruses HTLV-I and HTLV-II. These two retroviruses belong to the *oncovirus* subfamily. An oncovirus produces tumors in the host and, like the AIDS virus, attacks the T-4 lymphocytes (specialized white blood cells that respond to foreign molecules). However, the AIDS virus is not oncogenic because it does not directly cause tumors in the infected individual. Tumors may arise as a result of the underlying immune deficiency, but are not the direct result of the virus itself.

HIV is also related to a group of viruses called *lentiviruses* that cause direct neurological effects. Not surprisingly then, as AIDS became more prevalent physicians found that patients showed signs of neurological malfunctioning such as dementia, loss of motor coordination, and paralysis. This proved that the virus attacked brain as well as white blood cells. Although these facts suggest that the HIV virus might be a human form of a lentivirus, researchers are not yet convinced. There remain theories that the HIV virus is another subfamily of the retroviruses that has not yet been discovered.

## MOLECULAR ANATOMY OF A RETROVIRUS

As shown in Fig. 20.1, the free AIDS retrovirus consists of an RNA genome. *(No, that is not a typographical error. The retroviral genetic material here is really RNA, as in ribonucleic acid, not DNA!—CAM)* The RNA is wrapped within a coat of viral proteins, including a protein known as *reverse transcriptase.* This RNA–protein complex is surrounded by a membranous

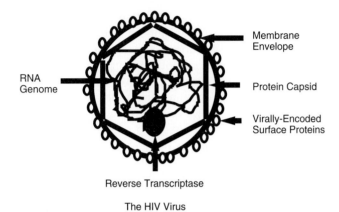

Reverse Transcriptase

The HIV Virus

FIGURE 20.1    Schematic diagram of an AIDS virus.

envelope carrying virally encoded proteins that give the virus its tissue specificity. Once the virus attaches to the cell, it enters the cytoplasm and the RNA–protein complex is released. The reverse transcriptase, which was encoded by the virus in the previous infection and included in the viral particle, allows the single-stranded viral RNA to be copied into double-stranded DNA. The mechanism of this reverse transcription is really complicated and will not be dwelled on here. *(Despite the wishes of one of our erstwhile editors, who in a previous incarnation was herself a retrovirologist.—RSH)* Suffice it to say, as drawn in Fig. 20.2, the reverse transcriptase converts the RNA molecule into a double-stranded RNA : DNA duplex. The RNA is removed by an enzyme called RNase H. The resulting single-stranded DNA molecule is then replicated into a double-stranded DNA molecule. A virally encoded protein, also carried within the mature virus and called *integrase,* is then capable of integrating that DNA into the chromosomes of the cell.

The HIV virus itself possesses three major open reading frames, or genes, known as *gag, env,* and *pol.* Once integrated into the chromosome, all three of these transcription units are transcribed into a single messenger RNA. Only the first one-third of this message, the part corresponding to the *gag* gene, is usually translated due to the presence of a stop codon at the end of this open reading frame. As shown in Fig. 20.3, the resulting Gag protein is then clipped into five smaller proteins, including four structural proteins and a protease (something that cleaves proteins). However, as shown in Fig. 20.4, in some of the translation events the ribosome can *slip* a single base just prior to that stop codon.

By slipping, or frameshifting, and bypassing the next stop codon the ribosome can encode a much larger polyprotein, thus encoding the *pol* and

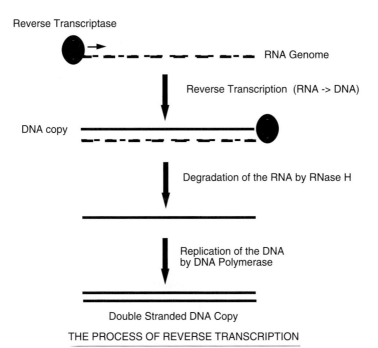

THE PROCESS OF REVERSE TRANSCRIPTION

FIGURE 20.2     Reverse transcription.

*env* genes. The necessary slippage events appear to occur in some 10% of translational events. As shown in Fig. 20.3, this much larger polyprotein is clipped into separate proteins by the virally encoded proteases. In the case of the *pol* transcription unit, these proteins include the reverse transcriptase, an RNA-processing enzyme called RNase H, and the viral integrase. The *env* gene encodes glycoproteins that will make up part of the envelope that surrounds the virus when it buds out of the cell. Once all of these proteins are produced, a structure is formed that consists of mature viral mRNA molecules complexed with structural proteins, integrase, and proteases. This RNA–protein complex is then wrapped in an envelope and extruded from the cell. The release of large numbers of such particles results ultimately in the death of the cell.

Think about the variations on the standard eukaryotic central dogma here. First, a genome made out of RNA. Second, messages that encode polyproteins that are then cleaved into individual gene products. Third, frameshifting by the ribosomes during translation as an obligate part of the viral life cycle. These rather deviant adaptations provide the molecular basis for some of the current treatments for AIDS. Some drugs that are used block reverse transcription, and the most promising new class of AIDS drugs called *protease inhibitors* prevent the cleavage of polyproteins.

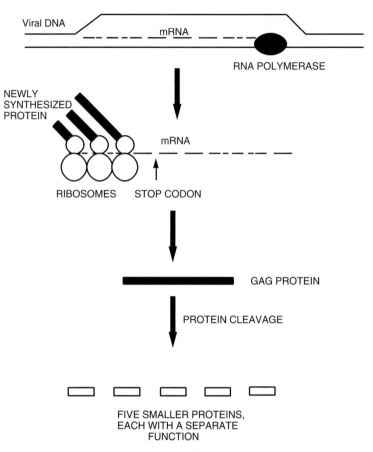

FIGURE 20.3    Polyprotein cleavage.

The HIV virus is surrounded by a protein coat that allows the virus to attach to cells such as lymphocytes and brain cells. Once attached to the cells, the virus is able to penetrate and infect. The protein coat that is exposed to the immune system is what stimulates antibodies to be produced. The virus genes that are responsible for the virus's protein coat are highly variable which causes the antigenic properties of the virus to vary as well. What this basically does is allow the virus to transform itself to the point where the original antibodies that were formed to protect the host against the infection are no longer effective. Thus, there are antibodies present, but they are unable to protect the infected individual against further infection or remove the virus from the bloodstream. The HIV virus is highly variable and adaptable; these two factors alone are what make AIDS such a mysterious and lethal disease. They also make creating a vaccine difficult, indeed.

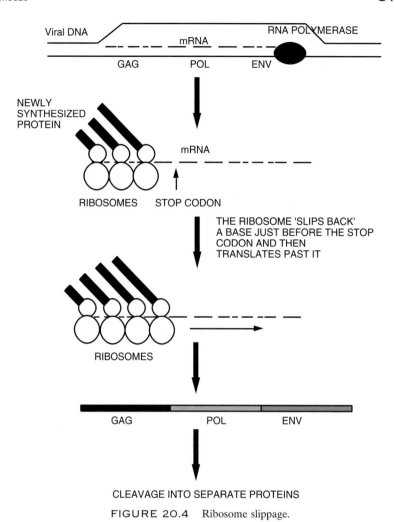

FIGURE 20.4   Ribosome slippage.

## HOW IS AIDS TRANSMITTED?

The AIDS virus is passed between people through infected blood and body fluids or by pregnant women who pass it on to the fetus. The major modes of transmission are (1) blood-to-blood transmission, such as by intravenous drug addicts sharing their needles; (2) unprotected sex with an infected partner; or (3) transmission from mother to fetus or infant. Blood transfusion used to be another source of transmission. However, because of excellent tests for the HIV virus in human blood, infections by transfusions are considered rare these days. The sharing of needles by IV drug abusers continues, apparently unabated, and represents the source of many new infections. The other major culprit is unprotected sex, i.e., sex without a condom.

Not all sexual activities are equally risky. Most researchers believe that unprotected anal intercourse represents the highest level of risk, with unprotected oral and vaginal intercourse also considered potentially dangerous but less risky. The high risk associated with anal intercourse reflects both the presence of the virus in the sperm of infected males and the high frequency of tearing in the anus and rectum as a consequence of this behavior. These small tears provide an avenue for the virus in the sperm to enter the bloodstream of the recipient. The prevalence of this behavior in some parts of the gay community account for much of the very rapid rise in infections among gay men. However, extensive education about safe sex within the gay community appears to be working, and the number of new AIDS cases in gay men is apparently no longer rising as fast as is observed for other risk groups.

Although unprotected oral and vaginal intercourse is considered less risky than anal intercourse, the chance of infection can be increased greatly if either party is also infected with one of several sexually transmitted diseases, such as syphilis or gonorrhea, which create open sores within the genitals. This complication may account for the fact that in areas of the world in which such infections are endemic, AIDS appears to be transmitted primarily by heterosexual intercourse. Nor can unprotected oral sex be considered "safe sex." Some workers argue that the presence of gingivitis, a chronic infection and inflammation of the gums involving bleeding around the gums, may create an entry site for the virus. *(Some folks estimate that nearly half of adult Americans have gingivitis.—RSH)* Basically, any wound or site of bleeding within the body provides an entry point for the virus, just as would a needle stick.

Please remember that you cannot tell whether someone is infected with the AIDS virus just by looking at them. There is no rule that says that someone who looks wholesome and unsophisticated, healthy and energetic, or young and rich (pick your favorite stereotypes) is free of this infection. It can reside in the body for a long time before it makes someone sick enough to make the apparent health status a warning signal. Even if the woman or man of your dreams has only slept with one other person, unprotected sex exposes you not only to his/her limited experience, but rather to the range of exposure of that his/her previous partner, plus anyone he/she slept with, plus anyone they slept with, and so on.

In one sense then the only truly safe sex can be considered intercourse between two completely monogamous individuals, BOTH of whom have NEVER had another partner (or engaged in any other high-risk activities such as sharing needles with others). Barring that, many researchers strongly recommend that condoms be used whenever possible, simply to prevent the transmission of bodily fluids. If you have questions about a specific activity or whether you need to be tested, call the National AIDS Hotline at 1-888-342-AIDS (or 1-888-342-2437) or your health-care professional.

*(This is an awful disease, and as said earlier, there is no way to tell if that great looking guy or woman you just met is infected. No way at all! So try to think about safe sex, please!—CAM If you are wondering if something you might have done might have put you at risk, please ask questions and get tested. There are many sites where you can be tested confidentially. Just call your student or county health service. There are even "home-test" kits that can be bought at the drug store! The phone number just given is a great place to start looking for information or to find a test site. If you test positive, get help. The treatment of AIDS has changed the face of the disease dramatically.—RSH)*

Once the virus does pass into an infected individual, however, it begins a rapid infection of the T cells of the immune system. At first the immune system can fight this battle, for months, sometimes even years. Eventually, however, the virus wins and the T-cell population begins to dwindle to a point where the immune system can no longer function. The result is AIDS.

## TREATMENT OF AIDS

A decade ago the only treatment for AIDS was to treat these infections as they arose. As the immune system began to fail, the sheer number of infections and their severity often killed people quite quickly. *(The first AIDS victim I knew, a graduate student at the big deal medical school in New York, died within only a few months of first becoming symptomatic.—RSH)*

In the intervening years, we have developed better drugs to fight the opportunistic infections, and better yet, we have also developed several sets of drugs that actually act to slow or prevent viral replication. As noted earlier, we have also identified drugs that inhibit the activity of the viral proteases. These findings have allowed the development of effective AIDS treatment protocols that seem likely to greatly extend the lifespan of many, if not most, infected individuals.

## ARE SOME PEOPLE GENETICALLY RESISTANT TO INFECTION BY HIV?

For many diseases, there exist genetic polymorphisms in our population that alter our susceptibility to these diseases. We have already seen how variations in human DNA sequences can change proteins in ways that cause susceptibility to disease. In the case of cystic fibrosis, there can be major problems from bacterial infections that are quite rare in normal individuals.

In the cases of genetic diseases that directly impair the immune response, such as severe combined immune deficiency, it seems obvious that a genetic defect that reduces the effectiveness of the immune system would increase susceptibility to infection. Thus, we see that the existence of differences in the human genome can affect the occurrence and progression of infections. What we haven't really discussed in much detail is the possibility that some genotypes may actually increase the body's ability to fight disease. However, researchers have identified a variety of ways in which altered blood cell proteins confer resistance to infection or give infected individuals a better chance of surviving the disease.

With great caution do we tell you that the repertoire of human genetic variations has conferred a potentially similar kind of defense on a very small number of people. In the last decade, scientists have found a very small number of people who survive HIV infection and people that appear to be resistant to infection. At least some of these people are resistant because of a mutation in a human gene that encodes a component of the attachment or "docking" protein complex that the virus uses to attach to the cell. People lacking or carrying structural variants of these proteins appear to be immune to infection. *(Again, not all mutations are deleterious.—RSH)*

It is unclear just how much of an advantage there is for heterozygous carriers of some of the known mutations, but it is clear that they do not share the level of protection of the homozygotes. The frequency of protective mutations is low enough that you might think that their discovery is not much help. However, the discovery that HIV must use these docking proteins to get into and infect the cell has led researchers to suggest the development of drugs aimed at blocking this point of entry into the cell (imagine a drug that blocked the attachment site), as well as assisting in the development of effective vaccines.

Why did we express caution about giving you this news, this rather exciting ray of hope after all of the bad news we have heard about AIDS? Because most of you do NOT have the protective mutations that have been identified so far; because it is unclear just how much protection comes from some of the mutations that have been identified so far; because there is a lot that is not yet understood about human resistance or susceptibility to infection with the HIV virus or the progression of the disease once infection has started; and because we do NOT want you to kid yourself into thinking that the existence of such mutations is a license to engage in risky behaviors.

## BRIGHT LIGHTS AT THE END
## OF THE TUNNEL

As stated earlier, our understanding of the molecular biology of the AIDS virus has created several new treatment modalities. Some researchers believe that a combination of three currently available anti-AIDS drugs has changed the course of the disease. However, AIDS research is still in its infancy and better drugs and perhaps a vaccine will come along. They just can't come soon enough. For right now, our best weapon in the fight against AIDS is information. You can't catch it from hugging, kissing, or helping. You cannot catch it from donating blood, but you can catch it from sharing needles and having unprotected sex.

# PRENATAL DIAGNOSIS

This section will continue the concept of prenatal testing and diagnosis introduced in Section III. The various tests and the disorders detected by them will be discussed along with the potential for gene therapy. This section will also focus on the problems inherent in human gene therapy and the approaches available to solve those problems.

# 21

# METHODS OF
# PRENATAL DIAGNOSIS

Much of what has been said previously focuses on providing information to parents regarding the genetic health of (1) children they might be planning in the future and (2) unborn fetuses some weeks or months past conception. This chapter focuses on the available tools for prenatal diagnosis. These tools are divided into two groups: noninvasive or minimally invasive tests such as ultrasound and maternal blood tests and invasive tests such as *amniocentesis* and *chorionic villi sampling* (CVS). Each type of test will be discussed in some detail, with attention to the disorders that can be detected by each test. We will also try to describe the tests themselves in enough detail to allow you to understand just what is involved in each procedure.

As these procedures are discussed, it is important to keep in mind just why we do them (see Boxes 21.1 and 21.2). Perhaps their primary value is to provide peace of mind to couples who fall into one or another so-called "risk groups," such as older mothers or parents who have already had one or more children with a genetic anomaly. In such cases, the test can often reassure the parents that the fetus developing inside the mother is in fact healthy, at least as far as we can assay it. However, simply because these couples fall into known risk groups, there obviously will be cases where the answer will not be a happy one (i.e., where the fetus does have an extra copy of chromosome 21 or bears some other significant hereditary defect). In such cases, this information is crucial in giving parents the choice to either prepare for a child who may have very special needs or perhaps to terminate the pregnancy. *(Frankly, Scott and I hate this choice. In a better world all mothers would have a third option: being able to cure or rescue*

BOX 21.1 WHEN TO SEEK HELP: REASONS TO SEEK
THE ADVICE OF A MEDICAL GENETICIST

1. Advanced maternal age.
2. Previously affected child.
3. Presence of a chromosomal rearrangement in one parent.
4. Parents are known or are likely to be carriers of a genetic defect.
5. Family history of a neural tube defect.

*the genetic defect either* in utero *or shortly after birth. That option currently doesn't really exist, but as we will discuss in the next (whew! almost there) chapter, its day may be coming soon.—CAM)*

A second reason for doing some of these tests is large-scale or *population screening.* Some of the simpler less invasive tests, such as a blood test for $\alpha$-fetoprotein (known as AFP; see later), are quick, minimally invasive, and inexpensive enough that they are routinely done on most pregnant women in many technologically advanced countries. In the case of AFP, a low concentration of this fetally produced protein in the mother's blood is a possible indicator for Down syndrome, whereas a very high value might be indicative of a neural tube defect (see Chapter 18). This test alone is not sufficient to diagnose either disorder, as there are too many "false positives," but a high or low value would justify more accurate and perhaps more invasive tests. As methods for identification of those few fetal proteins, and even some cells, that make in into the mother's bloodstream get faster and better in parallel with improvements in our assay techniques, we suspect that such large-scale screenings for additional disorders will become more and more common.

The third reason for doing such tests is simply that parents sometimes ask for them. Even though a given couple may not fit into a known risk group, they may have had a friend who gave birth to a child with this or that defect or read a newspaper or magazine article that raised concerns that could be addressed through such testing. There are also other, perhaps more questionable, reasons for seeking prenatal diagnosis. For example, most medical geneticists seem to have at least one story to tell of couples who sought out testing for purposes solely of determining the baby's sex. For one reason or another, these parents specifically want a child of one sex to the point that they might choose to terminate a pregnancy should the fetus be of the "unwanted" sex. Such an outcome raises the serious ethical conundrum of doing a test whose outcome might lead to the termination of a healthy pregnancy only on the grounds of sex.

Actually, testing for sex is sometimes done in cases of severe sex-linked disorders for which no specific test is yet available. The concept is that in cases where the mother is a known or obligate carrier of an X-linked disorder, the parents may decide that if they only have daughters, then they will not have to face the 50% chance of having a child with that disease. As we get better at testing for more and more traits, this problem is going to get more serious. At some point we are going to need to ask ourselves, "Just what traits are we going to test for and how we will deal with the outcomes of those tests?" In doing so we are going to need to balance the "value" of the test (as defined in a host of ways) against whatever "risk" that test (and its results) may pose to the mother or to the fetus. So before we can continue this discussion, we need to discuss the tests themselves.

## NONINVASIVE TESTS

The most common form of maternal blood test is used to assay the concentration of a fetal protein called $\alpha$-fetoprotein (AFP) in the mother's blood. This is a major blood protein in the fetus and a small amount of it leaks out into the mothers blood supply. As shown in Fig. 21.1 a concentration of AFP that is more than three times the median value for the general population is a strong indicator of spina bifida or other disorders known as *neural tube defects.* Such a tentative diagnosis can be, and sadly often is, quickly confirmed by ultrasound examination (sonar imaging of the fetus in the mother's uterus).

### MATERNAL BLOOD TESTS

Please note that a value of two times the normal value is obtained for both normal and neural tube defect-bearing babies. The requirement for further, more definitive tests, such as ultrasound, is underscored by the fact there are pieces of information that could help reassign some of the "false positives" into the negative category. For instance, if the age of the fetus had been underestimated, then an observed value (which changes with the age of the fetus) might be a normal value for the correct age but appears abnormal simply due to the age error. Similarly, if the value is high but there is a "multiple" pregnancy, the interpretation of the value is altered by that knowledge; a higher AFP value is expected if the product of two or three babies rather than one is measured. Because other very rare genetic anomalies can also lead to elevated values, knowledge that such a genetic anomaly runs in a family would assist in the correct interpretation of the test values.

There is also a worrisome possibility of false negatives: those cases in which the maternal blood AFP level for a fetus with a neural tube defect

BOX 21.2 WHERE TO SEEK HELP: PEOPLE AND
ORGANIZATIONS WHO DO MEDICAL GENETICS

### MEDICAL GENETICS SPECIALIST

These physicians have an MD degree plus specialty training in an area such as pediatrics or internal medicine, PLUS fellowship subspecialty training in the diagnosis and treatment of genetic disorders, birth defects, and other types of malformations. As more and more is learned about the genetic causes of human health problems and effects of teratogenic agents, the role of the medical geneticist is becoming increasingly important and specialized. Your family physician can give you some information about genetics, but there is a limit to how much training can take place on any given specialty topic during the education of a family practitioner, a pediatrician, or an internal medicine specialist. If you have questions about a complex, severe, or rare medical condition in yourself, your child, or some other family member and are trying to make major decisions about having additional children after a child with a birth defect and/or genetic disease has been born or decisions about continuing a pregnancy, having a genetic test done, or undertaking surgery or other major intervention because of a genetic disease, you may find yourself wanting to see a medical geneticist.

A medical geneticist's extra training includes not only information on genetics and birth defects, but also training in techniques and resources for sorting out some very complex health puzzles with underlying genetic and environmental causes. You might go to a medical geneticist because you are concerned about whether you have been exposed to teratogens that can cause birth defects, to find out more about a birth defect even if that kind of birth defect doesn't seem to run in your family, or because you have had many miscarriages and want to find out why. The problem that has you wondering does not necessarily even have to be something severe and life-threatening to take you to the medical geneticist's doorstep, and the question that takes you there may not be about a baby or a pregnancy. The medical genetics folks solve medical mysteries that don't get brought to them until the patient is an adult.

So as you can see, the medical geneticist really deals with prenatal, pediatric, and adult situations, and they deal not only with genetic disorders but also with birth defects and other situations in which the genetic origins may not be obvious to you. You don't necessarily have to have a referral from another doctor, but you should consult your health plan on this one as they might require that referral.

## GENETIC COUNSELOR

The other half of the medical genetics team is the genetic counselor. These health care specialists have an MS degree in genetic counseling. Their education has trained them to be able to assist you with the medical, psychological, and social repercussions of whatever it is that took you to the medical geneticist in the first place. The genetic counselor works with the medical geneticist to help determine the origins of a trait that runs in your family or the causes of a birth defect or the probability that something present in your relatives could turn up in your children.

Part of the genetic counselor's job is to help educate you, to be sure that you understand any tests that are offered, and the kinds of results that could come out of the tests. If test results are obtained, the genetic counselor explains the results and deals with questions and concerns you may have about what you have found out. The purpose of the genetic counselor is not to tell you whether you should have a genetic test done or whether you should have a child or continue a pregnancy. The purpose of the genetic counselor is to be sure that you are armed with all of the information that you need so that you can make the best, most informed, decision possible for you and your particular circumstances. As the amount of genetic information available increases and the complexity of the choices for testing or dealing with test results increases, the role of the medical genetics team becomes increasingly important.

## CLINICAL LABORATORIES

Many of the tests requested by your doctor, whether it is your family practitioner or the medical geneticist, will be done by clinical laboratories in your own hospital or near by. However, some very specialized testing, such as screening for mutations in a particular gene, may be done at only a few places in the country. If information on the gene responsible for your genetic disease is recent information, the testing available may be done on a research basis only and not available from clinical laboratories.

Mutation screening of a given gene may spend several years being developed in the laboratory setting before becoming available through a commercial clinical lab. If that is the case, you might have to agree to participate in the research to be able to get the testing done or you might be told that they cannot guarantee you that they will be able to give you results.

Once optimal screening processes are worked out in the research labs and once the patent lawyers finish squabbling, the test will become available commercially. However, even at that point there are no guarantees that you will find out what you want to know, as even finding out that you have the mutation can still leave you with many unanswered questions.

## RESEARCH LABORATORIES

All of the genetic tests that become available to patients start out by being developed by researchers working in laboratories. The research lab may be at a university, in a hospital, at a pharmaceutical company, or at a research institute like the Salk Institute. Such labs are headed by MDs or PhDs who design and direct the research conducted by a team of researchers. Medical genetics research includes a broad array of topics such as searches for genes, studies of how mutations are caused, investigation of how chromosomes are duplicated and distributed to the correct daughter cells, research on animal models of human diseases, and testing of improved methods for diagnosing genetic anomalies. Because the testing techniques first emerge out of research done in these labs, they are the only source of such testing available before commercial development of a test, and because they are research operations and not commercial labs, they are not usually set up to handle things in the same way a commercial lab might. The result is limited availability of testing.

## ORGANIZATIONS

There are many other places that you can turn to for information. For many different diseases, organizations have developed that raise funds for research, provide support groups, and provide information about the disease. One example is The Foundation Fighting Blindness, which supports research, carries on educational programs, has local chapters throughout the country, and holds national meetings attended by patients, family members, care givers, and educators who want to understand more about retinitis pigmentosa and related diseases that can cause blindness. By doing an Internet search, other organizations of this kind can be found such as the Alliance of Genetic Support Groups, the March of Dimes, and the National Organization for Rare Disorders. Some individuals who are trying to take an active role in communication about disorders in their families have established Web pages that present information or reach out to others with similar problems. There are several ways to locate an organization that provides information or support relative to a particular disease. The Internet address listed below is one good place to start. Often your doctor's office will have information about such organizations. If they don't, try checking with a medical geneticist or other specialist that sees a lot of cases of the disease in question. Even just looking under Social Service Organizations in the yellow pages of the phone book for a large city can connect you with a variety of organizations that can help you get information or support. Just looking in the phone book for a town with a population of 100,000 yielded organizations that can help relative to things such as cancer, lung diseases, multiple sclerosis, kidney disease, sickle cell anemia, epilepsy, and birth defects (the March of Dimes).

### PUBLICATIONS

We frequently end up getting information about genetics from the 10 o'clock news, from the newspaper, or from news magazines. Reporters provide us with sound bites and science writers provide us with a few paragraphs or pages on a gene that has been mapped or cloned. However, if the topic is a disease present in your family or is affecting someone you care about, you may want to know much more than you can get out of a headline from *The New York Times.* Libraries will offer some books, but you may need access to a university library if you want many details. If you have access to the Internet or to a university library, you may be able to tie into databases that have information about genetic diseases. A list of information resources on genetics, genetic testing, dysmorphology, rare disorders, cancer genetics, teratogens, support groups, and careers in human genetics can be found on the Internet at http://www.kumc.edu/gec/.

This is an information resource maintained by a genetic counselor and we expect information found here to be of high quality. However, we must offer a BIG caution about the information sources you use. Just as the caliber of information in books and magazines and newspapers can vary greatly (why wouldn't you believe Elvis was kidnapped by space aliens), information found in the press and on the Internet can vary greatly in accuracy and may not always come from experts. Although the scientific literature undergoes what we call peer review, a process by which each paper published must be approved by other scientists who review the content of the paper, many publications and most Internet sources are not protected by peer review. An exception to this is the many peer-reviewed scientific journals that now make their publications available on the Net. You need to be concerned about all of your sources of information and confirm this information with your doctor or genetic counselor if you are going to use the information to help you make decisions about your health care.

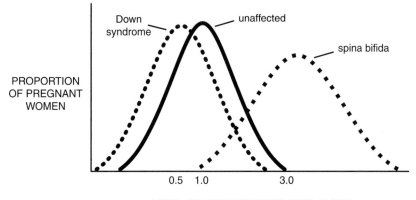

LEVEL OF AFP IN THE MOTHER'S BLOOD
(1.0 is the normal value)

FIGURE 21.1    AFP levels.

falls within the high normal range. Indeed, the best estimate is that 20% of such cases are missed because of AFP levels that overlap with those of normal fetuses. However, as discussed later, when AFP levels are measured in amniotic fluid taken from the mother's uterus and the examination is coupled with ultrasound studies, virtually all cases can be detected.

As shown in Fig. 21.1, the maternal blood AFP concentration is reduced in fetuses with Down syndrome and in those fetuses with some other autosomal trisomies as well. However, note that the distributions for normal and Down syndrome fetuses overlap far more substantially than for neural tube defects. Thus a low AFP level, like maternal age, can only be considered a risk factor used to determine whether more accurate, but also more invasive, testing should be pursued.

Because the number of false positives obtained by assaying AFP alone can be quite high, a better test has recently evolved. This test, called the "triple screen," measures two other chemicals in the maternal blood stream, human chorionic gonadotropin (hCG) and estriol (E3), as well as AFP. The combined levels of these three components of the mother's blood predict the presence of a Down syndrome fetus in 60–70% of the cases and show a much lower rate of false positives. The triple screen is now in wide use as a screening tool for Down syndrome, as well as several other fetal anomalies, in women 35 and over.

**ULTRASOUND**

Ultrasound studies are becoming more and more common in the medical management of pregnancies. Numerous studies clearly show that this test poses no risks to fetus or to mother and yet can provide extremely valuable information with respect to a variety of tests for neural tube defects, limb deformities, or some types of heart disease, as well as some disorders of the kidney and gut. High-resolution sonograms done later in pregnancy can also detect disorders such as cleft lip and cleft palate. Because of their safety and recent advances in technology, such tests are growing in their acceptance and usage. Moreover, prospective parents love the opportunity to "see the baby" months before delivery. *(Indeed, I note wearily, the first couple of pages of many children's photo albums are occupied by various sonograph pictures!—RSH)*

## FETAL CELL ISOLATION

This technique is far from common, but, yes, a small number of fetal cells can be isolated from the mother's blood using a machine known as a cell sorter. Once such cells are in hand they can be tested quickly for the presence or absence of a Y chromosome by straightforward molecular biology. Using the PCR technique discussed previously, it may soon be possible to look for Y chromosome sequences in unseparated maternal blood. Even a single XY fetal cell in a given sample of maternal blood should be sufficient to provide a positive signal. There are also *in situ* hybridization methods using fluorescent probes that can visualize individual chromosomes as bright dots in the isolated fetal cells. These methods are known as fluorescent *in situ* hybridization (FISH). A cell from a normal fetus will show two dots using a probe for chromosome 21, whereas a Down syndrome fetus will show three such dots. Techniques like this are currently used mostly in some specialized circumstances to answer very specific questions that aren't adequately answered with some of the simpler tests, but as technology improves that could change quite quickly.

## MORE INVASIVE TESTS

OK, think about this situation. Mom is 37 years old and the AFP level is on the low end of normal. There is clearly a risk here that this woman is carrying a fetus bearing an autosomal trisomy. To be certain, and we need to be certain, we need access to a reasonable number of fetal cells. Currently, there are two standard methods for getting these cells: amniocentesis and CVS. Both techniques can be done after the 10th to 12th week of pregnancy and both provide the necessary cells for a variety of genetic tests, most notably karyotyping.

### AMNIOCENTESIS

This test is diagramed in Fig. 21.2. Basically, using ultrasound as a guide, the doctor inserts a needle through the abdomen into the uterus and re-

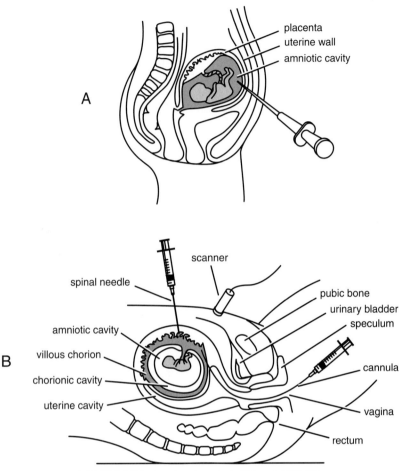

FIGURE 21.2　Amniocentesis (A) and CVS (B).

moves a small amount of fluid (known as amniotic fluid) that surrounds and cushions the developing fetus. Although many texts still say that amniocentesis should be done in the 15th to 16th week of pregnancy, improvements in the technique and in ultrasonography now permit the test to be done much earlier, often at 10–12 weeks. The doctor will tell you that the needle is small and that you will only feel a bit of pressure as the needle enters the abdomen. *(Nonetheless, my wife, who has been through this twice, describes the procedure as "someone sticking a turkey baster through my belly button." You can decide for yourself which version you trust.—RSH)* The doctor can see the tip of the needle, as it looks like a very bright star on the ultrasound. As the doctor is also able to see where the fetus is on the ultrasound screen, it is possible to aim so as to miss the fetus, and the chances that the needle will actually hit or damage the fetus are minimal. For these reasons, this procedure is now considered quite safe; the risk of inducing miscarriage is estimated to be less than 0.5% at centers routinely performing amniocentesis. *(You might realize that the reason why physicians now routinely encourage a pregnant woman over 35 to undergo amniocentesis is that her age-dependent risk of an autosomal trisomy begins to exceed the risk of miscarriage from the procedure alone. The disparity of the two risks will continue to increase as the mother gets older.—RSH)*

The amniotic fluid withdrawn by this method provides a rich source of both fetal cells and fetal proteins. The cells can be cultured using newer "microdrop" methods, and metaphase spreads suitable for karyotyping can be obtained in a few days to a week. Cells can be even more quickly analyzed by FISH assays, although such techniques provide information only about the specific chromosomes for which probes were available or used. Sufficient cells are available for both DNA analysis and biochemical testing. The use of each of these sorts of analyses will be deferred a paragraph or so while we consider the other means for getting fetal cells, CVS.

## CHORIONIC VILLUS SAMPLING

This test, which can be performed as early as the 9th week of pregnancy, is diagramed in Fig. 21.3B. Basically, the doctor inserts a flexible needle, known as a cannula, through the center of the cervix and into the uterus. The doctor then removes a small amount of tissue from a fetal tissue known as the chorionic villi, which is tissue that will go on to form the placenta. This tissue divides mitotically very actively, and thus metaphase cells for karyotyping can be obtained quickly. The tissue can also be subjected to DNA analysis.

A decade ago the major advantage to the use of chorionic villus sampling (CVS) was that it could be performed much earlier in pregnancy than amniocentesis. However, the use of CVS seems to have declined in recent years for several reasons. First, amniocentesis, which until some years ago could only be done at 16 weeks, has been improved to the point that it can now be performed before the 12th week. Thus CVS can be done about 3 weeks sooner than the current earliest time point for amniocentesis, offering only a minimal advantage in terms of early detection. Second, some workers report a lower success rate of karyotyping with CVS. Finally, there were troubling reports, which are currently in some dispute, that CVS might induce an elevated frequency of limb anomalies.

For these reasons, CVS is now used far less often than amniocentesis. Nonetheless, both methods provide the same three things: metaphase cells for karyotyping, fetal DNA for DNA analysis, and cells and enzymes for

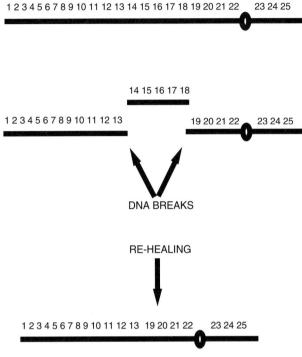

FIGURE 21.3    Schematic diagram of a deletion.

biochemical studies. The second half of this chapter will consider each of these types of assays.

## ANALYSIS OF FETAL CELLS

### KARYOTYPING

A picture of a normal human karyotype is presented in Chapter 9. Such figures are obtained by taking dividing cells from the fetal sample and lysing them (breaking them open) on glass slides so that the individual mitotic chromosomes spread out in a loose field. After the slides are stained with various dyes, the resulting clusters of chromosomes from each cell are photographed and examined. For the last four decades, folks have spent much time and money to build computer systems that will automate karyotyping, but the "old-fashioned" by-hand method still works best. Skilled cytogenetic technicians begin with large photographs of each metaphase spread. Then they carefully cut the picture of each chromosome out of the photograph and match each pair of chromosomes side by side on a piece of mounting paper. *(Having done only 20 such human karyotypes while taking a human cytogenetics lab course in graduate school, I can tell you that the folks who do this are truly gifted at pattern recognition and are among the hardest working people that I know.—RSH)*

Some disorders, such as autosomal trisomies, and sex chromosome anomalies, such as Turner (XO) and Klinefelter syndromes (XXY), are easily picked up by this method. *(Even novices like myself can identify them.— RSH)* However, a good karyotype can also recognize more subtle aberrations, such as deficiencies or duplications for small regions of the genome, translocations of material between chromosomes, inversions of the material on a given chromosome, and the fusion of two ends of a given chromosome to form a ring. Although such anomalies are not common in humans, they do occur and they often have phenotypic consequences. For example, being heterozygous for a deletion of material on the short arm of chromosome 5, which is to say possessing only one copy of that material, results in a disorder known as the *cri-du-chat* (or "cry of the cat") syndrome. This syndrome, recognizable because such infants mew like kittens when they cry, results in severe mental retardation and craniofacial anomalies. We have also noted previously the role of deletions in causing the Prader–Willi and Angelman syndromes. A partial list of other human chromosome rearrangements and their effects is presented in Box 21.3 (also see Fig. 21.3). *(Note that if the right culture conditions are used, the presence of "fragile sites," such as those exhibited by fragile X patients, can also be detected by karyotyping.—RSH)*

One might also note that karyotyping sometimes reveals abnormalities, such as an XYY karyotype, whose effects are not clearly understood or defined. Other abnormalities may be found such as balanced or reciprocal translocations that may not affect the health of the child, but might well affect his/her ability to produce children (Figs. 21.4 and 21.5). Some care is required in explaining such outcomes to the prospective parents. This point also applies when karyotyping anyone. *(As noted earlier, I took a lab course in human cytogenetics as a grad student. During the course of the class, we all did our own karyotpyes. Although 19 of us came up with normal karyotypes, one student discovered that she was heterozygous for a reciprocal translocation between two of her autosomes. Even though she was normal phenotypically, there was a risk that this anomaly might reduce the likelihood of her producing normal children.—RSH)*

It should be obvious that karyotyping will also, by default, tell you the sex of the fetus. Curiously, pairs of prospective parents differ in whether they want to be told what the karyotype reveals in this respect. Parents often say that they prefer to be "surprised" in the delivery room regarding the sex of the baby and thus ask that the report of the results of a normal

## BOX 21.3 REARRANGEMENTS AND EFFECTS

Sometimes damage to our cell's DNA is repaired in such a way as to change the structure of chromosomes. These changes are called *rearrangements*. The major classes of rearrangements are listed below.

### DELETIONS

As shown in Fig. 21.3, a deletion results from the loss of a piece of DNA from a given chromosome arm. Deletions can extend over large or small regions of the chromosome arm, but cannot remove either the centromere or the telomere. To be visible as a change in the banding pattern, a deletion must remove more than 4,000 kb of DNA. Thus, cytologically visible deletions may remove a number of contiguous genes. As a result, heterozygosity for a visible deletion can often result in multiple congenital defects. One such example is the *cri-du-chat* syndrome that arises as a consequence of deficiencies for part of the short arm of chromosome 5. We have described deletions within and encompassing the DMD gene. We have also mentioned the role of deficiencies in the genetics of Prader–Willi and Angelman syndromes.

### TRANSLOCATIONS

Translocations result from the interchange of material between two different chromosomes. Human cytogeneticists divide translocations into three major classes: reciprocal translocations (Fig. 21.4A); centric fusions, which are also known as Robertsonian translocations (Fig. 21.4B); and insertional translocations (Fig. 21.4C). The cases denoted in Fig. 21.4 are considered to be examples of *balanced rearrangements*. In each case the cell carries the normal copy number for each chromosome arm. Individuals carrying balanced rearrangements in the population are found at a frequency of approximately 1/1,000 live births and are usually phenotypically normal. However, during gametogenesis the meiotic segregation of these translocation heterozygotes can often produce aneuploid gametes bearing large deficiencies and/or duplications for the chromosomes involved in the translocation event. An example of a segregation event in a cell heterozygous for a reciprocal translocation involving chromosomes 13 and 21 is shown in Fig. 21.5. Note that the abnormal meiotic segregation events produce aneuploid gametes that can result in miscarriages or in the production of aneuploid offspring. When such individuals are karyotyped, the abnormal karyotypes are denoted as *unbalanced rearrangements*. We have already discussed the role of such unbalanced rearrangements in the etiology of Down syndrome in Chapter 9.

The role of somatically arising translocations in the activation of oncogenes and the subsequent promotion of human cancers were also discussed in Chapter 17. (The interested reader might also look at the role of X autosome translocation in the cloning the DMD gene, described in Chapter 16, and the problem of X inactivation in a female heterozygous for an X autosome translocation, described in Chapter 7.)

## RINGS

If a chromosome suffers two breaks, one near the end of each arm, the two new ends can fuse together to form a ring chromosomes. Although ring chromosomes are occasionally observed, they are quite unstable and thus are often seen in only a fraction of the patient's cells.

karyotype be limited to "everything looks fine." However, some parents want the advance information on this point. One even finds the occasional case of a split decision on this point. *(In the case of our first child, I wanted to know the sex after the amniocentesis and my wife preferred not to know. So when the genetic counselor called me with the result [my wife wasn't reachable], I asked and was told that the baby was a girl. When I finally tracked down my wife hours later, I respected her wishes by saying only, "46 chromosomes—the baby is fine." As we started to hang up, my wife paused and said, "You know the sex, don't you?" When I averred that I did, but, per our agreement, would not tell her, she laughed and said, "Good, just don't tell anyone." As soon as the I put the receiver back in its cradle, the phone rang again, with my wife demanding to know the result after all. Seems that if I knew, then she had to know.—RSH)*

As noted earlier, FISH analysis does offer an alternative to some aspects of karyotyping. For example, it is a much quicker way to look for a specific autosomal trisomy. Newer technology, called chromosome painting, allows the use of sequences that will hybridize all along the length of a specific chromosome, effectively painting the entire chromosome in the color of the fluorescent dye in the probe. In theory, such a technique could be used in sorting out complicated cases in which a simple observation of the banding pattern of the chromosomes wasn't enough information to allow correct assignment of which chromosomes or pieces of chromosomes were involved in a rearrangement. Once the fetus is identified as having a complex chromosomal rearrangement, can you imagine why anyone would go on to ask just which bits of chromosomes are involved? What repercussions to the child might there be if an extra bit of DNA tacked onto the end of

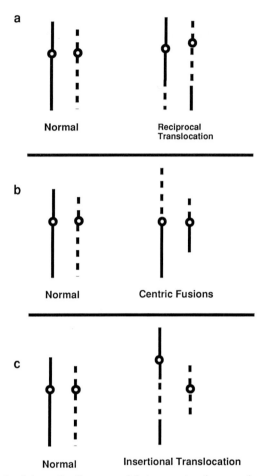

FIGURE 21.4    Schematic diagrams of heterozygotes for a reciprocal translocation (A), a centric fusion (B), and an insertional translocation (C).

chromosome 9 is a piece of the Y chromosome? What difference would it make if you knew instead that the extra DNA came from chromosome 18 and could be expected to act as a partial trisomy with the expectation that the baby might be born alive but then die in the first year of life? In some cases a simple karyotype is enough to identify the additional piece of DNA, but FISH (the use of a specific probe from one specific chromosomal region) or chromosome painting increases the chance that we will be able to get a specific answer to our questions about the baby's karyotype.

Nonetheless, because only a small number of chromosome regions can be examined in a given assay, its utility is somewhat more limited. However, this technique has the advantage that cells can be reliably scored by computer-based assays. As more probes become available and as the auto-

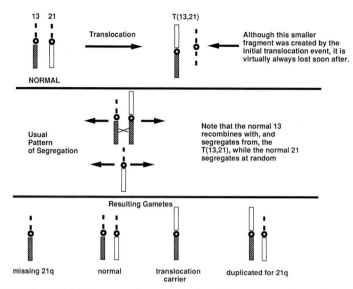

FIGURE 21.5    Meiotic segregation patterns of a cell heterozygous for a reciprocal translocation.

mation of this technique improves, it may well find greater utility. One could imagine automated analysis of hybridizations designed to look at specific regions for all 23 pairs of chromosomes. Even FISH, however, is limited to looking at relatively large regions of chromosomes. If one wants to look at individual genes, or mutations therein, one needs to take the next step down, looking at the DNA.

## DNA ANALYSIS

In a real sense, this section is simply a summary of the last 10 chapters. Suppose a couple's first child had Duchenne muscular dystrophy (DMD) and then came to you asking about the DMD status of their second, as an unborn child. Karyotyping might provide some reassurance since a female fetus is very unlikely to be affected. If the fetus is male, the answer is less clear cut as the odds of being affected are 50%. However, if you have fetal cells and know the nature of the DMD mutation borne by the first child, it is straightforward to determine whether this child carries the mutation. Now you can provide truly useful information to the parents. Very similar things can be done in the cases of fragile X and Huntington disease. Tests are being developed rapidly for a whole host of other disorders as well. Even if the gene is not yet cloned, a closely linked DNA marker can sometimes be used to diagnose the genetic state of the fetus if DNA from other family members is also tested. There is nothing special about fetal DNA, at least in terms of its chemical properties; any of the tests described so far can be used to assay the genetic state of a fetus.

It usually will make a big difference if the primary genetic defect has already been determined for other family members, affected members, and carriers before tackling prenatal diagnosis. For many, the test for a known mutation can be fast. However, as stated earlier, many genes are quite large. An open-ended search for an unknown mutation *somewhere* within one of these genes can take some time, precious time that you do not want to spend during the period in which prenatal testing takes place. For some genes, such as CF or DMD, tests have been developed that can identify a large number of known mutations plus many deletions. For other genes, however, the development of testing has not advanced as far, and it could take too much time to determine the primary mutation on the time scale of the prenatal test.

Thus, if you are concerned about a genetic defect and want to include mutation screening as part of your family planning, it will all work out much better if you start asking questions before you are pregnant. You may be told that a standard test is in place that can do everything you need done at the time of the prenatal test, or you may be told that you qualify for some type of "pre-implantation" testing. However, depending on what the gene is and what the genetic defect is, beginning your inquiries ahead of time might give you important choices that might not be available if you wait. Of course there are many cases (new mutations or recessive diseases that you don't realize are lurking in the genomes of both you and your spouse) that you cannot protect against until the first child with a problem is born to you. Even then, asking your relatives questions about the family medical history can sometimes offer a warning. If you have a cousin with CF and your spouse had a great uncle with CF you should be asking yourself whether you want to talk to a member of a medical genetics team before the first pregnancy. Just think, sometimes the answer will be the happiest one of all—that you are not carriers. However, if that is not the answer, being informed can save you from later saying, "If only I had asked."

This ability to do genetic testing is not a panacea. Indeed, the facility of such tests becomes most worrisome in terms of the material we have presented on the inheritance of complex traits. One worries about people in the near future attempting to test fetuses for DNA markers associated with traits such as mental illness, obesity, and sexual orientation. It is perhaps a rather personal set of prejudices, but we draw a distinction here between testing for traits an individual *will* express and which *will* greatly diminish their quality of and length of life, such as DMD or fragile X, and those traits that they *might* express and whose effects on their quality of life are harder to assess.

Aren't there diseases for which the responsible genes are neither mapped nor cloned? Yes, there are, and these disorders can often be diagnosed by biochemical or enzymatic assays performed on fetal tissues. Such tests are briefly described in the next section.

## BIOCHEMICAL ASSAYS

There are over 100 metabolic disorders that can be diagnosed by laboratory analyses of cells or fluid obtained by amniocentesis or CVS. Most of these disorders are rare, but many of them are sufficiently heritable that testing in families with an affected relative is worthwhile. These disorders include Tay-Sachs disease, galactosemia, and Maple-Syrup Urine disease. Such tests are not commonly performed and indeed are used only when a specific family's genetic history requires them. In addition, the measurement of AFP levels in amniotic fluid on a 15- to 16-week fetus in conjunction with ultrasound analysis can identify nearly all cases of neural tube defects.

Our ability to assess the genetic health of a fetus is impressive and our capabilities expand daily. To the extent that "truth is good," "knowledge is power," and informed choice is better than uninformed choice, then advances in the technologies of prenatal diagnosis can greatly improve our quality of life by providing us with better information from which to make better choices. However, just like all medical or scientific tools, there is a side effect or a down side: Each test becomes a hurdle that developing fetus must pass. We briefly consider the effects of such hurdles in ending this chapter.

## THE CONDITIONAL PREGNANCY

Years ago, women attempting to bear children waited to seek a pregnancy test until their second missed period. They then went in for the so-called "rabbit test" and a positive result was often sufficient to warrant announcing the "happy news." These days, the home tests available at most supermarkets or drug stores are accurate on the first day of the first missed period. However, earlier knowledge, coupled with increases in prenatal diagnostic techniques, has not always resulted in earlier acceptances or announcements of impending births. Rather, women are increasingly aware that, on average, one-sixth of human pregnancies identified by such tests will result in miscarriage before the end of the 12th week. Perhaps not surprisingly, at least some couples are then waiting to accept or announce the pregnancy either until after the end of the first trimester or until they have seen a normal developing fetus on a sonogram.

Couples are also becoming increasingly more guarded and concerned about genetic disorders as well. Some of these couples feel that if there is a chance that a negative result will lead them to terminate the pregnancy, then it is better if such matters are kept private. Some women say that they also consciously try to avoid accepting or "bonding" with the pregnancy for fear of becoming attached only to have to lose the pregnancy pending the adverse result of one or another test. Clearly, as our technology gets better, the number of disorders that can be analyzed will increase dramatically. One cannot help but wonder just how "conditional" pregnancy can become.

# 22

## POTENTIAL FOR
## GENE THERAPY

*Will you make me some magic with your own two hands?*
*Could you build an emerald city with these grains of sand?*
*Can you give me something I can take home?*
*I can do that.*

—*J. Steinman**

In the last chapter we noted our unhappiness with the choices faced by a couple bearing a child with a genetic defect. In a better world, we would like to be able to offer parents a third choice, namely a cure. In essence, we would like to be able to either add back a functional gene to the muscles of children with muscular dystrophy or replace the abnormal FMR1 gene with a normal copy of that gene in male children carrying the fragile X syndrome. This is clearly a worthwhile goal and, in many ways, is the crescendo of our idealistic objectives in doing genetic analysis.

The last decade has made enormous progress in this effort. Several approaches exist for cloning replacement genes and for getting them back into human cells, both cells from the patient that are grown in the lab and cells still residing in the patient. In the case of cells being grown in a laboratory culture, multiple methods of chemical and viral transformation allow us to put genes back into cells. These viral transporters or vectors are usually viruses that have the capacity to carry their DNA into a specific cell type and insert that DNA stably into the nuclear DNA of the cell. Such viral delivery systems are currently the primary method for inserting genes into cells residing in the patient. These advances have made gene therapy a reality.

This chapter will consider briefly the problems inherent in human gene therapy and the approaches available to solve those problems. Before we do so, we need to briefly discuss several general issues, such as (1) in just which tissues do you want to insert or replace a given gene; (2) what do you actually want to put into the cell; (3) just how big is the gene you need to insert or replace; (4) can you identify patients early enough in human development to correct the defect; and (5) are you willing to consider going into the germline to correct a deleterious mutation and prevent it from being transmitted to the next generation?

## TARGETING A SPECIFIC TISSUE

The question of which tissue needs to be "fixed" is a critical one, mostly because of access, but also largely because of biology. Cells from certain cell types, such as blood, skin, and muscles grow very well in culture. Cells could be harvested from the patient or from a suitable donor source, and genes could be inserted into these cells in the laboratory and the "transformed" cells reintroduced into the organ of choice. For other tissues, such as the lung or the brain, such options are not yet available. Moreover, if the gene in question is required in many cell types, then the difficulties inherent in gene therapy currently seem insurmountable. However, consider a disorder like muscular dystrophy. The deleterious effects of these genes are largely limited to muscle. Researchers are currently exploring methods in which muscle progenitor cells carrying gene constructs that produce high levels of dystrophin are produced in the lab and then introduced into muscles of affected children. Because these cells can be incorporated into the muscle, there is a possibility of rescuing the DMD defect. Realize that the original mutation is not being changed here; the mutant gene is simply bypassed by producing the protein from another source. The cell really doesn't care where the dystrophin comes from, it just needs it!

What about tissues that can't yet be grown in culture and reintroduced into the body, such as lung or brain cells? Such tissues present a serious problem. One needs to envision some kind of transporter, such as a crippled virus, that could specifically bring the rescuing gene construct into that cell type and insert the normal gene into the nuclei of a large fraction of cells in that tissue. One can imagine someone with cystic fibrosis inhaling a suspension of viruses that could, like the mythical Trojan horse, disguise a copy of the CF gene as some everyday ordinary virus that would carry that normal CF gene into the cells of the lungs and insert it into the nuclei of these cells. Other internal tissues, such as the brain or heart, pose more difficult, but perhaps not insurmountable, problems. A friend of Scott's, Dr. Leslie Leinwand of the University of Colorado, has developed nonviral techniques for inserting genes into the heart cells of living rats, and investigators all over the country are exploiting the possibilities of creating or finding viruses that specificially target various tissues. Keep in mind, however, that our immune systems have had a lot of practice at keeping viruses away from our cells so efforts to develop viral delivery systems will have to find ways to evade the immune system.

One of the most interesting targets for tissue-specific gene therapy is cancer cells. As diagramed in Fig. 22.1, several investigators have come up with potentially promising techniques for inserting genes encoding intracellular poisons into cells. Some of these suicide genes are rather nifty, as they are only activated in the presence of the protein products of certain oncogenes, proteins normally found in high abundance only in rapidly growing tumor cells. Such techniques may constitute the so-called "magic bullets" that are the holy grail of cancer research—a clever means of

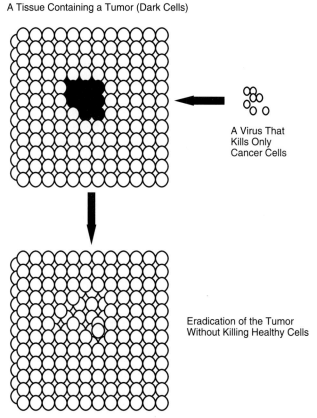

FIGURE 22.1    A virus that specifically kills tumor cells might be able to eradicate tumors without harming the surrounding tissue.

tricking tumor cells into killing themselves. Along a similar vein, researchers have already had promising results with developing viruses that can replicate in and then lyse tumor cells but cannot damage normal cells. For example, the ability of a type of human virus called adenovirus to replicate in and destroy human cells depends on a protein called E1B. The function of the E1B protein is to bind to and inactivate the cellular p53 protein. Mutants of adenovirus that fail to produce E1B protein cannot replicate in normal cells, but they can replicate quite nicely in cells that lack p53, as do many cancers. Thus infection with such mutants may provide an excellent way to destroy cancerous cells that are defective for p53 while leaving normal tissue intact.

*[What about cases of autosomal dominant disorders, cases where the mutation creates a poisonous protein? In such cases putting a normal or "good" gene back into the patient's cells may not solve the problem. Think about it: A cell is making a protein that is causing damage to that cell. You can't just replace the gene or gene product, and you can't fix the problem by sending a magic bullet to kill the cell. The problem here is in fact that the mutant form of the protein is killing the cell. Some clever approaches to solving the autosomal dominant problems are being tried. Using one approach, instead of inserting a gene that will express the message that makes the gene product, the cell is given the ability to produce a short piece of RNA that is complementary to a small part of the gene transcript that contains the mutation. This short transcript, called an antisense RNA, will base pair with the gene transcript (yes, RNA is usually single stranded but you can get double-stranded sections of RNA) and keep it from being used to make protein. If the "mutant" antisense RNA pairs with the normal transcript, there is not a long enough stretch of continuously matched base pairing to keep the two strands together efficiently. The normal transcript ends up dissociating from the mutant antisense RNA and becomes available for use in translation to produce the normal protein product. The mutant antisense RNA forms a much more stable structure with the RNA that exactly matches its sequence so that many copies of the mutant RNA stay in a double-stranded state (or end up degraded in many cases), resulting in the mutant RNA not getting used for translation. The utility of such techniques remains to be seen, but we want you to know that the possibility is there.—JER and RSH]*

## HOW MANY CELLS NEED TO BE FIXED

Even if we knew what we wanted to put into the cell, and we could get proper targeting of the gene to the right cell type, there is the question of just how many cells need to acquire the rescuing DNA in order to ameliorate the phenotype. Do most, all, or just a few cells need to be "transformed"? On the flip side of such concerns, one might also worry about "fixing things" too well. What if your newly introduced gene produces too much product?

Might that be as bad as a deficiency of the product? What if other cells, in addition to your target cells, picked up the newly introduced gene? Might there be a potential for damaging misexpression of the gene? Moreover, what if the cells you are targeting turn over with some regularity and are replaced by new "untransformed" cells? Just how often are you willing or able to repeat this technique? Unless you can transform the stem cell population, this is going to be a serious problem. Realize that if a viral vector is used, repeated treatments might lead to immunity against the virus itself. This problem is going to take some ingenuity to solve! None of these issues are trivial problems. Each of these concerns will need to be addressed as gene therapy protocols are worked out.

## THE PROBLEM OF CORRECTING BIG GENES

The discussion of the DMD gene leads us into the next general concern: the size of the gene you need to replace. Giant genes like DMD are very difficult, if not impossible, to introduce whole into cells. Very large pieces of DNA are mechanically fragile and do not fit into the viral protein coats that we would like to use to stabilize and deliver the large DNA fragment. There are just no currently available means to get them in. So in these cases geneticists must resort to building so-called "mini-genes" in which the introns have been removed and the regulatory regions have been pared down to the bare essentials. Although such genes are sometimes tricky to build, it can usually be done. Even so, in the case of the DMD gene, the stripped-down version is still quite large. Still, these are problems that can be solved. More serious problems might accrue if the gene you want to insert either needs to be modified in some way, a la imprinting, or needs to be in a specific place in the genome. These problems haven't been faced yet, so let's defer them to the next edition. *(What next edition? Who said anything about a next edition? I'm tired! I need a trip to Maui and my MBA completed before I type another word.—CAM . . . Yeah, what next edition? Who said anything about a next edition? We don't know if anyone will buy this one.—the editor)*

## CAN YOU CORRECT THE PROBLEM IN TIME?

Okay, so you think you can identify the target tissue, you know what you want to put into the cell, and you have a vector to get your gene there, but can you do it in time? As noted in the chapter on Down syndrome, folks have identified a single gene whose triplication may explain much of the pathology of this disorder. Great! Suppose we could build a vector that could go into a cell, integrate itself into one of those three copies of a gene, and kill that gene. Techniques for this process, called gene disruption, are well worked out in yeast, flies, mice, and human cells grown in culture. Indeed, it is this technique that allows people to make mouse mutants that mimic human genetic diseases.

Describing how disruption is done in detail is beyond the scope of this book. In brief, a mutant or, better yet, an internally deleted copy of the gene is brought into the cell by a vector. As shown in Fig. 22.2, this mutant copy of the gene recombines with the normal gene. In doing so, it inserts itself into the chromosome while popping out the normal homolog, which is then lost. In practice, this takes a lot of whiz-bang genetic engineering and some very clever uses of genetic markers and selection schemes, but it is doable. As noted below, it is done frequently in other mammals, most notably mice. Researchers are also working on techniques for simply injecting short stretches of mutant or normal DNA into a cell and allowing these sequences to recombine with a resident allele. All of this is possible.

However, at least for Down syndrome patients, such possibilities are fraught with problems even if we could work out the technology of gene disruption. First, how do you knock out *only one* copy of the gene? Think about it. While the whiz-bang genetic engineering is tossing the extra un- wanted copy of some gene on chromosome 21 into the molecular garage, how do you keep the other two copies from being removed at the same time? Well, no problem, you point out, all the cells that knock out more than one copy will probably die. Yeah, but these are brain cells you are

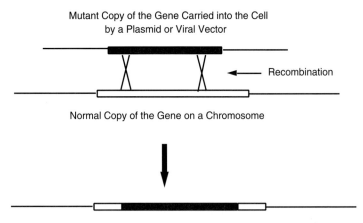

FIGURE 22.2    Gene disruption. A mutant copy of a given gene is brought into the cell by a viral or plasmid vector. The mutant sequences are inserted into the "good" chromosomal sequence by recombination. The vector DNA is eventually eliminated. Thus a "good" copy has been replaced by a "bad" or "mutant" copy. Realize that the same trick might also allow one to replace mutant genes in the chromosome with normal copies.

worried about here. Wholesale cell death is not an acceptable casualty. Even if you could solve the "just kill one and only one copy of the gene problem" with genetic "smart bombs," there is a far more serious problem. By the time you have a live-born baby with Down syndrome, much of the damage has already been done. Some, or maybe much, of that damage may be irreversible. So we just need to treat earlier. How early? Imagine the difficulties inherent in doing gene therapy in a fetus growing in the mother's uterus. Moreover, how early is early enough? Could some damage be done even before you can diagnose the defect? Perhaps, perhaps not. Our point here is to get you to realize that not all genetic diseases may be treatable by gene therapy or by gene disruption. The best hope is for those diseases with later onset where we have easy access to the affected tissues. *(This is not to say that techniques of fetal gene replacement might not be worked out for some diseases. They probably will, but they are unlikely to be the first gene therapy problems to be solved successfully.—RSH )*

Thus, in many cases, it may be more feasible to work on treating these disorders with drugs rather than with gene therapy. One might imagine, for example, working to design a drug that inhibited the overexpressed protein that appears to produce many of the defects associated with Down syndrome. It may be possible to devise a dosage for this inhibitor that basically titrates that protein back to normal activity. Again, the cell couldn't care less what the genes do as long as the resulting protein products are right!

## EARLY ATTEMPTS AT GENE THERAPY

For some disorders, such as cancer and muscular dystrophy, gene therapy may be a true miracle, as is the case of a genetic disorder called ADA (due to homozygosity for mutations that cause a deficiency for a protein called adenosine deaminase). The absence of this enzyme in white blood cells causes the death of an important type of white blood cell called T cells. The result is a severe and fatal immunodeficiency called severe combined immunodeficiency (SCID). A number of affected children have been given transfusions of white blood stem cells taken from their umbilical cord blood that had been transformed with a normal copy of the ADA gene, i.e., the stem cells were taken from the cord blood, transformed in the lab, and reintroduced into the newborn children. Amazingly, the success rate, at least among the small number of children treated, seems to be quite high. ADA gene replacement stands as the hallmark success of gene therapy. Why? Because all the conditions were met. A single gene needed to be replaced, the gene was a workable size, the primary defect was limited to an accessible tissue, and stem cells could be obtained and reintroduced.

There is a hint of another putative success story with gene therapy for sickle cell anemia. As noted earlier, sickle cell anemia is a common blood disorder, especially among black Americans, that results from a mutation in one of these genes (the $\beta$-globin gene) encoding hemoglobin. The defective protein can cause red cells to be deformed and thus to block small capillaries. Eric Kmiec and colleagues injected blood precursor cells from a patient homozygous for the sickle cell mutation with a short stretch of chemically modified DNA containing the correct DNA sequences required to replace the mutation. Following this injection a fraction of the blood precursor cells actually replaced the mutant base pair in their hemoglobin gene with the correct base pair from the donor. To the best of our knowledge, such corrected cells have not yet been placed back into a patient, but the possibility is certainly there.

Although a few trials of gene rescue protocols have been tried for other diseases, such as CF or muscular dystrophy, there are as yet no similar success stories. Both of these two diseases have in common the fact that correcting only a single tissue (the lung epithelium in CF and skeletal muscle in muscular dystrophy) would greatly ameliorate the severity of the disorder in many patients. In addition, target tissues are accessible for both DMD and CF. One could imagine direct injections of either viruses bearing the DMD gene or transformed muscle precursor cells into muscle or inhalation of a suspension of CFTR gene-bearing viruses into the lung. Both diseases are being tackled in this fashion. The problems are just much harder. Still, the people who work on these two diseases in particular, and others as well, have good ideas to get around the current difficulties. Success seems likely in at least some cases.

Even if all of these problems can be solved in the next decade or so, we might also want to think for a moment about the potential misuses of gene therapy. Just what things are to be corrected? One could imagine parents of a body destined to be of intermediate height requesting injections of growth hormone. More frighteningly, after one of his lectures on the genetics of sexual orientation, Scott has had several students ask him publicly and privately whether he imagined that homosexuality might eventually be "cured" by gene therapy. Pointing out that homosexuality was not an illness did not appease the questioners. We as a society will need to think long and hard about just what differences are a normal part of the variation within our species and which are severe enough medical problems to warrant correction. These are not going to be easy decisions and we are unclear as to who is going to make them.

## SHOULD GENE THERAPY BE LIMITED TO SOMATIC CELLS?

This brings us to the last issue: Is it possible to quit dealing with mutations in each generation by going into the germline and correcting the defect once and for all? This is a very touchy issue these days. There are many people *(and we are among them—RSH and CAM)* who have few, if any, ethical problems with gene replacement or disruption in somatic cells to correct a serious medical ailment, but who get queasy at the mere thought of germline gene replacement. To some extent it's a moot issue. Because of the continuous occurrence of new mutations, we will never rid the human gene pool of deleterious variation, even if we were able to correct each person, even heterozygous carriers, at each generation. Moreover, the cost and scale of such an effort might be unimaginable. Still, for some disorders, such as Huntington disease, there might be some real value to this approach. As also noted in many such cases, other techniques, such as embryo selection, can be used to create the same result. As long as an individual does not pass on a deleterious gene to their offspring, it is immaterial whether the germline defect is corrected.

Still, people talk a lot about germline gene therapy and replacement. Why? Well, the answer is partly because we can and partly because it lets us fix the problem once without having to fix it again with each succeeding generation. For over a decade or so we have been able to do germline DNA replacement in the mouse. The basic technique requires a cell line known as embryonic stem (ES) cells that can be easily grown and manipulated in culture. However, as shown in Fig. 22.3, these cells can also be injected into early mouse embryos that were created by *in vitro* fertilization. Once injected, these cells are incorporated into the developing embryo and become part of the developing mouse fetus. These donor cells contribute to the various tissues in the mouse and thus are frequently incorporated into the germline of that mouse. If the ES cells are transformed with a given gene prior to being injected into the embryo and the donor ES cells get into the germline, gametes carrying the exogenous gene will be produced by the resulting mouse. Voila! Germline gene therapy! The resulting progeny mice (those created by gametes bearing the inserted gene) are known as *transgenic mice. (There are a couple of other approaches to this problem, such as inserting genes directly into fertilized eggs or early embryonic cells, but the general idea is the same.—RSH )*

Using somewhat similar techniques, researchers have produced transgenic pigs, goats, rats, rabbits, cows, and sheep. There is nothing truly specific to mice here. Some transgenic animals prove useful for producing large quantities of valuable human proteins in their milk, a great aid in producing these proteins for therapeutic purposes. Similarly, using the gene disruption protocols mentioned earlier, scientists have "knocked out" the mouse genes corresponding to important genes in human. Thus a mouse deficient for the mouse homologs of the ataxia telangiectasia or the p53 gene have been produced, just to name two of a very large number of examples. These mutant mice provide invaluable "mouse models" for human diseases on which new treatments and therapies can be tested. Indeed,

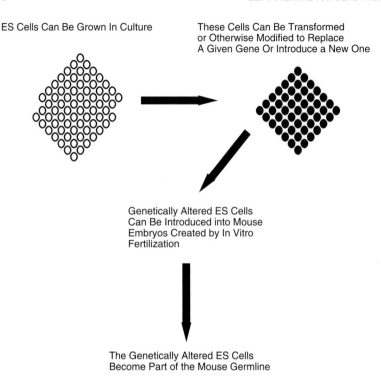

FIGURE 22.3    Use of mouse ES cells to introduce mutations into the germlines of mice.

the number of useful mouse models of human disease seems to increase almost daily.

There are times, however, when mouse models are not as useful as we might hope. Knocking out the DMD gene in mice does not create a mouse with muscular dystrophy. Perhaps mice just don't live long enough for dystrophin to be crucial. Similarly, knocking out the mouse homolog of the CFTR gene doesn't properly mimic cystic fibrosis. Mice are a good model organism, but they aren't human beings. For these reasons, researchers talk about doing targeted gene knockouts in larger mammals, such as dogs, and pigs. Right now such work is limited by the absence of equivalents to mouse ES cells for animals other than mice, but progress in addressing this problem appears to be rapid. Indeed, we suspect that by the year 2000, knockout mutants in higher mammals such as pigs, sheep, and dogs will be common. We are not aware of anyone talking about doing such things in monkeys yet, but we suspect that it won't be long before someone does it. One can easily imagine the sorts of ethical objections to these experiments proposed by various groups of people, including people concerned about the rights of animals and those concerned with the release of genetically altered organisms into the biosphere.

The concerns of such people are aggravated by talk of using such transgenic animals on a large scale for agricultural purposes. Plant researchers have already brought such things as the Flavor-Saver tomato, a genetically altered tomato that stays ripe longer, to the marketplace. People are working hard to introduce genes into plants that increase the yield or prevent infection. Genes that prevent plants from being consumed by various pests are also in the works, as well as genes that extend the shelf life of various products. People are also at work on introducing genes that will allow the growth of bigger and leaner meat-producing animals, as well as increasing milk production. It's all possible and it is all being done.

The techniques are probably generally applicable to most mammals, *including humans,* and there's the rub. We may not yet have a burning need for this technology, but we very likely could easily develop it. If we wanted to make transgenic humans, we probably could. To the best of our knowledge, no one has tried making a transgenic human. Indeed, such experiments are currently forbidden by the National Institutes of Health. So we haven't, but we could. There are many reasons to suggest that we still know far too little to go mucking about with our germlines. Again, we probably shouldn't, but we could, and it is exactly that nagging "but we could" idea that alarms people. Human experience suggests that whatever we *can* do, we often *will* do. There are some exceptions. At any time in the last 40 or so years we could have annihilated life on earth by nuclear war, but we didn't. Perhaps we won't mess up on the human genome either. Perhaps.

The problem is that sooner or later someone will come along with a reason. A few years ago at a conference, Scott was sitting around in the evening chatting with some colleagues when this issue came up. *(Scott, you might at this point wish to point out that it was late at night and that you guys were sitting in a bar.—CAM)* At first people just joked about making photosynthetic humans or toes that were capable of nitrogen fixation. Silly ideas such as skin cells making fluorescent proteins, making us visible in the dark, were also bandied about with laughter. However, it didn't take long for someone to get serious. What if we could introduce a gene for an enzyme that allowed us to break down cellulose or carried with it resistance to some toxin or some disease? The conversation came to a chilling halt as soon as someone realized just what we were talking about: "improving" human beings. One cannot imagine a scarier idea. The ideas, however, will keep coming, and sooner or later someone is going to come up with an idea they think is really good, good enough to be worth the inherent risk. When they do get that idea, they'll probably try to get permission to try it, partly because they can convince someone it's a good idea or convince some company that it's a money-making idea and partly they will try it for the most human of all reasons: because they can. . . .

## ONE LAST THOUGHT ABOUT THE SCIENCE

We began this book discussing Mendel's laws and their biological bases. For the ensuing 19 chapters, we have systematically broken those laws one by one. Imprinting, linkage, sex linkage, nondisjunction, mutation, and X inactivation, these are all egregious violations of Mendel's laws. Nonetheless, it is exactly such violations that taught us how Mendelian inheritance works.

We learned from studying and especially cherishing the exceptions. *(An even better maxim in life than in science.—RSH)* We keep trying to watch things work and when something didn't work as expected, or didn't work at all, we jumped on it, beat on it, caressed it, and dissected it until we understood why. This is the way of science. To quote from a poem about bird feeding from one of Scott's heroes, the scientist–poet Loren Eiseley:*

> I went back to the kitchen
> 　and they came in the old way.
> 　Like man they have problems
> 　like man, what worked once, may work again
> This is the root of magic and science
> 　life's response to uncertainties
> 　if a thing works, you try it
> 　and try it once more, and again
> 　until you are absolutely sure
> 　it will never work, then try it once more.
> 　That is magic and science
> 　and animals and people
> 　live or die by the uncertainties

It has been an astounding journey. A journey that now leaves all of us on the precipice of being able to effectively tinker with our own heredity. That information will be available to each of you in the oncoming years. As said earlier, we strongly urge you to use it to benefit yourselves and your family. On that note, follow us into the epilogue and we will try to tie up the loose ends.

* Reprinted with permission of Scribner, a division of Simon & Schuster from *Notes of an Alchemist* by Loren Eiseley. Copyright © 1972 by Loren Eiseley.

# Epilogue:

# Fears, Faith, and Fantasies

*Although the nightmare should be over, now some of the terrors are still intact.*

—*Jim Steinman**

We can't leave you here without trying to take one more quick look in our crystal ball. If it hasn't occurred to you already, realize now that you stand on the threshold of a revolution that will change the lives of our species forever. We now have the capacity to assess a very tiny amount of the meaning in our genes. We will get better at this particular trick, much better. We will also quickly develop the skills necessary to modify our genomes, surely in our somatic cells, possibly in our germlines as well. There are truly many potential benefits of this technology, but there are also some very big potential pitfalls. As an example, we want to digress into a bit of a history lesson. Specifically, we want to talk about a subject called the *science of eugenics*. Eugenics is the selective breeding of the human population for purposes of improving the quality of the human race.

## THE AMERICAN EUGENICS MOVEMENT

Indeed, we want to talk about government-sponsored eugenics programs in the not too distant past, programs that made laws that deprived people of their ability to be parents. We're going to talk about governments that made laws about sterilization, incarceration, and even about people's right to be alive. The eugenics programs we are talking about happened here in the United States! And it wasn't monsters who carried these programs out. It was done by good upstanding people, people who were considered pillars of their community. They supposedly did these things in the name of God and the public good, and that's what's very scary.

In 1903, an organization called the American Breeders Association was formed. The idea of the association was to bring Mendelian ideas to the United States. Much of what they did dealt with horses, mice, and other animals, but they also began to follow up some rather theoretical work on human breeding begun by a man named Galton in England. Shortly after the formation of the American Breeders Association, the American eugenics movement began. This movement evolved as a federally funded agency located at Cold Spring Harbor, New York, called the eugenics record office. It was run by a Harvard professor named Charles Davenport. Davenport was considered to be one of the great liberal minds of his day. The agency was interested in collecting data about the human population. They trained people to go out into the rest of the country and find pedigrees so they could gather evidence about how certain human traits or "diseases" were transmitted. There was a lot of money in this work in terms of what the government was spending and the amount they were willing to pay their record keepers for finding appropriate families.

One particular bit of data they collected that stands out was two pedigrees that "showed" that seafaringness is an X-linked trait. How could it *not* appear to be X linked at a time when virtually all sailors were men? There were pedigrees for idiocy, silliness, nomadism (the love of wandering), vagrancy, criminality, and more. Remember, people were paid to go out and create these pedigrees. They went to prisons, mental institutions, and anywhere they could. As far as anyone can tell, it was all fraudulent. Money was tough and no one was checking. These people often brought back pedigrees that didn't even have names on them.

Fueled by such information, the American eugenics movement quickly built up real steam. Eugenics booths and education programs were set up in county fairs and schools all over the country. Some brochures for the movement urged people to "wipe out idiocy, insanity, imbecility, epilepsy, and create a race of human thoroughbreds such as the world has never seen." These views on heredity seemed to fit with general common sense. People knew that some of these traits or behaviors did tend to run in families and that certain traits tended to occur in some families more so than in others. In this sense, the American eugenics office was providing the so-called "evidence" to buttress well-established prejudices.

Unfortunately, much of these data were used as justification for new laws. Several states passed laws prohibiting people with certain traits from marrying. It seemed reasonable to many that one way to make a better society was to simply prevent marriages that were predicted to produce certain categories of "defective" progeny; state governments passed laws that idiots, criminals, and epileptics couldn't get married. Eugenics-based marriage laws quickly became the norm in our growing country.

*(When I was about 16, I came across a section on marriage laws in an almanac and found that about seven states still have laws proscribing epileptics from marrying. I have epilepsy. These laws are/were fossils of the American eugenics movement. They were based on a now discredited idea that epilepsy is associated with insanity and imbecility and that epileptics are dangerous. These rather odd views are themselves probably remnants of Medieval beliefs suggesting that seizures were an exposition of demonic possession. The science was bad, but the laws were made anyway.—RSH)*
Based on ideas of racial superiority, buttressed by the eugenics movement, and fueled by prevailing racial prejudices, 34 states also passed laws making marriage illegal for people of different races, the so-called *antimiscegenation laws.* People worried and talked openly about the so-called dangers of "racial degeneracy." Please try to remember these things were not just happening in Nazi Germany; they were happening here.

Soon the law would take marriage laws even further. In 1907, Indiana passed the first law requiring involuntary sterilization. It mandated that people with certain traits, including epilepsy, be sterilized; soon other states would follow suit with similar laws. By the 1930s, more than 30 states had passed laws of mandatory sterilization for an incredibly large number of traits. Between the 1920s and the 1940s, it is estimated that 30 to 35,000 people were sterilized. This number is very likely to be a very gross *underestimate* since not all cases were reported. The people in the eugenics movement were deadly serious and were backing up their politics with knives.

Things got even worse around the 1920s and 1930s. Life got hard for people, and prosperity's infinite view was changing. Immigration was on the rise from all parts of the world. People here worried that some of these new arrivals were genetically inferior and that these genetically inferior people were bringing undesirable genetic traits into this country. Much of the testimony that helped the passage of the Immigration Restriction Act of 1934 was centered on arguments that high fractions of immigrants from certain countries were "feebleminded." Indeed, a progenitor of the IQ test was administered to newly arriving immigrants and suggested an enormous frequency of feeblemindedness among people from certain countries. People felt that such individuals should be denied entrance into this country because they were genetically inferior. However, the curious thing about these IQ tests was that they were administered in *English* to people who had just gotten to the United States and spoke not a word in English. Still these data served as the basis for one of the most restrictive immigration acts in history, which stayed on the books until 1960.

As painfully crazy as all of this must seem to you, it is important to realize that these laws had wide backing throughout American society, even at the highest levels. There was a landmark case on involuntary sterilization that went to the United States Supreme Court in 1922. The case was decided by none other than Oliver Wendel Holmes, known then as an independent vital force for social reform and difference. Holmes was really a terrific, kind, and intelligent man in many ways, but let us quote from the decision in which Holmes and his court upheld the rights of states to sterilize supposedly genetically inferior individuals against their will:

> We have seen more than once that the public welfare may call upon the best citizens for their lives. It would be strange then, if it could not call upon those of us who already sap the strength of the state for these lesser sacrifices, often not to be felt as such by those concerned, in order to prevent our being swamped with incompetence. It is better for all the world if instead of waiting to execute the degenerate offspring for crime or to let them starve from imbecility, societies can prevent the genetically unfit to continuing their own kind. The principles that sustain the compulsory vaccination are broad enough to cover the cutting of fallopian tubes. If we are willing to ask the best of us that they lay down their lives in the defense of their country, in the defense of liberty, and in the defense of freedom, why can not we ask from the weakest of us to voluntarily deprive themselves of the right to reproduce children.

Again, these laws were passed and supported not by monsters, but by very good people. They did what they did in the name of right. That's what terrifies us — that these good people, acting for the good of reason, with the full support of the church and of the state, were able to do so much evil to so many with so few voices being raised.

In the end, the wave of immigration changed this society. Humans finally began to realize, after 20 or 30 years, that immigrants from various parts of the world are more or less the same everywhere and that everyone had an enormous amount to contribute here. Common experience belied the messages of the eugenics movement. More importantly, real genetics was blooming as a science. People were getting an idea of what genetics could and could not do. Good scientists were trying to do serious human genetics, were discovering that nothing was as simple as the eugenics people said, and were also discovering that the eugenics movement pedigree data could not be replicated. So by the end of World War II, most of this went away. All that is left are what is in the history books and some fossilized laws about marriage in those seven states.

## COULD IT HAPPEN AGAIN?

A few years ago, when Scott was a professor at a medical school, he heard a very famous physician give a lecture to a medical school class on human genetics. This fellow started his lecture by saying,

I want you to understand that genetics is not just an important course — it will be the most important course you are ever going to take, because everything you see is going to be genetic. Now, I see you don't believe that. You probably think that at some point you are going to be in an emergency room dealing with a gun shot wound, plugged by a police officer in the middle of a bad drug deal in the middle of south Bronx. You think that is not genetics, but you are wrong because I will argue to you that it was that person's genes that led them to be dealing drugs in the first place.

This professor is not alone in his views.

Consider a passage from a book called *On Human Nature* written by Edward O. Wilson:*

The question is no longer whether human social behavior is genetically determined. It is to what extent. The accumulated evidence for a large hereditary component is more detailed and compelling than most persons, including even geneticists, realize. I will go further, it is decisive.

Wilson argued that genetic differences in terms of behavior and capability were decisive, that we are what we are entirely because of genes. Not because of who raised you, not because of what you believe, or because of how hard you worked, but all because of what was written down in your genes.

There are many serious people who seem to agree with this point of view, or at least are broadly sympathetic to it. The worry here is that many of these people base those views on good science. They interpolate broadly from data available on mental illness and sexual orientation that once we understand everything, that everything will be genetic. Therein lies the worry; in previous times, the investigators were working on fraudulent data, they didn't have good tools, and they didn't know how to figure things out, but now the tools are getting better everyday. We can figure out who is going to be born with what disease and how tall, etc. We will have better and better insights into what genetic predispositions exist. This time it will be real science.

Realize that these problems will get worse as the Human Genome Project continues. We have already talked about mapping genes for diseases, mental illness, behavioral traits, etc. We have also talked mapping complex traits. More frighteningly, we are building the first rough tools for gene replacement. We will get better at such things. It is actually imaginable that we may be able to *cure* muscular dystrophy and cystic fibrosis within a decade. People talk about injecting functional tumor suppressor or "guard" genes into tumor cells to cure cancer. None of this is science fiction anymore. Right now the would-be "gene doctors" are hamstrung by a lack of good delivery systems.

However, as our understanding of human viruses improves, so will our ability to target specific genes into the nuclei of specific cells. The same

* From *On Human Nature* by Edward O. Wilson. Copyright © 1978 by the President and Fellows of Harvard College. Reprinted by permission of Harvard University Press.

can be said for our ability to specifically target tumor cells. As we learn more about the proteins that commit a cell to malignant growth, we can identify those proteins, or their absence, as targets for gene therapy. The technology will rise to meet the challenge; these things will happen! As this technology comes into being, we are going to need to seriously discuss how to use it.

## MAYBE WE CAN DO A BETTER JOB THIS TIME

It will fall to your generation to figure out how to control this technology because the human genome is going to be dumped into your laps. We are paying more money than you can ever imagine for complete sequencing and complete mapping of the human genome for techniques that will allow the genes required for virtually anything you want to be quickly mapped and identified. In the not too distant future, a vast amount of information will be in your hands whether you like it or not. If you don't figure out some way to control this technology, then people like those quoted earlier will figure out a way.

The good news is that people, and not just scientists, are talking very seriously about this problem. The press publishes stories on the new genetics frequently and issues are discussed on talk shows, in college classrooms, and lectured on from pulpits. Some states have already passed laws against genetic discrimination and others are pending. There is much resistance to such efforts, some of it, not surprisingly, from insurance companies. There are also people who are crying loudly for us to just stop all of this research, to turn the clock back. Legislators in Washington spend a lot of time trying to set limits on just what research the federal government will fund. To what degree are human embryos acceptable as experimental organisms? Just what safeguards need to be in place before someone takes a stab at gene therapy?

Many of our colleagues view these discussions as unwanted and unnecessary intrusions into the progress of science. Their view is that they know what needs to be done and that they should be left alone to do it. We suspect that the leaders in the American eugenics movement made similar claims. However, we view the whole tapestry of this discussion, including the extremists on both ends, as very healthy. The more we as a society discuss the ethical implications of this new genetics, the better off we shall be.

Let us make our point with one more example. Back in 1945 in New Mexico, the first atomic bomb was detonated near a town called Almagordo. As the mushroom cloud billowed over the white sands, a physicist who led part of the project turned to his colleagues in the control room and said, "Oh my God, what have we done?" It was way too late for that. The sibling of that bomb was already on its way to an air base in the Philippines. A short while later a Japanese city would be vaporized, and then another. It was too late to wonder what the scientists who built those bombs had done. The whole point of this epilogue, and indeed this book, is to stimulate more discussion, enough discussion so that no matter how this all comes out, we won't find ourselves like that physicist, worrying about ethics just a little bit too late.

And so ends our book. We hope you enjoyed reading it. We certainly enjoyed writing it. *(Well, most of the time. . . .—RSH and CAM)* Indeed, we think of the chapters really as letters from us to you. If you get the chance, write to us in care of the publisher or send Scott an e-mail (rshawley@ucdavis.edu) and let us know what you think. Good luck.

**Catherine**                    **Scott**

# APPENDIX

# SUGGESTED ADDITIONAL READINGS

\*    indicates material easily accessible to any reader
\*\*    reading these articles may require some training in the biological sciences
\*\*\*    intended for those with advanced training or specialists in the field

### SUGGESTED READINGS RELATIVE TO TOPICS IN SECTION 1: THE BASICS OF HEREDITY

*For those who want to trace and construct the pedigree of their own family:*

*The Geneologist's Companion and Sourcebook* by Emily Croom.\*

"Recommendations for Standardized Human Pedigree Nomenclature" by R. L. Bennet in *American Journal of Human Genetics* **56,** 745–752, 1995.\*\*

*For more about Gregor Mendel:*

*Gregor Mendel: The First Geneticist* by Vitzslav Orel.\*

*For a different view of genetics that reaches back 1,500 years:*

"Perspectives on Genetics in China" by J.-L. Fu and colleagues in *Annual Review of Genetics* **29,** 1–18, 1995.\*\*

*For more about chromosomes, including color pictures of "painted" chromosomes:*

"Chromosomal Bar Codes Produced by Multicolor Fluorescence *in Situ* Hybridization with Multiple YAC Clones and Whole Chromosome Painting

Probes" by C. Lengauer and colleagues in *Human Molecular Genetics* **2,** 505–512, 1993.***

"Cytogenetic Approches to Genome Mapping" by J. C. Hozier and L. M. Davis in *Analytical Biochemistry* **200,** 205–217, 1992.***

"Chromosomal Control of Meiotic Cell Division" by K. S. McKim and R. S. Hawley in *Science* **270,** 1595–1601, 1995.**

"Direct Evidence of a Role for Heterochromatin in Meiotic Chromosome Segregation" by A. F. Dernburg and colleagues in *Cell* **86,** 135–146, 1996.***

"Early Studies on Human Chromosomes" by D. G. Harnden in *Bioessays* **18,** 162–168, 1996.**

"Unresolvable Endings: Defective Telomeres and Failed Separation" by R. S. Hawley in *Science* **275,** 1441–1443, 1997.**

***For more about the unraveling of the structure of DNA:***
*The Double Helix* by James Watson.*

*The Eighth Day of Creation* by Horace Freeman Judson.*

***For more information on any aspect of replication, transcription, or translation:***
*The Molecular Biology of the Cell* by B. Alberts and others.*

*The Molecular Biology of the Gene* by J. D. Watson.*

*Genetics and Molecular Biology* by R. Schleif.*

***For more on the regulation of gene expression:***
"Regulation of Gene Expression" by N. Rosenthal in *New England Journal of Medicine* **331,** 931–933, 1994.**

"Transcription Factor Mutations and Disease" by D. S. Lachtman in *New England Journal of Medicine* **334,** 28–33, 1996.**

***For more information on mutations and the resulting characteristics:***
"Evolution Evolving: New Findings Suggest Mutation Is More Complicated than Anyone Thought" by T. Beardsley in *Scientific American* **277,** 15, 18, 1997.*

"Identical Mutations and Phenotypic Variation" by U. Wolf in *Human Genetics* **100,** 305–321, 1997.**

*For information on mitochondrial (nonchromosomal) genetics:*
"Mitochondrial Disorders" by M. Zeviani and C. Antozzi in *Molecular Human Reproduction* **3**, 133, 1997.\*\*\*

"Mitochondrial DNA and Disease" by A. Suomalainen in *Annals of Medicine* **29**, 235, 1997.\*\*\*

## SUGGESTED READINGS RELATIVE TO TOPICS IN SECTION 2: HOW GENES DETERMINE OUR SEX

*For more on sex determination:*
"Male Sex Determination: Current Concepts of Male Sexual Differentiation" by M. L. Gustafson and P. K. Donahoe in *Annual Review of Medicine* **45**, 505, 1994.\*\*

"Mutational Analysis of SRY in XY Females" by J. R. Hawkins in *Human Mutation* **2**, 347, 1993.\*\*\*

"A Clinician Looks at Androgen Resistance" by R. Balducci and colleagues in *Steroids* **61**, 205, 1996.\*\*

"Sex Determination in Humans" by A. J. Schafer and P. N. Goodfellow in *Bioessays* **18**, 955, 1996.\*\*

*For more on X inactivation (sometimes also called Lyonization):*
"Some Milestones in the History of X-Chromosome Inactivation" by M. F. Lyon in *Annual Review of Genetics* **26**, 17, 1992.\*\*

*For more on sexual identity and sexual reassignment:*
"Man and Woman/Boy and Girl" by J. Money and A. A. Ehrhardt.\*

*Sexual Signatures: On Being a Man or Woman* by J. Money and P. Tucker.\*

"Sex Reassignment at Birth: Long-Term Review and Clinical Implications" by M. Diamond and H. K. Sigmundson in *Archives of Pediatrics and Adolescent Medicine* **151**(3), 298–304, 1997.\*\*\*

"The True Story of John/Joan" by J. Coppolino in *Rolling Stones* **775**, 54–92, 1997.\*

On the Web: "Hermaphrodites with Attitude," http://www.holonet.net/isna/HWA/Winter94-95/Winter94-95.html.\*

*For more on sexual orientation:*

"Evidence for a Biological Influence in Male Homosexuality" by S. LeVay and D. H. Hamer in *Scientific American* **270**, 44–49, 1994.*

"Development and Familiality of Sexual Orientation in Females" by A. M. Pattatucci and D. H. Hamer in *Behavioral Genetics* **25**, 407–420, 1995.***

"Population and Familial Association between the D4 Dopamine Receptor Gene and Measures of Novelty Seeking" by J. Benjamin and colleagues in *Nature Genetics* **12**, 81–84, 1996.***

"A Linkage between DNA Markers on the X Chromosome and Male Sexual Orientation" by D. H. Hamer and colleagues in *Science* **261**, 321–327, 1993.***

"Linkage between Sexual Orientation and Chromosome Xq28 in Males but Not in Females" by S. Hu and colleagues in *Nature Genetics* **11**, 248–256, 1995.***

"Portrait of a Gene Guy" by Robert Pool in *Discover* **18**, 50–57, 1997.*

### SUGGESTED READINGS RELATIVE TO TOPICS IN SECTION 3: WHEN MEIOSIS OR MENDELIAN INHERITANCE FAILS

*For more information on the causes of nondisjunction:*

"Direct Proof through Non-disjunction that the Sex-Linked Genes of Drosophila Are Bourne by the X Chromosome" by C. B. Bridges in *Science* **40**, 107–109, 1914.*

"Meiotic Nondisjunction Does the Two-Step" by T. Orr-Weaver in *Nature Genetics* **14**, 374–376, 1996.**

"Recombination and Nondisjunction in Flies and Humans" by K. Koehler and others in *Human Molecular Genetics* **5**, 1495–1504, 1996.**

*For more information on Down syndrome and trisomies:*

"Down Syndrome Genetics: Unravelling a Multifactorial Disorder" by D. Hernandez and E. M. Fisher in *Human Molecular Genetics* **5**, 1411–1416, 1996.**

"Functional Screening of 2 Mb of Human Chromosome 21q22.2 in Transgenic Mice Implicates *minibrain* in Learning Defects Associated with Down Syndrome" by D. J. Smith and others in *Nature Genetics* **16**, 28–36, 1997.***

"Simple Minded Mice from 'in Vivo' Libraries" by I. Kola in *Nature Genetics* **16,** 8–9, 1997.**

"Ethical and Legal Issues Regarding Selective Abortion of Fetuses with Down Syndrome" by N. M. Glover and S. J. Glover in *Mental Retardation* **34,** 207–214, 1996.**

*For the story of the hunt for the gene for Huntington disease:*
*Mapping Fate* by Alice Wexler.*

*For more information on the role of imprinting in the phenotypic expression of Turner syndrome:*
"A Father's Imprint on His Daughter's Thinking" by P. McGuffin and J. Scourfield in *Nature* **387,** 652–653, 1997.**

"Evidence from Turner's Syndrome of an Imprinted X-Linked Locus Affecting Cognitive Function" by D. H. Skuse and others in *Nature* **387,** 705–708, 1997.***

"Daddy's Little Girl" by M. D. Lemonick in *Time* **149** (June 2), 50, 1997.*

*For more information on imprinting and methylation of DNA:*
"Proposed Mechanism of Inheritance and Expression of the Human Fragile-X Syndrome of Mental Retardation" by C. D. Laird in *Genetics* **117,** 587–599, 1987.***

"Parental Imprinting and Human Disease" by M. Lalande in *Annual Review of Genetics* **30,** 173–195, 1996.**

"Genomic Imprinting: Potential Function and Mechanisms Revealed by the Prader-Willi and Angelman Syndromes" by C. C. Glenn and colleagues in *Molecular Human Reproduction* **3,** 321–32, 1997.***

"X-Chromosome Activity: Impact of Imprinting and Chromatin Structure" by R. V. Jamieson and colleagues in *International Journal of Developmental Biology* **40,** 1065–1080, 1996.***

"Imprint Switch Mechanism Indicated by Mutations in Prader-Willi and Angelman Syndromes" by G. Kelsey and W. Reik in *Bioessays* **19,** 361–365, 1997.**

"Formation of Methylation Patterns in the Mammalian Genome" by M. S. Turker and T. H. Bestor in *Mutation Research* **386,** 119–130, 1997.***

*For information about the kinds of scientific breakthroughs that have won Nobel prizes:*

*The Nobel Prize Winners: Physiology or Medicine* edited by Frank N. Magill.*


## SUGGESTED READINGS RELATIVE TO TOPICS IN
## SECTION 4: HUMAN GENES

*For information on the Human Genome Project and other genome projects:*

"The Human Genome Initiative: A Statement of Need" by J. D. Watson in *Hospital Practice* **26,** 69–73, 1991.*

"Positional Cloning Moves from Perditional to Traditional" by F. S. Collins in *Nature Genetics* **9,** 347–350, 1995.**

"Biodiversity, Genomes, and DNA Sequence Databases" by D. D. Leipe in *Current Opinion in Genetics and Development* **6,** 686–691, 1996.**

"The Complete Genome Sequence of *Escherichia coli* K-12" by F. R. Blattner and colleagues in *Science* **277,** 1453–1474, 1997.***

"Codon Usage and Genome Evolution" by P. M. Sharp and G. Matassi in *Current Opinion in Genetics and Development* **4,** 851–860, **1994.**\***

"Third World Participation in Genome Projects" by S. D. Pena in *Trends in Biotechnology* **14,** 74–77, 1996.**

*Other resources with information on the Human Genome Project and cloning:*

National Institutes of Health
Office of Recombinant DNA Activities
6000 Executive Blvd., Suite 302, MSC 7010
Bethesda, MD 20892-7052

National Human Genome Research Institute
Office of Communications
9000 Rockville Pike, 31 Center Drive
Bethesda, MD 20892

The National Center for Biotechnology Information
http://www.ncbi.nlm.nih.gov/

*For more information on cloning genes:*

"Biotechnology: An Introduction to Recombinant DNA Technology and Product Availability" by R. P. Evans and M. Witcher in *Therapeutic Drug Monitoring* **15,** 514–520, 1993.**

"From Cloning to Commercial Realization: Human alpha Interferon" by E. Baron and S. Nerula in *Critical Reviews in Biotechnology* **10,** 179–190, 1990.**

"YACs, BACs, PACs, and MACs: Artificial Chromosomes as Research Tools" by A. P. Monaco and colleagues in *Trends in Biotechnology* **12,** 280–286, 1994.**

"Microdissection and Microcloning of Human Chromosome Regions in Genome and Genetic Disease Analysis" by F. T. Kao in *Bioessays* **15,** 141–146, 1993.**

*For more information on cloning organisms:*

"Writing and Reading about Dolly" by E. F. Keller and J. Ahouse in *Bioessays* **19,** 740–742, 1997.**

"Cloning Humans?" by M. Johnson in *Bioessays* **19,** 737–739, 1997.**

*For more on restriction fragment length polymorphisms and PCR:*

"Chromosome Mapping with DNA Markers" by R. White and J.-M. Lalouel in *Scientific American* **258,** 40–48, 1988.*

"The Unusual Origin of the Polymerase Chain Reaction" by K. B. Mullis in *Scientific American* **262,** 56–61, 1990.*

*For more on cystic fibrosis:*

"Cystic Fibrosis: From the Gene to the Dream" by M. Buchwald in *Clinical and Investigative Medicine [Medecine Clinique et Experimentale]* **19,** 304–310, 1996.**

"Genotype and Phenotype in Cystic Fibrosis" by L. C. Tsui and P. Durie in *Hospital Practice* (Office Edition) **32,** 115–118, 1997.**

*For more information on muscular dystrophy:*

"Dystrophin-Associated Proteins and the Muscular Dystrophies" by R. H. Brown, Jr., in *Annual Review of Medicine* **48,** 457–466, 1997.**

"When You Remember Me," a film about a disabled boy who fights for patients' rights in institutional settings.*

*For more information on neurofibromatosis:*

"Genetic and Epigenetic Mechanisms in the Pathogenesis of Neurofibromatosis Type I" by L. J. Metheny and colleagues in *Journal of Neuropathology and Experimental Neurology* **54**, 753–760, 1995.***

"Neurofibromatosis Type 1: An Update and Review for the Primary Pediatrician" by Y. Goldberg and colleagues in *Clinical Pediatrics* **35**, 545–561, 1996.***

"The Diagnostic Evaluation and Multidisciplinary Management of Neurofibromatosis 1 and Neurofibromatosis 2" by D. H. Gutmann and colleagues in *Journal of the American Medical Association* **278**, 51–57, 1997.***

*For more information on cancer:*

"Tying It All Together: Epigenetics, Genetics, Cell Cycle and Cancer" by S. B. Baylin in *Science* **277**, 1948–1949, 1997.**

"The Genetic Basis of Cancer" by W. K. Cavanee and R. L. White in *Scientific American* **272**, 72–79, 1995.*

"Antioncogenes and Human Cancer" by A. Knudson in the *Proceedings of the National Academy of Sciences* **90**, 10914–10921, 1993.**

*For more information on breast cancer:*

"Inherited Breast and Ovarian Cancer" by C. Szabo and M. King in *Human Molecular Genetics* **4**, 1811–1817, 1995.**

"Linkage of Early-Onset Familial Breast Cancer to Chromosome 17q21" by J. Hall *et al.* in *Science* **250**, 1684–1689, 1990.**

*To look up more details on any specific disease you might be curious about:*

*The Metabolic and Molecular Basis of Inherited Disease* by Charles Scriver, Arthur L. Beaudet, William S. Sly, and David Valle.**

On the Internet: http://www3.ncbi.nlm.nih.gov/Omim/.***

## SUGGESTED READINGS RELATIVE TO TOPICS IN SECTION 5: INTERACTIONS OF GENES AND THE ENVIRONMENT

*For more information on behavioral genetics:*

"Recent Developments in Human Behavioral Genetics: Past Accomplishments and Future Directions" by S. L. Sherman and colleagues in *American Journal of Human Genetics* **60**, 1265–1275, 1997.**

"Understanding the Genetic Construction of Behaviour" by R. J. Greenspan in *Scientific American* **272**, 72–78, 1995.*

"Crime and Punishment: Meeting on Genes and Behavior Gets Only Slightly Violent" by T. Beardsley in *Scientific American* **273**, 19, 22, 1995.*

"Genetic Aspects of the Mechanisms of Learning" by W. Ponomarenko and N. G. Kamyshev in *Neuroscience and Behavioral Physiology* **27**, 245–249, 1997.***

"Genes and Aggressiveness: Behavioral Genetics" by L. H. Tecott and S. H. Barondes in *Current Biology* **6**, 238–240, 1996.*

"Genetics, Antisocial Personality, and Criminal Responsibility" by S. H. Dinwiddie in *Bulletin of the American Academy of Psychiatry and the Law* **24**, 95–108, 1996.**

"Behavioral Genetics" in *Why Aren't Black Holes Black?* by Robert M. Hazen and Maxine Singer, p. 264, 1997.*

"A Gene for Nothing" by R. Sapolsky in *Discover* **18**, 40–49, 1997.*

**For more information on aggression and a mutation at the MAOA gene:**
"X-Linked Borderline Mental Retardation with Prominent Behavioral Disturbance: Phenotype, Genetic Localization, and Evidence for Disturbed Monoamine Metabolism" by H. G. Brunner and others in *American Journal of Human Genetics* **52**, 1032–1039, 1993.***

"Abnormal Behavior Associated with a Point Mutation in the Structural Gene for Monoamine Oxidase A" by H. G. Brunner and others in *Science* **262**, 578–580, 1993.***

"MAOA Deficiency and Abnormal Behaviour: Perspectives on an Association" by H. G. Brunner in *Ciba Foundation Symposium* **194**, 55, 1996.**

**For more information on inherited psychiatric disorders:**
"Genetic Mechanisms in Childhood Psychiatric Disorders" by P. J. Lombroso and colleagues in *Journal of the American Academy of Child and Adolescent Psychiatry* **33**, 921–938, 1994.**

"Understanding the Genetic Basis of Mood Disorders: Where Do We Stand?" by V. I. Reus and N. B. Freimer in *American Journal of Human Genetics* **60**, 1283–1288, 1997.**

"Genetic Linkage and Bipolar Affective Disorder: Progress and Pitfalls" by M. Baron in *Molecular Psychiatry* **2,** 200–210, 1997.**

"Writing Amish Culture into Genes: Biological Reductionism in a Study of Manic Depression" by J. Floersch and colleagues in *Culture, Medicine and Psychiatry* **21,** 137–159, 1997.*

"The Molecular Genetics of Schizophrenia: Progress So Far" by G. Kirov and R. Murray in *Molecular Medicine Today* **3,** 124–130, 1997.**

"Implications of Genetic Research for Child Psychiatry" by M. Rutter in *Canadian Journal of Psychiatry [Revue Canadienne de Psychiatrie]* **42,** 569–576, 1997.**

"Personality and Psychopathology: Genetic Perspectives" by G. Carey and D. L. DiLalla in *Journal of Abnormal Psychology* **103,** 32–43, 1994.**

*For more information on AIDS:*

*AIDS in the World II,* edited by Jonathan Mann and Daniel Tarantola, Oxford University Press, New York, 1996.**

"High-Tech Assault on HIV: Gene Therapy" by P. Cotton in *Journal of the American Medical Association* **272,** 1235–1236, 1994.**

"In Search of AIDS Resistant Genes" by S. J. O'Brien and M. Dean in *Scientific American* **277,** 44–51, 1997.*

"New Strategies for Treating AIDS" by P. A. Sandstrom and T. M. Folks in *Bioessays* **18,** 343–346, 1996.**

### SUGGESTED READINGS RELATIVE TO TOPICS IN
### SECTION 6: PRENATAL DIAGNOSIS

*For more information on prenatal diagnosis and other genetic testing:*

"Prenatal Diagnosis — Why Is 35 a Magic Number?" by Susan and Stephen Pauker in *New England Journal of Medicine* **330,** 1151–1152, 1994.**

"Catching a Bad Gene in the Tiniest of Embryos" by Philip Elmer-Dewitt in *Time* **140,** 81–82, 1992.*

"Cystic Fibrosis and DNA Tests: Implications of Carrier Screening" by the Office of Technology Assessment of the U.S. Congress.**

"Cystic Fibrosis, Duchenne Muscular Dystrophy and Preimplantation Genetic Diagnosis" by J. Liu in *Human Reproductive Update* **2**, 531–539, 1996.***

"The Genetic Testing of Children for Cancer Susceptibility: Ethical, Legal, and Social Issues" by A. F. Patenaude in *Behavioral Sciences and the Law* **14**, 393–410, 1996.*

"Legislating to Preserve Women's Autonomy during Pregnancy" by I. L. Feitshans in *Medicine and Law* **14**, 397–412, 1995.*

"Legal and Regulatory Issues Surrounding Carrier Testing" by R. A. Charo in *Clinical Obstetrics and Gynecology* **36**, 568–597, 1993.**

"Electric Genes: Current Flow in DNA Could Lead to Faster Genetic Testing" by D. Paterson in *Scientific American* **272**, 33–34, 1995.*

***For more on the use of genetic markers in forensics:***
"Population Genetics in the Forensic DNA Debate" by B. S. Weir in the *Proceedings of the National Academy of Sciences* **89**, 11654–11659, 1992.**

"High profile: The Simpson Case Raises the Issue of DNA Reliability" by J. Horgan in *Scientific American* **271**, 33, 36, 1994.*

***For more information on gene therapy:***
*Gene Therapy: A Primer for Physicians* by Kenneth W. Culver.**

"A Brief History of Gene Therapy" by T. Friedmann in *Nature Genetics* **2**, 93–98, 1992.**

"Human Somatic Cell Gene Therapy" by A. Bank in *Bioessays* **18**, 999–1007, 1996.**

"Gene Therapy Strategies for Novel Cancer Therapeutics" by M. E. Rosenfeld and D. T. Curie in *Current Opinion in Oncology* **8**, 72–77, 1996.***

"Regulation of Somatic-Cell Therapy and Gene Therapy by the Food and Drug Administration" by D. A. Kessler and colleagues in *New England Journal of Medicine* **329**, 1169–1173, 1993.**

"Overcoming the Obstacles to Gene Therapy" by T. Friedmann in *Scientific American* **276**, 96–101, 1997.*

"What Cloning Means for Gene Therapy" by S. Mirsky and J. Rennie in *Scientific American* **276**, 122–123, 1997.*

"Gene Therapy for the Nervous System" by D. Y. Ho and R. M. Sapolsky in *Scientific American* **276,** 116–120, 1997.*

"Nonviral Strategies for Gene Therapy" by P. L. Felgner in *Scientific American* **276,** 102–106, 1997.*

"Gene Therapy for Cancer" by R. M. Blaese in *Scientific American* **276,** 111–115, 1997.*

"Patently obvious: Want to do gene therapy? Ask Sandoz" by T. Beardsley in *Scientific American* **273,** 45, 1995.*

"Review: Molecular Biology of Transgenic Animals" by A. L. Boyd and D. Samid in *Journal of Animal Science* **71,** 1–9, 1993.*

"The Use of SCID (Severe Combined Immunodeficient) Mice in Biotechnology and as a Model for Human Disease" by J. S. Sandhu and colleagues in *Critical Reviews in Biotechnology* **16,** 95–118, 1996.**

### SUGGESTED READINGS RELATIVE TO TOPICS IN THE EPILOGUE

*For more on eugenics:*
"Medicine against Society: Lessons from the Third Reich" by J. A. Baron-dessin in the *Journal of the American Medical Association* **276,** 1657–1661, 1996.*

"Redrafted Chinese Law Remains Eugenic [editorial]" by M. Bobrow in *Journal of Medical Genetics* **32,** 409, 1995.*

*For many other suggested readings on the ethical side of the Human Genome Project:*
"ELSI Bibliography: Ethical, Legal, and Social Implications of the Human Genome Project" by Michael S. Yelsey and Michael R. J. Roth, U.S. Department of Energy, Office of Energy Research, Washington, DC, 1993.*

*For more about the legal side of gene cloning:*
"On Gene Patenting" by B. Healy in *New England Journal of Medicine* **327,** 664–668, 1992.*

*For insights from a geneticist who did follow-up studies on Japanese survivors of the atomic bombs:*
*Physician to the Gene Pool* by James V. Neel.*

*For more information on the ethical, legal, and social problems created by genetic technologies:*

"Summary Statement of the Asilomar Conference on Recombinant DNA Molecules" by P. Berg and colleagues in *Proceedings of the National Academy of Sciences* **72,** 1981–1984, 1975.**

"Asilomar Conference on Recombinant DNA Molecules" by P. Berg in *Science* **188,** 991–994, 1975.**

"Genetic Discrimination and the Employment Provisions of the Americans with Disabilities Act: Emerging Legal, Empirical, and Policy Implications" by P. D. Blanck and M. W. Marti in *Behavioral Sciences and the Law* **14,** 411–432, 1996.*

"The Human Genome Project: View from the National Institutes of Health" by L. Fink and F. S. Collins in *Journal of the American Medical Women's Association* **52,** 4–7, 1997.*

"Human Genome Research and the Public Interest: Progress Notes from an American Science Policy Experiment" by E. T. Juengst in *American Journal of Human Genetics* **54,** 121–128, 1994.**

"Intellectual Property Issues in Genomics" by R. S. Eisenberg in *Trends in Biotechnology* **14,** 302–307, 1996.**

"Ethics, Genomics, and Information Retrieval" by K. W. Goodman in *Computers in Biology and Medicine* **26,** 223–229, 1996.**

"Genetic Predisposition to Cancer: Issues to Consider" by D. Malkin and B. M. Knoppers in *Seminars in Cancer Biology* **7,** 49–53, 1996.**

# Glossary

α-**Fetoprotein (AFP)** a protein secreted by a developing fetus. The level of AFP in the mother's blood can be a useful tool for prenatal diagnosis of Down syndrome and neural tube defects (see Chapter 21)

**Acquired immunodeficiency syndrome** a disease caused by infection with the HIV virus that attacks the body's immune system (see Chapter 20)

**Adenine** one of the bases that make up DNA and RNA, usually abbreviated by the symbol A (see Chapter 3)

**Agarose gel electrophoresis** a process that separates DNA molecules according to size (see Chapter 12)

**Allele** different forms of a gene are called alleles. Alleles of a given gene differ in terms of their DNA sequence. Homo- or heterozygosity for some alleles results in differences in the phenotype, but other alleles have a different sequence without causing any change in the phenotype (see Chapters 1 and 2)

**American Eugenics Movement** a disastrous 20th century political movement in this country to "improve" the American population by enforced sterilization, stringent laws against immigration, and laws against interracial marriage (see Chapter 23)

**Amniocentesis** a technique used for prenatal diagnosis of a large number of genetic abnormalities. A needle is used to remove a small amount of amniotic fluid that surrounds the developing fetus. This fluid contains a sufficient number of fetal cells to allow chromosome analysis (karyotyping), DNA analysis, and tests for a variety of biochemical processes (see Chapter 21)

**Amnion** one of the membranes that surround the developing fetus. The amnion is composed of cells derived from the developing embryo and is thus genetically identical to the embryo (see Chapter 21)

**Amniotic fluid** fluid that fills the amnion and provides a cushion for the developing fetus (see Chapter 21)

**Anaphase** the stage in mitosis when sister chromatids separate and move to opposite poles. The term anaphase I is used to refer to the stage in the first meiotic division when homologous chromosomes separate and begin to move towards opposite poles. The term anaphase II is used to refer to the stage in the second meiotic division when sister chromatids separate and move to opposite poles (see Chapter 4)

**Anaphase I** anaphase of the first meiotic division (see Chapter 4)

**Anaphase II** anaphase of the second meiotic division (see Chapter 4)

**Androgen insensitivity syndrome (AIS)** a genetic disorder caused by loss of function in the X chromosomal gene (AR) that encodes the testosterone receptor. XY males carrying such mutations develop as externally normal females who are not fertile. Sometimes referred to as Tfm (testicular feminization) (see Chapter 6)

**Androgen receptor (AR) gene** gene that encodes the testosterone receptor (see Chapters 6 and 10)

**Anencephaly** an abnormality of the fetus involving widespread failures of brain formation. No brain structures above the midbrain are formed. These babies do not survive early infancy (see Chapter 21)

**Aneuploid** possessing an incorrect number of chromosomes. The chromosome constitution of a normal human being is 46 chromosomes, composed of 22 pairs of homologous chromosomes and the sex chromosome pair (XX or XY). Any deviation from this chromosome composition, such as possessing 45 or 47 (or more) chromosomes, is considered to be aneuploid. Normal human gametes possess 23 chromosomes, one from each pair. Gametes with more

or fewer chromosomes than 23 are aneuploid (see Chapters 4, 6, and 9)

**Aneuploidy** the state of being aneuploid (see Chapters 4, 6, and especially 9)

**Angelman syndrome** a genetic disorder caused by possessing only a paternally inherited copy of a region in band q11 on chromosome 15. This disorder is also known as the "happy puppet" syndrome. Symptoms include a jerky gait, a happy demeanor, severe developmental delays, absent speech, and often seizures. This disorder indicates the requirement for a maternally derived copy of this region (see Chapter 11 and Prader-Willi syndrome)

**Anticipation** a phenomenon in which the severity of a trait increases (or age at which onset of symptoms occurs decreases) with succeeding generations in a given family (see Chapter 10)

**Anticodon** the three base sequence on a tRNA molecule that undergoes Watson–Crick base pairing with the three base codon on the mRNA. A tRNA with a particular anticodon always carries the same amino acid as other tRNAs with that same anticodon (see Chapter 3)

**Antimiscengenation laws** laws prohibiting interracial marriages (see Chapter 23)

**Ataxia telangiectasia (AT)** a human repair deficiency syndrome inherited as an autosomal recessive disorder. AT occurs in individuals lacking a normal allele of the ATM gene. This gene encodes a protein required for a DNA damage sensitive checkpoint. Children with AT display a high frequency of both neurological disorders and certain types of cancers (see Chapter 17)

**Autosomal dominant** a pattern of inheritance displayed by dominant mutations on the autosomes. As described in Chapter 5, the most notable features of this type of inheritance are that affected individuals always have an affected parent, children of both sexes are equally likely to display the trait, the trait can be passed on by parents of either sex, and about half of the progeny of an affected individual are expected to express the trait (see Chapter 5)

**Autosomal recessive** a pattern of inheritance displayed by recessive mutations on the autosomes. As described in Chapter 5, the most notable features of this type of inheritance are that affected individuals often do not have an affected parent, children of both sexes are equally likely to display the trait, and about one-quarter of the children produced by the mating of two people that are heterozygous for an autosomal recessive mutation will be affected (see Chapter 5)

**Autosome** chromosomes 1–22 (or all the chromosomes except the X and the Y chromosome). Those chromosomes that are present in an identical pair in both sexes (see Chapters 2 and 4)

**β-Galactosidase** the enzyme produced by the *LacZ* gene in *E. coli* bacteria. This enzyme plays a critical role in the ability of these cells to metabolize the sugar lactose (see Chapter 3)

**Balanced rearrangement** a chromosomal aberration in which parts of two nonhomologous chromosomes have been interchanged, but no material has been lost or gained (see Chapter 21, Table 1); this may also be referred to as a balanced translocation

**Banding** a process by which metaphase chromosomes on a glass slide are treated with chemical dyes. Some regions along the chromosomal arm take up more dye and other regions take up less dye, resulting in a highly reproducible banding pattern (see Chapter 4)

**Barr body** a cytologically visible mass in the nucleus corresponding to the inactive X chromosome. Nuclei from XY males do not display a Barr body, but nuclei from XX females possess one (see Chapter 7)

**Base pair** Watson–Crick-paired nucleotides in a double-stranded DNA molecule (A–T or G–C) (see Chapter 3)

**Becker muscular dystrophy** a less severe type of muscular dystrophy mutation caused by mutations in the DMD gene (see Chapter 16)

**Benign** referring to a tumor that is neither invasive nor metastatic (see Chapter 17)

**Bipolar affective disorder** a mood disorder characterized by a period of manic behavior and periods of major depression (see Chapter 18)

**Bivalent** two paired homologous chromosomes, usually physically connected by the sites where recombination has taken place, during the first division of meiosis (see Chapter 4)

**Bloom syndrome** an autosomal recessive disorder characterized by a growth deficiency, a predisposition to various kinds of cancer, and chromosomal instability (see Chapter 17)

**Cancer** a genetic disease resulting in uncontrolled cell division. Beginning with a single cell, this rapid division results in the formation of one or, in the cases of metastatic cancers, many tumors. Cancer results from mutations in the cellular genes that control cell division and/or in the genes whose products carry out DNA repair (see Chapter 17)

**cDNA**   a DNA copy of a messenger RNA molecule. The production of a cDNA allows one to clone a DNA sequence corresponding to a given mRNA molecule and thus to study that molecule (see Chapter 12)

**Cell cycle**   the real "circle of life." The term cell cycle describes the passage of a cell from one mitotic division to the next. During the course of duplication of a single cell and separation of the two resulting daughter cells, a cell passes through a growth period (G1), a period in which its DNA is replicated (S), and a second growth period (G2) before reentering mitosis (M). Most cells in the body are off the cell cycle track and are sitting in a permanent "parking lot" called G0 (see Chapters 3, 4, and 17)

**Cell cycle regulator**   a protein that regulates the cell cycle. Mutations in these genes are leading causes of inherited and sporadic cancers (see Chapter 17)

**Centriole**   a microtubular structure within the centrosome (see Chapter 4)

**Centromere**   the site on the chromosome that pulls the chromosomes toward the poles of the spindle during mitosis and meiosis. A set of proteins bound to the centromeric DNA sequences creates a structure called the *kinetochore*. Kinetochores attach themselves to microtubules, and "motor proteins" within the kinetochore actually pull the chromosomes to the poles (see Chapter 4)

**Centrosome**   a structure found at the poles of mitotic spindles in both sexes and at the poles of meiotic spindles in males only. Centrosomes serve to organize the microtubule fibers into spindles (see Chapter 4)

**CF**   *see* Cystic fibrosis

**Checkpoint**   a point in the cell cycle where the cell assesses its ability to continue the cell cycle. There are, for example, DNA damage-sensitive checkpoints in G1 and G2 that are activated by damage to DNA. The activation of one of these checkpoints will halt the cell cycle until that damage is resolved (see Chapter 17)

**Chiasma**   a recombination event that is visible by microscopy at diplotene and diakinesis (i.e., during late prophase of meiosis I) (see Chapter 4)

**Chiasmata**   more than one chiasma

**Chimera**   an organism that is a mixture of two genetically different types of cells. For example, if a new mutation happens in a very early embryo, so that the resulting baby has the mutation in some cells but not in others, the baby would be considered chimeric. If the subject of discussion is cloned DNA or genes, chimera can mean a cloned piece of DNA that contains DNA from more than one DNA source. An example would be a clone in which part of the cloned insert comes from chromosome 1 and part of the insert comes from chromosome 5 (see Chapter 6)

**Chorion**   an extraembryonic membrane, also derived from cells of the early embryo, that will go on to become the placenta. Because the chorion and the fetus are both derived from cells of the early embryo, they are genetically identical. Thus genetic analysis of cells from the chorionic villi (structures in the chorion) provides information about the genetic composition of the embryo (see Chapter 21)

**Chorionic villus sampling**   a technique used for prenatal diagnosis of a large number of genetic abnormalities. Basically, the doctor inserts a flexible needle, known as a cannula, through the center of the cervix and into the uterus. The doctor then removes a small amount of tissue from the chorionic villi, a tissue that will go on to form the placenta. This tissue divides mitotically very actively, and thus metaphase cells for karyotyping can be obtained quickly by letting the small number of cells from the tissue sample grow in a culture dish until there are enough cells present to carry out the test. The tissue can also be subjected to DNA analysis (see Chapter 21)

**Chromatid**   either of the two full-length DNA molecules produced by replicating a eukaryotic chromosome. Thus prior to replication each chromosome is composed of only one chromatid. Following replication, each chromosome is composed of two identical chromatids (denoted *sister chromatids*). At mitosis, or in meiosis II, the two sister chromatids separate from each other and move to opposite poles (see Chapter 4)

**Chromatin**   the mixture of DNA and proteins that comprises a chromosome (see Chapter 3)

**Chromosome**   a DNA molecule, and associated proteins, that is capable of stable meiotic and mitotic segregation. Chromosomes carry a single centromere. Most chromosomes consist of linear DNA molecules and thus also possess two telomeres at their ends. (The exceptions to the requirement for two telomeres are rare examples of ring chromosomes.) Prior to replication, each chromosome is composed of only one chromatid. Following replication, each chromosome is composed of two identical chromatids (denoted *sister chromatids*) (see Chapter 4)

**Chromosome walking**   the process of obtaining sequentially overlapping clones along a given region of a chromosome. Chromosome walking becomes an essential tool to clone specific genes within the human genome once tightly linked DNA polymorphisms have been identified by linkage analysis (see Chapters 12–14)

**Cleft palate**   a congenital malformation of the upper part of the mouth. It may or may not be associated with a cleft lip (see Chapter 18)

**Cloning**   unfortunately for every biologist, this term has multiple meanings. In the parlance of horror movies, journalists, and politicians, the term means creating genetically identical copies (clones) of a living organism. In the jargon of cell biologists, cloning means isolating populations of genetically identical cells derived from a single progenitor cell. Geneticists and molecular biologists use the term to mean introducing a given piece of DNA into a host organism (usually the bacteria *E. coli*) and then allowing the replication of that host to make copies of that DNA molecule. This process of gene cloning requires inserting the DNA molecule to be cloned into a cloning vector and the introduction of that resulting molecule into the host organism (transformation). Cloning an organism gets you additional copies of the organism. Cloning a piece of DNA gets you copies of the piece of DNA (see Chapter 12)

**Cloning vector**   a DNA molecule such as a plasmid or a virus that can be used to "clone" other DNA molecules in a given host cell. The cloning vector provides a set of essential functions that are not present on the targeted piece of DNA that is being cloned. Cloning vectors must be able to replicate in the host cell. Plasmids used as cloning vectors also carry a gene, usually a gene conferring resistance to an antibiotic, that allows the scientist to select only host cells carrying the plasmid. They must also possess restriction enzyme cut sites that allow other DNA molecules to be inserted into the vector (see Chapter 12)

**Codominant** this term (or codominance) refers to cases where the phenotype of the heterozygote is distinguished easily from either homozygote. Excellent examples can be found for genetic markers such as RFLPs or microsatellite repeat polymorphisms (see Chapter 13), which are considered to be codominant. Another example is the gene for the ABO blood group. In this case individuals heterozygous for both the A and the B alleles express both the A and the B surface antigens on their blood cells rather than just A or just B. (Note, however, that not all alleles of a given gene are codominant. In the case of the ABO blood group, the O allele is fully recessive. Thus, AO individuals are phenotypically identical to AA individuals.) The term codominant has also been extended to describe dominant mutations that exert a stronger phenotype in homozygotes than in heterozygotes

**Codon** three contiguous bases in an mRNA molecule that specify the addition of a specific amino acid to the growing amino acid chain as the ribosome moves along the mRNA (see Chapter 3)

**Complementation test** an incredibly elegant genetic test that allows a ge-

neticist to determine whether two independently occurring recessive mutations are in the same or different genes (see Box 5.1 in Chapter 5). The concept of this test is simple. If the genetic defects in two different organisms are the same defect, such as if both parents have defects in gene A, then the copy of gene A provided by one parent cannot complement or compensate for the defect in A provided by the other parent and the offspring will have two defective copies of A and will be affected. However, if the defect in one parent is in gene A and the defect in the other parent is in gene Q, then the A-defective parent can provide a "good" copy of Q to compensate for the Q defect, and the Q-defective parent can provide a "good" copy of gene A to complement or compensate for the A defect, so that the offspring will be heterozygous for both A and Q and be unaffected

**Compound heterozygote** an individual heterozygous for two different loss-of-function mutations in the same gene (e.g., m1/m2, where both m1 and m2 are mutations in the same gene). Such an individual will likely express the phenotype characteristic of the less severe of the two possible homozygotes, m1/m1 or m2/m2 (see Chapter 14)

**Consanguinity** marriages between genetically related individuals. Also known as "inbreeding" (see Chapter 14 and many daytime talk shows)

**Contiguous gene syndromes** cases where a single deficiency or deletion of a region of DNA removes several essential genes and thus produces a complex set of phenotypes are often referred to a contiguous gene syndromes because they are presumed to reflect the simultaneous loss of two or more closely linked genes (see Chapter 16)

**Coupling relationship** consider the case of an individual who is heterozygous at both of two closely linked genes A and B. Thus their genotype is AaBb. The term coupling relationship refers to arrangement of alleles of the A and B genes on the same homolog. The genotype AB/ab is not in the same coupling relationship as the genotype Ab/aB (see Chapter 12–14)

*Cri-du-chat* **(cry of the cat)** a hereditary disorder caused by deletion of part of the short arm of chromosome 5 from one of the two copies of chromosome 5. The syndrome is characterized by head and facial anomalies and severe mental retardation. The syndrome is normally heralded by the fact that *cri-du-chat* babies make a high-pitched "cat-like" cry. Most affected children die in early childhood but a few cases survive to early adulthood and are severely mentally retarded.

**Crossing over** the process of genetic *recombination* (see Chapter 4)

**Crossover** a *recombination* event (see Chapter 4)

**Cystic fibrosis (CF)** a fairly common hereditary disorder characterized by autosomal recessive inheritance. Mutations in the CF gene prevent the production of a functional version of an essential protein known as the cystic fibrosis transporter (CFTR). This protein is essential for transporting certain types of molecules (ions) across the membranes of cells in a number of tissues, including the lung, pancreas, and testes. Symptoms for CF homozygotes (or compound heterozygotes) include chronic respira-

tory infection, pancreatic insufficiency, and male sterility. The exact defects depend on the nature of the mutant alleles. Heterozygotes are common (approximately 1/20 individuals) in some populations (see Chaper 15)

**Cytoplasm** material between the outside of the nuclear membrane and the inside of the membrane surrounding the cell. Protein production and energy production both occur in the cytoplasm. There are also a number of cellular organelles, such as the mitochondria and the golgi apparatus, found floating in the cytoplasm (see Chapter 3)

**Cytosine** one of the bases that make up DNA and RNA, usually abbreviated by the symbol C (see Chapter 3)

**Deletion (also referred to as a deficiency)** a type of mutation that removes one or more contiguous base pairs within a region of DNA. The term can be used to refer to the removal of one or more bases within a gene, but it can also refer to much larger aberrations that remove hundreds, thousands, or millions of bases, thus removing an entire gene or group of linked genes (see Chapters 3, 5, 15, 16, and 21)

**Denature** to separate a double-stranded DNA molecule into two separate strands (see Chapter 12)

**Dentatorubral-pallidoluysian atrophy (DRPLA)** a form of hereditary nervous system deterioration, including epilepsy and dementia among the symptoms, that is caused by a trinucleotide repeat expansion (see Chapter 10)

**Dicentric** a chromosomal rearrangement in which a given chromosome now possesses two centromeres. Although such chromosomes should be very unstable, often times one of the two centromeres appears to be inactivated (see Box 21.1 and Chapter 21)

**Differential splicing** the processing of a single transcript into different final mRNA products by including some exons and excluding others in some versions of the transcript and including a different combination of the exons in other versions of the transcript. Differential splicing is sometimes used to make slightly different proteins in different tissue types, but sometimes multiple different transcripts (and the resulting different proteins) will be found in the same tissue together (see Chapter 16)

**Diplo-** used in reference to gametes carrying two copies of the same chromosome (e.g., "diplo-X ova" would mean an egg carrying two copies of the X chromosome) (see Chapters 4, 6, and 9)

**Diploid** carrying two copies of each chromosome (except the sex chromosomes for which either two Xs or an X and a Y will be present in a diploid cell). Ploidy (whether or not a cell is diploid) is determined by counting the number of chromosomes, whether or not they have replicated, and *not* the number of *chromatids*. Thus a human cell remains diploid during both G1 and G2 of the cell cycle, despite the fact that each chromosome is composed of one chromatid at G1 and two chromatids at G2. Listen very carefully here, as nothing has caused more confusion among students. When you are determining whether a cell is diploid, *haploid* (one copy of each chromosome), or *triploid* (three copies of each chromosome), you need to count chromo-

somes not chromatids! Here is the key to it all: to count chromosomes, count the centromeres, not the numbers of chromatids or "arms" that are present on the chromosome (see Chapter 4)

**Diplotene/diakinesis** the stage in male meiosis when paired homologs begin to repel each other. At this stage they are held together only by the recombination events that occurred earlier during meiosis (see Chapter 4)

**DMD** *see* Duchenne muscular dystrophy

**DM protein kinase (DMPK)** the protein whose absence results in myotonic dystrophy (DM). Mutations leading to DM are the result of triplet repeat expansion (see Chapter 10)

**DNA (deoxyribonucleic acid)** the stuff of life. DNA, the hereditary material, is made up of the four types of molecules called bases, named adenine, cytosine, guanine, and thymine (abbreviated A, C, G, and T), that are linked together by a sugar (deoxyribose)-phosphate backbone. In our cells, DNA is normally found as a *double helix* composed of two DNA strands. The two strands are arranged with their bases pointing inward like steps on a ladder and with the sugar-phosphate backbones on the outside. The two sides of this ladder are twisted around each other to form the characteristic "double helix." The bases on each are connected to bases on the opposite strand by weak chemical bonds, and A is always paired with T and G is always paired with C. Thus the two strands are not identical, but rather complementary in sequence. The sequence of these "Watson–Crick base pairs" makes up the alphabet by which DNA encodes all of the information needed to make the components of the cell and carry out the functions of the cell. Contiguous stretches along a DNA molecule correspond to genes (see Chapter 3 and the rest of the book)

**DNA–DNA hybridization** the process of allowing two single-stranded DNA molecules (one Watson strand and one Crick strand) with *complementary* DNA sequences to rejoin (reanneal) to form a double-stranded DNA molecule. This term is most often used to describe cases where one of the DNA molecules is bound to a filter or a glass slide and the other is a labeled probe molecule, but this process can also take place between two DNA molecules that are both in liquid (see Chapters 12 and 13).

**DNA ligation** the act of sealing two double-stranded DNA molecules together at their ends (see Chapter 12)

**DNA polymerase** an enzyme capable of replicating a DNA molecule (see Chapters 3 and 13)

**DNA repair** the process of repairing chemical damage done to a DNA molecule. A number of different types of damage can occur within a DNA molecule, including double strand breaks, single strand breaks, misincorporated bases, and chemically modified bases. Different classes of DNA repair systems exist to repair these various types of damage. Mutations in genes that encode DNA repair proteins can lead to hypersensitivity to some DNA-damaging agents and in some cases to cancer. Diseases such as ataxia telangiectasia, xeroderma pigmentosa, and Bloom syndrome result from homozygosity for mutations in DNA repair genes (see Chapter 17)

**DNA replication** the process of replicating a double-stranded DNA molecule (see Chapter 3 and 13)

**Dominant** a form of inheritance in which only one "defective" copy of the gene needs to be present for the phenotype to be manifested. It is also possible to talk about a dominant allele. Consider the case of an Aa heterozygote that exhibits the same phenotype as an AA homozygote. In this case the A allele is considered to be dominant and the a allele is said to be recessive (see Chapters 1 and 2)

**Down syndrome (trisomy 21)** a genetic disorder arising as a consequence of possessing three copies of chromosome 21. Children with this disorder exhibit characteristic facial anomalies as well as some degree of mental retardation and other health problems (see Chapter 9)

*Drosophila melanogaster* the common fruit fly. Genetic studies on this organism proved to be fundamental to much of our understanding of genetics (see Chapters 4 and 5)

**Duchenne muscular dystrophy (DMD)** a disorder characterized by muscle atrophy and wastage that is due to mutations, often deletions, in an X-linked gene that encodes a protein known as dystrophin. DMD displays a pattern of inheritance characteristic of a sex-linked recessive mutation (see Chapter 16)

**Duplication** a chromosome rearrangement that "duplicates" a given region of DNA. Consider the case ABCDEFGHIJK, where the letters A–K indicate genes along a chromosome. A duplication of the BCDEF region might have the sequence ABCDEFBCDEFGHIJK. This is called a *tandem duplication.* One also recovers cases where the duplicated sequences are found in an inverted orientation (i.e., ABCDE-

FFEDCBGHIJK) (see Box 21.1 and Chapter 21)

**Dystrophin** the protein product of the DMD gene. The complete absence of this protein leads to Duchenne muscular dystrophy. However, some types of DMD gene mutations, which apparently allow the production of some amount of partially functional dystrophin, can lead to a less severe type of dystrophy known as Becker muscular dystrophy (BMD) (see Chapter 16)

**Dystrophy** wastage or atrophy of the muscles, as in Duchenne muscular dystrophy (DMD; see Chapter 16) and myotonic dystrophy (DM; see Chapter 10)

**Edward syndrome** *see* Trisomy 18 (see Chapter 9)

**Electrophoresis** for the purposes of this book, the term electrophoresis refers to the separation of RNA or DNA molecules by using an electric current to propel these molecules through a gel. Under these conditions smaller molecules move faster than larger molecules. Outside of nucleic acid molecular biology the term has much broader meanings (see Chapter 12)

**Embryo selection** the technique of selecting genetically "normal" embryos from a population of embryos created by *in vitro* fertilization and implanting those embryos into the mother. This allows parents who carry deleterious mutations in their germlines (such as the Huntington disease mutation) to ensure the production of children that do not inherit that mutation (see Chapters 10, 21, and 22)

***Escherichia coli*** a bacteria commonly used as a host cell for cloning. Moreover, studies of the genetics of *E. coli*

served to elucidate a number of important genetic principles (see Chapters 3 and 12)

**Euchromatin** the region of a chromosome that is rich in active genes, as opposed to heterochromatin, from which few genes are actively expressed (see Chapters 4 and 6)

**Eugenics** the "science" of "improving" the human race by selective breeding and enforced sterilization (see Chapter 3)

**Eugenics records office** a U.S. Government-sponsored office in New York that existed during much of the first half of this century. This office directed research into human breeding and provided recommendations to the government

**Eukaryote** an organism whose cells possess a nucleus, as opposed to prokaryotes, such as bacteria, in which the chromosome is found in the cytoplasm. Humans are eukaryotes (see Chapter 3)

**Exon** a stretch of a eukaryotic gene that is included in the final processed mRNA molecule, as opposed to intron sequences, which are spliced out during mRNA processing (see Chapter 3). Although it seems backward to give the name introns to the parts that are excised and the name exons to the parts that are left in, the nomenclature becomes logical if you think of exons as part of the completed RNA that is exported from the nucleus and introns as the part that get cut out and left in the nucleus.

**Facultative heterochromatin** chromosomes or regions of heterochromatin that can exist as heterochromatin (i.e., highly condensed and transcriptionally inactive) but are not always so. An inactivated X chromosome, or

Barr body, is an excellent example of facultative heterochromatin (see Chapter 7)

**Familial adenomatous polyposis**   an inherited form of colon cancer (see Chapter 17)

**Fanconi's anemia**   an autosomal recessive disorder characterized by bone marrow failure, predisposition to cancer, and hypersensitivity to certain DNA-damaging agents (see Chapter 17)

**Fluorescence** *in situ* **hybridization (FISH)**   a technique used to map cloned sequences of DNA onto metaphase chromosomes using DNA–DNA hybridization. Fluorescently labeled "probe" DNA made from the cloned sequence is hybridized to denatured metaphase chromosomes spread out on glass slides (see Chapter 12)

**FMR1**   the gene defined by the fragile X mutation. This gene is required for proper development of the face, brain, and testes. Triplet repeat expansion in this gene leads to both fragile X syndrome and a cytologically detectable fragile site (see Chapter 10)

**Founder effect**   the presence of a given allele at a higher frequency within a genetically isolated population where many of the current members of the population are descended from one or more of the small number of individuals that "founded" that population and who carried that allele (see Chapter 14)

**Fragile site**   a site on a metaphase chromosome where breaks or stretching are often observed. Many, if not most, fragile sites are thought to be the result of triplet expansion mutations (see Chapter 10)

**Fragile X E (FRAXE)**   another fragile site on the X chromosome that arises as a consequence of large expansions of a GCC triplet. Individuals who carry such mutations often display mild mental retardation (see Chapter 10)

**Frame shift mutation**   a mutation that alters the reading frame of a message (see Fig. 3.10 in Chapter 3). As an example, consider mutations that insert (or delete) 1 or 2 bp. As a result of such mutations, the normal sequence is read in the correct frame up until the codon bearing the insertion or deletion. However, the insertion or deletion of one or two bases shifts the reading frame of the rest of the message so that a completely different string of amino acids is produced. In many cases, the frame will shift so that one of the new codons to be read will be a stop codon. In other cases, the new reading frame will result in reading of codons beyond the point at which the "real" stop codon is located. In either case, it is common for a frame shift to alter the length of the protein produced in addition to causing major changes in the amino acid sequence beyond the point of the mutation. In reality, a frame shift can be created by the insertion or deletion of any number of bases that is not a multiple of three.

**G0 (G zero)**   the stage in the cell cycle occupied by cells that are no longer dividing. Cells in this stage contain unreplicated and greatly decondensed chromosomes that are actively transcribing mRNAs. G0 cells comprise most of the cells in our body (see Chapter 17)

**G1 (G one)**   the stage in the cell cycle that follows mitosis but precedes replication. Each chromosome is composed of a single chromatid. During this stage the decondensation of the chromosomes provides an opportunity for gene expression. Cells in G1 must pass through a point in the cell cycle called START in order to continue cell division. Passage through START is under very tight genetic control (see Chapter 17)

**G2 (G two)**   the stage in the cell cycle that follows replication (S phase) but precedes mitosis. Cells in G2 have replicated their DNA and thus each chromosome is composed of two sister chromatids. During this stage the decondensation of the chromosomes provides a second opportunity for gene expression. Cells in G1 must pass through one or more "check points" in the cell cycle in order to continue cell division (see Chapter 17)

**Gamete**   a sperm or an egg (see Chapters 1, 2, and 4)

**Gene**   geneticists use the term gene to mean a number of things. Perhaps the best definition is that a gene is a region of DNA that can encode a protein. (This definition is somewhat problematic in that there are some genes, such as those that encode tRNAs and rRNAs, that encode important RNA products that are not themselves translated into proteins. Nonetheless, it is still a pretty good working definition.) More broadly, one could define a gene as a region of DNA that is transcribed and for which that transcript contributes to some aspect of the phenotype of that organism. This definition approximates Mendel's definition of a gene as a unit of heredity. There are other definitions, but these two usually suffice. *(I often wonder if chemists have as much trouble defining atoms as geneticists do agreeing on a definition for a gene.—RHS)*

**Genetic background**   this term refers to all the other genes in a person's genome whose expression might affect the expression of a given gene or set of genes under study. Genetic background is probably as critically important as it is difficult to assess. *(Right now it is used primarily as an excuse for ignoring data that are incongruent with some scientists' favorite explanation for a given process or for ignoring the problem of environmental contribution to a phenotype.— RSH)* (see Chapter 5)

**Genetic code**   the mapping between the set of 64 possible three-base codons and the amino acids (or stop sequences) specified by each of those codons (see Table 3.1)

**Genome**   the complete DNA content of an organism (see Chapter 3)

**Genotype**   the allelic composition of a given individual for one or more genes (e.g., the genotype of one person might be Aa rather than the AA or aa genotypes found in other individuals or the genotype under consideration might be something complex such as AaBBccFi $m_1$ Fi $m_2$ T t$Y^+Y_{fn}Z_3Z_5$) (see Chapter 2)

**Gonadal sex**   whether an individual possesses ovaries or testes (see Chapter 6)

**Guanine**   one of the bases that make up DNA and RNA, usually abbreviated by the symbol G (see Chapter 3)

**Haploid**   possessing only one complete copy of the genome. For example, a human sperm bearing 23 chromosomes, one from each pair, is haploid (see Chapters 2 and 4)

**Hemizygote**   an individual carrying only a single copy of a given gene or chromosomal region. For example, males are hemizygotes for any gene that is only present on the X chromosome (i.e., for the vast majority of their X chromosomal genes, excepting only those genes that are also present on the Y chromosome). Individuals heterozygous for an autosomal deletion are also sometimes referred to as hemizygotes in terms of the genes removed by that deficiency (see Chapters 4 and 5)

**Hemizygous**   the state of being a hemizygote (see Chapters 4 and 5)

**Hemoglobin**   the protein in red blood cells that carries oxygen (see Chapters 4 and 20, Box 20.1)

**Hereditary nonpolyposis colon cancer**   an inherited form of colon cancer due to a mutation in a gene involved in DNA repair (see Chapter 17)

**Heritability**   an estimate of the extent to which the variation in phenotype for some trait within a population is determined by genetic variation. Expressed as a number from 0 to 1.0, the higher the heritability the greater the role that genotype plays in determining that trait in that population at that point in its history. Comparisons of heritability values between two populations are essentially meaningless and provide no information about genetic differences between those populations (see Chapter 18, especially Box 18.1)

**Heterochromatin**   the region of a chromosome that contains few active genes, is rich in highly repeated simple sequence DNA (so-called satellite DNAs), and is usually very highly condensed during interphase. As opposed to gene-rich euchromatin, which composes most of the arms of our chromosomes, heterochromatin is usually found near the centromeres and telomeres (see Chapters 4 and 6)

**Heterozygote**   an individual or organism that possesses two different alleles for a given gene (see Chapter 2)

**Heterozygote advantage**   a condition where a heterozygote survives better than either homozygote. Sickle cell anemia may exemplify such a case in that the same mutant allele of the hemoglobin gene that causes sickle cell disease in homozygotes confers resistance to malaria in heterozygotes. Thus in areas in which malaria is epidemic, the heterozygote may do better than either homozygote

**Heterozygous**   the state of being a heterozygote (see Chapter 2)

**Holandric**   the pattern of inheritance exhibited by mutations in genes located only on the Y chromosome (see Chapter 5)

**Homozygote**   an individual or organism that possesses two identical alleles for a given gene (see Chapter 2)

**Homozygous**   the state of being a homozygote

**Hormone**   a chemical messenger secreted by one cell type that induces a change in the state or behavior of one or more other groups of cells or tissues (e.g., testosterone and estrogen)

**Host cell**   a bacterial, yeast, or animal cell used to propagate clones, infectious agents such as viruses, or plasmids carrying elements such as drug resistance genes. Host cells used to make copies of clones, plasmids, or bacterial viruses are bacterial cells that have special properties that allow growth under laboratory conditions but limit the ability of the cell to grow outside of the laboratory. Some clones, called yeast artificial chromosomes (YACs) are produced in yeast cells. In some cases, human or animal cells in culture are used to host a clone that produces a protein whose functions are studied relative to that cell type (e.g., putting a clone of the CFTR gene back into a culture of lung cells from someone with cystic fibrosis to demonstrate that replacement of the CFTR gene can fix the defect in salt transport by the cells) (see Chapters 12 and 22)

**Human immunodeficiency virus (HIV)** the virus that causes AIDS (see Chapter 20)

**Imprinting** a germline process that "presets" or predetermines the potential of a transmitted gene or chromosome to be active or inactive without changing the actual sequence of As, Gs, Cs, and Ts in the DNA. Cases of imprinting that are known occur in the embryo. For example, in kangaroo females, and in the extraembryonic membranes of XX human fetuses, it is always the paternally derived X chromosome that is chosen to be inactivated. Imprinting presumably reflects a modification of the DNA or proteins in such a way as to preset activity in the embryo. However imprinting occurs, it is "erasable" in the germline of the next generation (see Chapter 11)

**Independent assortment** the observation that for the heterozygote AaBb, where the A and B genes are unlinked, all four classes of gametes (AB, Ab, aB, ab) are produced at equal frequency. Thus whether a given gamete carries the A or a allele of the A gene does NOT influence which allele of the B gene (B or b) that same gamete will carry (see Chapter 2)

**Indifferent gonads** structures in the early embryo that can develop as either testes or ovaries (see Chapter 6)

*In situ* **hybridization** literally means "on site" or "in place" hybridization. When used to map cloned DNA sequences to chromosomes, this term refers to the hybridization of a labeled DNA probe to chromosome spreads affixed to a glass microscope slide (the DNA in the chromosomes has been denatured into separate strands). The resulting hybridization of the probe to the homologous region on the chromosomes results in the labeling of one site on a specific chromosome or two sites on a pair of homologs. The position of that signal can be identified by examining the banding pattern of the chromosome. When *in situ* hybridization is used to assess transcription activity, a labeled DNA probe containing only the coding strand of the DNA is hybridized to sections of cells from various tissues, again affixed to a glass microscope slide. The formation of DNA:mRNA duplex molecules here results in the labeling or staining of these cells, indicating that that gene is transcribed in those cells (see Chapter 12)

*In vitro* **fertilization** the process of creating zygotes in a test tube by combining sperm with eggs outside of the human body. Eggs are obtained by hormonally stimulating a woman's ovaries to produce ova. The resulting eggs are removed from the woman and then mixed in the laboratory with sperm obtained from the prospective father. The resulting zygotes are then allowed to go through several cell divisions to produce early embryos. One or two cells from these embryos can be removed safely, allowing the performance of a variety of PCR-based genetic tests. Once "healthy" embryos are identified, they can be introduced into the uterus of the prospective mother

**Integrase** an enzyme produced by a retrovirus that inserts a DNA copy of the retroviral genome into the cellular DNA (see Chapter 20)

**Intron** a stretch of an eukaryotic gene spliced out of the final processed mRNA molecule, as opposed to exon sequences that remain during mRNA processing (see Chapter 3). Although it seems backward to give the name introns to the parts that are excised and the name exons to the parts that are left in, the nomenclature becomes logical if you think of exons as part of the completed RNA that is exported from the nucleus and introns as the part that gets cut out and left in the nucleus

**Invasive** a property of tumors that denotes the ability of tumor cells to spread into the surrounding normal tissues (see Chapter 17)

**Inversion** a type of chromosome aberration in which some or all of a chromosome arm has been inverted with respect to the centromere. Consider the sequence ABCDEFGHIJoKLM-NOPQRST, where the symbol o denotes the centromere. An inversion that reverses the sequence of sites BCDEF would have the following structure AFEDCBGHIJoKLM-NOPQRST. This type of inversion, which does not include the centromere, is called a *paracentric inversion.* An inversion that reverses the sequence EFGHIJoKLMNO would have the structure ABCDONML-KoJIHGFEPQRST and be called a *pericentric inversion* (see Box 21.1 and Chapter 21)

**Jumping genes** genetic elements, more properly called transposons, that can move about the genome under certain conditions (for reasons that now elude us, we never discussed such

things in this book, but it's an important concept anyway)

**Junk DNA** DNA that carries out no useful functions for the host organism. The debate continues whether the large amount of noncoding sequence in the human genome qualifies as junk DNA or whether it only seems useless because we have not yet discovered what it does. Some of the noncoding sequences are stretches of sequence that are repeated. Some of them are repeated a few times, whereas others are repeated 1000, 50,000, or even a million times. As we learn more, we discover that many of these repeated sequences are actually doing things, but for many of them, no useful function has yet been found

**Karyotyping** the analysis of human chromosomes by viewing under a microscope. Metaphase cells are broken onto glass slides in such a way that all 46 chromosomes are well separated within a small area. The chromosomes are then stained in a way that leaves a characteristic pattern of bands that can be used to identify and distinguish the different chromosomes from each other. Individual chromosomes and chromosome pairs are then identified and scrutinized for errors in number (e.g., trisomy 13, 18, or 21, or monosomy for the X) or structure (chromosome aberrations such as large deletions or translocations) (see Chapters 4, 5, and 6)

**kb** *see* Kilobase (see Chapter 12)

**kilobase** one thousand bases (see Chapter 12)

**Kinetochore** the DNA–protein complex assembled at the *centromere* that allows chromosomes to attach to, and move along, the *microtubules* in the mitotic or meiotic spindle. The kinetochore is known to include proteins called *motor proteins* (kinesins and dyneins) that function to pull or push the chromosomes along the spindle (see Chapter 4)

**Klinefelter syndrome (XXY)** a genetic anomaly resulting from the presence of an extra X chromosome in the male. Males exhibiting this disorder are sterile because of testicular atrophy. They may also show some degree of external feminization, including breast development (see Chapter 6)

**Lactose metabolism** biochemical processes that allow the sugar lactose to be used as an energy source. This process requires the enzyme $\beta$-galactosidase (see Chapter 4)

*lacZ* **gene** the *lacZ* gene in *E. coli* bacteria produces the enzyme $\beta$-galactosidase. This enzyme plays a critical role in the ability of these cells to metabolize the sugar lactose. The understanding of the regulation of this gene was a pivotal step in our understanding of the mechanism by which genes are regulated. We frequently hear about this gene when we talk about cloning because of its presence in some cloning vectors (see Chapter 4)

**Lentivirus** a type of retrovirus (see Chapter 20)

**Leptotene** an early stage in prophase I in which the chromosomes become visible

**Lethal mutation** a mutation that creates a sufficiently severe defect as to be incompatible with life (see Chapter 7)

**Library** in genetic terms, a collection of clones representing part or all of a genome or, in the case of cDNA libraries, a population of DNA clones

corresponding to a set of mRNA molecules. It is also a place with many books, but we don't have time for that here (see Chapter 12)

**Ligase** an enzyme that can link two DNA molecules together at their ends (see Chapter 12)

**Linkage disequilibrium (also known as allelic association)** the association in a given population of individuals of a particular allele at one gene, or genetic marker, with a specific allele of some other gene or genetic marker (or in some cases a disease gene allele) at an unexpectedly high frequency. Consider the case of a disease-causing mutation and a separate genetic marker (A), both of which are segregating in a given population. If 90% of individuals in the population that carry the disease-causing mutation also carry allele A1 of genetic marker A, but only 10% of the general population has allele A1, then allele A1 and the disease-causing mutation are considered to be in linkage disequilibrium or allelic association with each other (see Chapter 14)

**Locus** a specific site or position on a chromosome as determined by recombinational mapping. Often used to refer to a gene at that position (see Chapter 14)

**LOD score** a statistical tool used in the determination of recombination frequencies in human populations. In brief, a LOD score is a measure of the probability of observing a given pedigree, or set of pedigrees, assuming that the map distance that has been identified as separating two markers (or genes) is really the actual distance between them. The acronym LOD is an abbreviation of "log of the odds." A LOD score is the logarithm base 10 of the odds ratio obtained by taking the probability of obtaining observed data for any recombination frequency (called Q or theta) and dividing that probability by the probability of obtaining the observed data if the recombination fraction is actually 0.5 (that is to say that the two markers are unlinked). For the human genome, a LOD score of 3.0 or greater is required to consider that the evidence for linkage between a marker and a gene is highly significant (likely to be true) (see Box 14.1)

**Loss of heterozygosity (LOH)** a genetic event often observed in cells heterozygous for some recessive mutation that causes the loss of the normal or wild-type allele. As such the cell is now either homozygous or hemizygous for the recessive allele and will likely express the mutant phenotype. LOH can occur by mutation or deletion of the normal allele as a result of mitotic recombination or by loss and reduplication of the entire chromosome that carries the mutant allele. LOH is thought to play a pivotal role in the development of many, if not most, types of cancers (see Chapter 17)

**Machado–Joseph disease** an autosomal dominant neurologic disorder caused by a (CAG)n trinucleotide expansion in a gene located on 14q (see Chapter 10)

**Malignant** a characteristic of a tumor, indicating the ability of cells of that tumor to spread to other sites of the body and colonize new tumors and/or the ability of that tumor to invade surrounding normal tissues (i.e., a tumor that is metastatic and/or invasive) (see Chapter 17)

**Map unit** a measurement of recombination frequency. One map unit equals one percent recombination (see Chapters 4 and 14)

**Meiosis** the process by which haploid gametes are produced. Basically, homologous chromosomes pair and recombine during meiotic prophase. At anaphase I of the first meiotic division, entire homologs move to opposite poles of the spindle. The first meiotic division, *meiosis I,* in humans thus creates two *haploid* daughter cells, each with 23 chromosomes. (Realize that whether a cell is haploid or diploid depends on the number of chromosomes it has, not the number of chromatids.) Each product of meiosis I has 23 chromosomes (as determined by counting the number of centromeres), as compared to the diploid number of 46, so it is haploid. The second meiotic division, *meiosis II,* is then essentially a haploid mitosis, creating four haploid daughter cells, each with 23 chromosomes (see Chapter 4)

**Meiosis I** the first meiotic division (see Chapter 4)

**Meiosis II** the second meiotic division (see Chapter 4)

**Melanoma** a type of skin cancer (see Chapter 17)

**MERRF** *see* Myoclonus epilepsy with ragged red fibers

**Metaphase** in *mitosis,* the stage in human cell division where all 46 chromosomes are individually balanced between the two poles of the spindle, with one sister chromatid attached to each pole of the spindle. In *meiosis I,* the stage at which each pair of homologous chromosomes (bivalents) are balanced between the two poles of the spindle, with a centromere attached to each pole. This stage is referred to as metaphase I. In *meiosis*

*II,* the stage at which all 23 chromosomes are individually balanced between the two poles of the spindle, with one sister chromatid attached to each pole of the spindle. This stage is referred to as *metaphase II* (see Chapter 4)

**Metaphase I** the metaphase of first meiotic division (see Chapter 4)

**Metaphase plate** an imaginary line across the equator of the spindle where chromosomes are normally aligned (see Chapter 4)

**Metastasis** the process of a tumor spreading to new sites in the body. A characteristic of malignant tumors (see Chapter 17)

**Microsatellite** a region of genomic DNA composed of dinucleotide or trinucleotide repeats (e.g., CACACACACACACACACACA). These sequences are dispersed widely throughout the genome and are extremely polymorphic in terms of the number of copies of the repeating unit. As such, they have proven incredibly valuable as genetic markers for mapping human genes (see Chapters 13 and 14)

**Microsatellite repeat polymorphism (MSR)** a specific allele of a given microsatellite (see Chapters 13 and 14)

**Microtubules** fibers that make up the meiotic and mitotic spindles. Chromosomes attach to these fibers and move along them as the chromosomes migrate to the poles (see Chapter 4)

**Mismatch repair** a process of repair of damage to DNA caused by misincorporation of the wrong base during replication or by chemical damage to one or more bases. Defects in this process play a role in the origin of some types of cancers (see Chapter 17)

**Missense mutation** a mutation that changes a codon in such a way as to direct the incorporation of a different amino acid at that site in the protein (see Chapter 3 and Figure 3.8)

**Mitochondria** cellular organelles responsible for energy production. Mitochondria possess their own small circular DNA genomes. Mitochondria are thought to have arisen from bacterial parasites during the early stages of eukaryotic evolution (see Chapters 3 and 5)

**Mitochondrial inheritance** the pattern of inheritance exhibited by mutation in the genome of the mitochondria. In general, all or most children of affected mothers will be affected. Because sperm do not donate mitochondria to the zygotes, paternal transmission is never observed (see Chapter 5)

**Mitochondrial myopathy** a disease causing muscle weakness and fatigue resulting from a mutation in the mitochondrial genome (see Chapter 5)

**Mitosis** the process by which a cell divides to produce two identical daughter cells. Unlike meiosis, mitosis does not change the ploidy of a cell. The human cell starts out diploid, with 46 chromosomes (identifiable by the presence of 46 centromeres), and the two daughter cells both carry 46 chromosomes. In addition, chromosomes do not usually pair or recombine prior to mitosis (see Chapter 4)

**Monoamine oxidase A (MAOA)** an enzyme involved in the breakdown of neurotransmitters. The effects of a mutation in this gene on violent aggression are described in Chapter 19

**Monosomy** possessing only one copy of a given chromosome pair. Turner syndrome (45, XO) could also be described as monosomy for the X chromosome (see Chapters 4, 6, 7, and 9)

**Mosiacism** being composed of two different genotypes. Consider a case in which an error in the first mitotic division of an embryo resulted in an XX and XO daughter cell. The resulting baby would be composed of normal 46, XX cells and 45 XO cells. This girl would be a mosaic for Turner syndrome. 46XX females may also be considered mosaics with respect to X chromosomal genes because they carry a population of cells in which only their father's X is active and a population of cells in which only their mother's X is active (see Chapter 7)

**mRNA** an RNA copy of the coding strand of a gene. The mRNA molecule is transported to the *ribosomes* in the cytoplasm where it is translated into protein (see Chapter 3)

**Mullerian ducts** structures present in the early embryo of both sexes. Unless their development is inhibited by the presence of Mullerian inhibiting factor, the Mullerian ducts will go on to produce the cervix, uterus, and fallopian tubes (see Chapter 6)

**Mullerian inhibiting factor (MIF)** a hormone produced by the developing testes that inhibits the development of the Mullerian ducts (see Chapter 6)

**Multifactorial inheritance** a pattern of inheritance in which the final phenotype is determined both by multiple genes and by the interaction of alleles of those genes with the environment (see Chapter 18)

**Muscular dystrophy** *see* Duchenne muscular dystrophy and Becker muscular dystrophy

**Mutation** a stable and heritable change in the base sequence of a DNA molecule (see the entire book, but especially Chapter 3 for a discussion of the various types of mutations)

**Neural tube** a structure in the early embryo that eventually forms the brain and the spinal cord (see Chapter 21)

**Neural tube defects** errors in the formation or closure of the neural tube that result in disorders ranging from incomplete closure of the spine (spina bifida) to an absence of the brain (anencephaly) (see Chapter 21)

**Neurotransmitter** a chemical used by one nerve cell (neuron) to stimulate the activity of the next nerve cell in the pathway (see Chapter 19)

**Noncoding strand** some other sources use this term to denote the DNA strand actually copied by RNA polymerase. We refer to this strand as the *template strand* (see Chapter 3)

**Nondisjunction** in meiosis I, the segregation of two homologous chromosomes to the same pole. In meiosis II and mitosis, the separation of two sister chromatids followed by their migration to the same pole of the spindle. The end result of nondisjunction is the production of aneuploid daughter cells (see Chapters 4, 6, and 9)

**Nonsense mutations** mutations that convert a codon that specifies an amino acid into a stop codon. Such mutations cause a halt in translation and produce a truncated protein product (see Chapter 3 and Figure 3.9)

**Northern blotting** a tool for determining the presence or absence of a particular mRNA molecule among a population of RNA molecules obtained from a given set of cells or tissue. The process of transferring a population of RNA molecules that have been separated on a gel by electrophoresis onto a filter or membrane. The resulting Northern blot can then be hybridized to a given labeled DNA probe (see Chapter 12).

**Nucleotides**   the basic building blocks of nucleic acids, composed of a base (A, T, G, C, or U), a sugar, and a phosphate (see Chapter 3)

**Nucleus**   the structure in the cell that contains the genomic DNA. This is the site of transcription and replication (see Chapters 1–3)

**Nullo**   used in reference to gametes carrying no copies of a given chromosome (e.g., the term nullo-21 ova would mean an egg lacking chromosome 21) (see Chapters 4, 6, and 9)

**Oncovirus**   a virus that causes tumors (see Chapter 17)

**Oocyte**   a female meiotic cell (see Chapter 4)

**Open reading frame (ORF)**   a stretch of DNA that could encode a continuously translatable stretch of mRNA (i.e., a stretch of DNA whose transcript does not include a stop codon) (see Chapter 12)

**Operator**   a sequence in the regulatory region of the *lacZ* gene in *E. coli* to which the Lac repressor binds (see Chapter 3)

**Origin of replication**   a site on a chromosome or plasmid that is sufficient to initiate DNA replication at that site

**Pachytene**   the stage of prophase of meiosis I in which homologous chromosome pairing is complete (see Chapter 4)

**Patau syndrome**   *see* Trisomy 13 (see Chapter 9)

**Penetrance**   a measure of the degree to which a given genotype is correlated with the phenotype. Strongly penetrant mutations are almost always associated with a given phenotype (the so-called "mutant phenotype"). Less penetrant mutations may generate the "mutant" phenotype in only some characters (see Chapter 5)

**Phage**   a bacterial virus (see Chapter 12)

**Phage library**   a collection of phages bearing different DNA insertions derived from the genome of interest. For example, a human phage DNA library consists of a bacterial virus, each carrying a piece of human DNA inserted into the viral genome (see Chapter 12)

**Phenotype**   the actual physical manifestation of a trait that is determined by genes, e.g., height, hair color, or disease (see Chapters 2 and 5)

**Plasmid**   a circular DNA molecule, often used as a cloning vector, that can replicate in a bacterial or yeast cell (see Chapter 12 and cloning vector)

**Ploidy**   the number of haploid genomes possessed by a given cell. In ploidy, only chromosomes are counted, whether or not they are replicated (thus one never counts a pair of sister chromatids as two chromosomes). Cells with two copies of each chromosome are called *diploid*. Gametes with one copy of each chromosome are called *haploid*. In humans the diploid number is 46 and the haploid number of chromosomes is 23. Cells possessing three copies of each chromosome are *triploid*. *Triploidy* occurs reasonably frequently among human conceptions, but is incompatible with fetal normal development and results in spontaneous abortion. Cells carrying too few or too many copies of just one or more chromosomes are called *aneuploid.* Types of aneuploidy include trisomies and monosomies (see Chapters 4, 6, and 9)

**Polar body** one of the nonfunctional products of female meiosis (see Chapter 4)

**Polygenic inheritance** the type of inheritance observed when a phenotype is determined by two or more genes (see Chapters 5 and 18)

**Polymerase chain reaction (PCR)** the process of rapidly amplifying a defined region of DNA by sequential steps of denaturation and replication (see Chapter 13)

**Polymorphism** in this book, one of many alleles of a given genetic marker or gene. A gene or genetic marker with many variant forms is considered to be highly polymorphic (see Chapters 12, 13, and 14)

**Prader-Willi syndrome** a disorder of young children characterized by diminished fetal activity, obesity, muscular weakness, mental retardation, and short stature resulting from the absence of a paternally derived copy of genes on chromosome 15, either by transmission of a deletion-bearing chromosome by the father or by uniparental disomy for the maternally derived copy of chromosome 15. An example of the importance of imprinting (see Chapter 11 and Angelman syndrome)

**Premutation** a genetic change, as at the FMR1 gene, that creates a phenotype. Either a second mutation or an epigenetic change is required for that premutation to confer a phenotype (see Chapter 10)

**Primary sexual characteristics** gonads (testes, ovaries) and external genitalia (penis, scrotum, clitoris, vagina) (see Chapter 6)

**Proband** the individual that brought a family or kindred to the geneticist's attention (see Chapter 8)

**Probe** a DNA or, in some cases, an RNA molecule labeled by a radioactive or fluorescent tag that is used in DNA:DNA or DNA:RNA hybridization reactions (see Chapter 12)

**Prokaryote** an organism lacking a nucleus. The DNA of the genome is in the cytoplasm. Bacteria are prokaryotes (see Chapter 3)

**Promoter** the site at which a RNA polymerase binds (see Chapter 3)

**Pronucleus** one of the two gametic nuclei in a fertilized egg (see Chapter 4)

**Prophase** the first step in mitosis and meiosis I during which chromosome condensation occurs. In meiosis I, pairing and recombination also occur during prophase (see Chapter 4)

**Protease** an enzyme that cuts up other proteins. Most proteases are specific for a specific site on a protein or set of proteins. A virally encoded protease is essential for the replication of the HIV virus (see Chapter 20)

**Protease inhibitor** a chemical that can inhibit the activity of a specific protease (see Chapter 20)

**Protein** these molecules do most of the work of our cells. They carry out enzymatic reactions, produce energy, and serve as the "skeletons" and "muscles" of our cells. Each protein is composed of a linear array of amino acid building blocks. The sequence of those amino acid building blocks is encoded in the gene that produces that protein. It is not much of an exaggeration to say that the function of most of our genes is simply to encode the proteins that allow us to live (see Chapters 1–3)

**Proto-oncogene**   a gene normally required only in cells undergoing rapid division. Inappropriate expression of these genes can play an important role in tumor formation (see Chapter 17)

**Pseudo-autosomal region (PAR)**   a region of homology between the X and the Y chromosomes. Meiotic recombination in the PAR located at the tips of the short arms of the X and Y chromosomes is required for proper meiotic segregation of the X and Y chromosomes (see Chapters 6 and 7)

**Pure-breeding**   in reference to Mendel's pea experiments, a group of plants that produce only offspring like themselves (see Chapter 2)

**Rearrangement**   one of a class of mutations that alter the sequence of genes along a chromosome or interchanges genetic material between nonhomologous chromosomes. Such mutations include deficiencies (deletions), dicentrics, duplications, inversions, rings, translocations, and transpositions (see Box 21.1 and Chapter 21)

**Recessive**   a mutation is said to be recessive if its expression is "masked" by the normal or wild-type allele (i.e., it does not produce a phenotype when heterozygous with the normal allele). In other words, a mutation whose phenotype is only observed when the mutation is homo- or hemizygous (see Chapters 2 and 5)

**Reciprocal translocation**   *see* Translocation

**Recombination (also called crossing over and exchange)**   the process by which two homologous chromosomes exchange genetic material. This process requires the breakage and rejoining of two homologous DNA molecules followed by rehealing.

For two DNA molecules denoted ABCDEFG and abcdefg, the two products of recombination between C and D would be ABCdefg and abcDEFG. During *meiotic* recombination, paired homologs undergo multiple recombination events along their length. These exchanges play a critical role in ensuring segregation at anaphase I. During late prophase, recombination events are visible along the bivalent as sites called *chiasmata*. Map distances are measurements of the frequency with which recombination occurs between two genetic markers. Homologous recombination occurs much more rarely in mitotic cells. However, in mitosis, recombination between two homologs does *not* commit those chromosomes to segregate from each other at the next mitotic division (see Chapters 4, 14, and 17)

**Regulatory elements (REs)**   components of a eukaryotic gene that determine when and where that gene will be expressed. Such sequence include *enhancer* elements that promote transcription as well as elements that repress transcription. The function of these elements is mediated by the proteins that bind to them. These proteins, known as *transcription factors,* determine the accessibility of the gene to RNA polymerase and thus the efficiency with which it will be transcribed (see Chapter 3)

**Repair-deficient mutations**   mutations in genes whose protein products are required for one or more DNA repair processes. Such mutations impair the ability of the cell to repair damage to its DNA. At least some types of repair-deficient mutations in humans strongly predispose individuals who carry such mutations to develop cancer (see Chapter 17)

**Repressor** a protein that represses gene activity (see Chapter 3)

**Restriction enzyme** a protein that cleaves DNA molecules at specific sequences (e.g., AATTTA). Restriction enzymes are critical tools in the analysis of DNA molecules (see Chapter 12)

**Restriction fragment length polymorphism (RFLP)** a difference in the pattern of cleavage of a given DNA sequence by a restriction enzyme. These polymorphisms are critical tools for genetic mapping and reflect the presence or absence of mutations that alter or create the short sequences recognized by the restriction enzyme and at which cleavage occurs. Imagine a 3-kb DNA sequence that contains a "cut site" for enzyme A at a position 1 kb from the left end and a second site 2 kb from the left end. Digestion of this molecule with enzyme A will result in three pieces of DNA that are 1 kb in length. Now imagine a mutation that changed the cut site at 2 kb from the left end. There is now only a single cut site, located 1 kb from the left end. Digestion of this mutant DNA with enzyme A results in two products, one of which will be 2 kb in length (see Chapter 12)

**Retinoblastoma** a hereditary tumor of the eye (see Chapters 5 and 17)

**Retrovirus** a virus whose genome is composed of RNA (see Chapter 20)

**Reverse transcriptase** an enzyme that creates a complementary DNA copy of a RNA molecule. Reverse transcriptase is an essential component of retrovirus replication (see Chapter 20)

**Reversion** a mutation that reverses the phenotypic effects of a previously induced or isolated mutation in the same gene. Imagine a single base pair insertion that caused a frame shift mutation. The deletion of an adjacent base would restore the reading frame and *might* restore activity of the resulting protein. Some precise reversions actually reverse the original mutation, whereas others simply compensate for the effect of the first mutation. Reversion events are usually quite rare (see Chapter 10)

**Ribonucleic acid (RNA)** RNA is made up of the four chemical bases, adenine, cytosine, guanine, and thymidine (abbreviated A, C, G, and U), connected by a sugar (ribose) backbone. Most of the RNA in the cell is single stranded. RNA is produced by the transcription of DNA, and its synthesis follows the rules of Watson–Crick base pairing with the stipulation that U replaces T in RNA. For our purposes, there are four classes of RNAs: mRNAs that can be translated to produce proteins, tRNAs that assist in the process of translation, rRNAs that serve as structural components of the ribosome, and snRNAs that play critical roles in the processing of other RNA molecules (see Chapter 3 and the rest of the book)

**Ribosomal proteins** proteins that comprise the ribosome (see Chapter 3 and 7)

**Ribosomal RNA** *see* rRNA

**Ribosome** a structure located in the cytoplasm that carries out the process of *translation*. Ribosomes are composed of ribosomal proteins and structural RNAs (called rRNAs or ribosomal RNAs). Ribosomes attach to a mRNA molecule at the 5′ end and translate it sequentially by reading each *codon* (see Chapter 3)

**Ring**   a type of chromosome aberration in which the two ends of the DNA molecule have been fused to form a complete circle (see Box 21.1 in Chapter 21)

**RNA polymerase**   an enzyme composed of many protein subunits that carries out transcription through synthesis of an RNA "copy" of the coding strand by making an RNA strand complementary to the template DNA strand. RNA polymerase begins transcription by binding to the promoter region of the DNA located just 5′ of the point at which the transcript begins. There are several classes of RNA polymerases that carry out the transcription of different classes of genes (see Chapter 3)

**RNA processing**   the chemical modifications of an RNA transcript, including splicing out introns, that take place before the final mRNA transcript is exported to the cytoplasm for translation (see Chapter 3)

**RNase H**   an enzyme that degrades RNA in a RNA:DNA hybrid (see Chapter 20)

**rRNA**   a class of RNA molecules that serve as structural components of the ribosome. Several classes of rRNA molecules play different roles in determining the structure and the function of the ribosome.

**Secondary sex characteristics**   physical differences between the two sexes other than gonads and external genitalia. These include breast development, facial and body hair, and skeletal and muscular differences (see Chapter 6).

**Semiconservative replication**   the method by which a double-stranded DNA molecule is copied by making complementary copies of the two strands of the original DNA molecule. After replication is completed, each of the two new double-stranded DNA molecules consists of an old strand and a new strand (see Chapter 3 and 13)

**Sequence**   the order of base pairs in a DNA molecule (see Chapter 3)

**Sex chromosome**   the X and Y chromosomes (see Chapters 4, 6, and 7)

**Sex-limited trait**   a trait that can be expressed in only one sex (see Chapter 5)

**Sex-linked dominant**   a pattern of inheritance exhibited by dominant mutations of the X chromosome. All daughters of affected males are affected (see Chapter 5)

**Sex-linked recessive**   a pattern of inheritance exhibited by recessive mutations on the X chromosome. Unless the mother is also a carrier, all children of affected males are normal. Half the sons, but none of the daughters, are affected in matings of normal males to carrier females (see Chapter 5)

**Sexual identity**   a component of gender. The sex or set of sex roles with which an individual identifies. This is usually, but not always, congruent with biological sex (see Chapter 8)

**Sexual orientation**   the sex to which an individual is sexually attracted (see Chapter 8)

**Sex vesicle**   a structure in the nucleus of spermatocytes (the cells undergoing meiosis in males) in which pairing and recombination of the X and Y chromosomes take place (see Chapter 6)

**Sickle cell anemia**   a hereditary disorder caused by a mutation in the hemoglobin gene. Homozygosity for the sickle cell mutation causes a serious medical condition called sickle cell disease. However, heterozygosity for this mutation allows resistance to malaria (see Box 20.1)

**Sister chromatids** the two identical copies of a DNA molecule produced by semiconservative replication. Prior to mitotic anaphase, sister chromatids usually remain tightly connected along their lengths, especially at their centromeres. The separation of sister chromatids heralds the onset of anaphase in mitosis. During the first meiotic anaphase, sister chromatids separate along their arms, but the sister chromatids remain connected at the centromeres. Separation of sister centromeres occurs only at anaphase II (see Chapter 4)

**Spermatocyte** a meiotic cell in a male (see Chapter 6)

**S phase** the period in the cell cycle when DNA replication takes place (see Chapter 17)

**Spina bifida** a type of neural tube defect resulting in incomplete closure of part of the spinal column (see Chapter 18)

**Spinocerebellar ataxia type 1 (SCA 1)** a neurological disease whose symptoms begin in the fourth decade of life. The genetic basis of this disease is an expansion of a trinucleotide CAG repeat. Patients with early onset disease have a larger gene size than those patients with late onset disease (see Chapter 10)

**Splice site** the point in the sequence of the RNA molecule at which splicing takes place. Splice site sequences are found at the exon–intron boundaries and are essential for correct splicing of the transcript. The minimum feature of a donor splice site (at the 5′ end of an intron) is that the first two bases inside of the intron must be the

bases GT and that the last two bases (at the 3′ end of the intron) must be the bases AG. Additional sequences inside of the introns are also important to the splicing process, with sequences closest to the splice sites being the most important (see Chapter 4)

**Splicing** the process by which introns are removed from a primary transcript (the RNA copy of the coding strand of a gene) so that the exons are joined together in the mRNA molecule in the order that they are found on the chromosome (see Chapter 4)

**Start codon** the first AUG codon to be used by the ribosome when translating a message (see Chapter 3)

**Stop codon** the codons UAA, UGA, or UAG, which cause the termination of translation. Because there are no tRNA molecules with a complementary *anticodon*, these codons cannot be translated. When any of these three codons occur "in frame" on an mRNA molecule, synthesis of the protein stops and the mRNA and the protein are released from the ribosome (see Chapter 3)

**Structural protein** proteins that form the structures of which organelles, cells, tissues, and organs are made. Structural proteins can usually be thought of as distinct from enzymes that catalyze biochemical reactions. However, some proteins that may be thought of as part of a cell structure are also capable of carrying out an enzymatic or signaling function under some circumstances (see Chapter 1)

**Subtractive hybridization** the process of removing a specific set of DNA molecules from a larger population of DNA molecules by hybridization. Imagine that DNA sample A and

DNA sample B contain some sequences that are held in common and some sequences that differ between samples. Now further suppose that you only want sequences that are unique to sample A. Subtractive hybridization uses techniques that result only in sequences unique to sample A being left at the end of the experiment. So in this subtractive hybridization experiment, sample B DNA plus any DNA that hybridizes to sample B must get physically removed from the population of DNA molecules (see Chapter 15)

**Telomerase**   the enzyme that extends telomeric DNA by adding new telomere DNA onto the ends of the chromosome

**Telomere**   a region of repeated DNA sequences, and associated proteins, that is found at the end of eukaryotic chromosomes (see Chapter 4)

**Telophase**   the last stage during mitosis and meiosis II. The chromosomes have reached the poles and the nuclear envelope reforms around them (see Chapter 4)

**Template strand**   in *replication*, the DNA strand being copied by DNA polymerase. In *transcription*, the DNA strand being copied by RNA polymerase. Other books may also refer to this strand as the *noncoding strand* (see Chapter 3)

**Testis determining factor (TDF)**   the gene on the Y chromosome whose protein product induces the indifferent gonads to develop as testes. In the absence of TDF, the indifferent gonads develop as ovaries (see Chapter 6)

**Thymine**   one of the bases found in DNA. Thymine is not found in RNA, and usually abbreviated by the symbol "T" (see Chapter 3)

**Transcription**   the process by which the enzyme *RNA polymerase* makes an RNA molecule complementary to the sequence of the *template strand* of the DNA. Transcription begins with binding of the RNA polymerase to the promoter region 5′ to the region to be transcribed (see Chapter 3)

**Transcription factor**   other proteins, in addition to RNA polymerase, that bind to specific sequences in the DNA in the region around the promoter, especially 5′ to the promoter, in such a way as to influence whether transcription of the gene takes place and what the level of transcription will be. Some transcription factors induce elevated levels of transcription, some suppress transcription of the gene, and are present in some cell types but not others so that their binding to the DNA helps provide tissue specificity of expression of some genes (see Chapter 3)

**Transfer RNA**   *see* tRNA

**Transformation**   the process of introducing exogenous DNA into a cell (see Chapter 3 and 12)

**Transgenic mice**   mice bearing a foreign gene introduced into their genomes (see Chapter 12)

**Translation**   the process by which a ribosome makes a protein by "reading" the sequence of an mRNA codon by codon. To carry out translation, three steps must happen: initiation, elongation, and termination. Initiation occurs when the ribosome recognizes the AUG start codon and begins assembly of the protein chain with a methionine (the amino acid that corresponds to AUG). Elongation is the process by which the ribosome moves to each new codon along the mRNA chain, "reading" the next codon in line by finding and putting into place the tRNA whose "anticodon" recognition sequence is complementary to the codon being read. The amino acid carried by the tRNA whose "anticodon" can recognize the codon that is "up to bat" is then added to the growing protein chain. The mRNA then advances by three bases so that a new codon is in position to be read, and a new tRNA that recognizes that codon moves into position and its amino acid is added to the protein chain. This process of advancing along the mRNA chain, adding new amino acids to the protein, continues until termination occurs. Termination of translation results when one of the three stop codons, UAA, UAG, or UGA, is read by the ribosomal system, and translation is terminated with no insertion of an amino acid corresponding to those codons. Three kinds of RNA take part in the translation process: rRNA that is part of the ribosome, tRNA that recognizes the codons and carries the amino acid that "goes with" that codon, and the mRNA that dictates the order of the amino acids in the final protein (see Chapter 3)

**Translocation** a type of chromosome aberration in which material from one chromosome is moved onto another chromosome. In the case of *reciprocal translocation,* material is exchanged between two nonhomologous chromosomes in a reciprocal fashion. Imagine two nonhomologous chromosomes with the sequences ABCDEFGHloJKLMOP and 1234506789, where o donotes the centromeres. A reciprocal translocation might result in the formation of two different chromosomes with the sequences 1234FGHloJKLMOP and ABCDE5o6789. In the case of *inser-* *tional translocation,* a segment of one chromosome is excised from that chromosome and inserted into a site on a nonhomologous chromosome (cf. ABCD2345EFGHloJKLMOP) (see Chapter 21, Box 21.1)

**Triploid** a cell carrying three copies of each chromosome. Triploid conceptions do occur in humans, but are always inviable.

**Trisomy** possessing three copies of a given chromosome (see Chapter 9)

**Trisomy 13** the phenotype of this disorder includes severe malformations of the nervous system, as well as a host of other problems, including heart defects and facial anomalies. A large fraction of cases of trisomy 13 result in spontaneous miscarriage, and of live-born children it is virtually always lethal by the sixth month (see Chapter 9)

**Trisomy 18** this disorder is observed at a very low frequency (1 out of 8000 live births). The incidence at conception is much higher, but most of these embryos abort spontaneously. These usually have a characteristic set of malformations along with significant neurological deficiencies. There are usually severe cardiac problems and survival beyond 6 months is very rare (see Chapter 9)

**Trisomy 21** *see* Down syndrome

**tRNA** the RNA molecule that provides specificity to the translation process. Each type of tRNA molecule carries a specific anticodon complementary to the codon on the mRNA and an appropriate amino acid. By matching the anticodon and the codon, the appropriate amino acid is placed in the lengthening peptide chain (see Chapter 3)

**Tumor** a mass of rapidly and inappropriately dividing cells arising from a single progenitor cell (see Chapter 17)

**Tumor suppressor gene**   a gene whose protein product acts to prevent inappropriate cell division. Mutation "knockouts" of tumor suppressor genes are a key element in the etiology of many types of cancers (see Chapter 17)

**Turner syndrome**   a genetic disorder observed in 45 XO women. Turner females are sterile and may display a number of other phenotypes, including short stature (see Chapter 6, 7, and 11)

**Uracil**   a base found in RNA but not DNA, and usually abbreviated by the symbol "U" (see Chapter 3)

**Virus**   an infectious parasite composed of protein and nucleic acid that can replicate only when its genome is inserted into an appropriate host cell (see Chapter 20)

**Watson–Crick base pairing**   the other secret of life. Think about it, Watson–Crick base pairing (A–T, G–C in DNA and A–U, G–C in RNA) is the basis for EVERYTHING (i.e., replication, transcription, translation, recombination). Watson–Crick base pairing makes both replication and transcription accurate and possible. It is Watson–Crick base pairing between the mRNA and the tRNA that underlies translation (see Chapter 3)

**Wilm's tumor**   a hereditary form of kidney cancer (see Chapter 17)

**X autosome translocation**   a translocation involving the X chromosome and an autosome (see Chapters 7 and 15)

**Xeroderma pigmentosa**   a hereditary disease associated with a deficiency in DNA repair (see Chapter 17)

**XX(TDF) male**   a 45 XX individual in which the Y chromosomal TDF gene has been translocated onto the tip of one of the two X chromosomes. These individuals develop into sterile males (see Chapter 6)

**XY(del TDF) female**   a 46 XY individual bearing a Y chromosome lacking the TDF gene. Such individuals develop into sterile females (see Chapter 6)

**XYY male**   a male bearing an extra Y chromosome. Such men are taller than usual and are overrepresented in prison populations. However, there is no evidence that they are either more aggressive or more violent than their XY counterparts. If that strikes you as a contradiction, it should (see Chapter 19)

**Zygote**   the product of the fusion of the two pronuclei following fertilization

**Zygotene**   the stage of prophase of meiosis I during which chromosome pairing is accomplished (see Chapter 4)

# INDEX